D1618135

Digital Transformation of the Laboratory

Digital Transformation of the Laboratory

A Practical Guide to the Connected Lab

Edited by

Klemen Zupancic
Tea Pavlek
Jana Erjavec

The Editors

Klemen Zupancic
SciNote LLC
3000 Parmenter St.
53562 Middleton WI
United States

Tea Pavlek
SciNote LLC
3000 Parmenter St.
53562 Middleton WI
United States

Jana Erjavec
BioSistemika LLC
Koprska ulica 98
1000 Ljubljana
Slovenia

Cover Image: Tea Pavlek

Library of Congress Card No.:
applied for

British Library Cataloguing-in-Publication Data
A catalogue record for this book is available from the British Library.

Bibliographic information published by the Deutsche Nationalbibliothek
The Deutsche Nationalbibliothek lists this publication in the Deutsche Nationalbibliografie; detailed bibliographic data are available on the Internet at <http://dnb.d-nb.de>.

Print ISBN: 978-3-527-34719-3
ePDF ISBN: 978-3-527-82505-9
ePub ISBN: 978-3-527-82506-6
oBook ISBN: 978-3-527-82504-2

Typesetting SPi Global, Chennai, India
Printing and Binding CPI Group (UK) Ltd, Croydon, CR0 4YY

Printed on acid-free paper

C099641_050521

Contents

Preface

The subject of digital transformation is actually about you.

Your science, your everyday work environment, your partnerships and collaborations, and the impact of your work on the future of scientific progress.

Welcome to this book.

As a brilliant astronomer, Maria Mitchell once said, "mingle the starlight with your lives and you won't be fretted by trifles."

The greater meaning of digital transformation shifts the perspective toward the global scheme of things. The main evaluating factors behind the lab digitalization and digital transformation answer important questions: Are we improving the quality, efficiency, and the pace of innovation?

Lab digitalization is a people-driven initiative that aims to address the global challenges and provide solutions, backed by unquestionable integrity of traceable and reproducible scientific data.

At the moment, regardless of the laboratory type or size, people are struggling with the growing amount of generated data and leveraging its value. It is extremely challenging to organize data and keep everything traceable and reusable long term.

To address the challenge, modularity and flexibility are being incorporated on different levels of lab operations. Labs are becoming inviting spaces suitable for interdisciplinary partnerships in a digital, virtual, or personal environment. Data integrity initiatives and setup of new, digital systems prioritize integration of all tech solutions used in the lab for optimal performance. Through effective integration of tools, improved scientist-to-scientist interactions and intellectual contributions, and quality change management, lab digitalization places the human element at the very forefront of the overall progress toward the digital future.

This can be intimidating to some and exhilarating to others.

That is why this book is divided into modules: Inspiration, Knowledge Base, Practical, Case Studies, Continuous Improvement, and Vision of the Future. Each module covers different aspects of lab digitalization.

Inspiration

We start this book with an inspiring overview of lab evolution, new technologies, and new science being done. It will give you a complete overview of the subject of laboratories of the future and, hopefully, add to the vision of your own career in science and technology.

Knowledge Base

Knowledge Base section focuses on crucial terms to understand. It will give you a solid basis of knowledge that you will be able to apply further on as your lab grows and evolves.

Practical

The Practical chapters give you examples and guidance on defining your lab's digitalization strategy.

Case Studies

We present different case studies and expert comments on the subject of going from paper to digital. You will be able to read how different laboratories and professionals approached the subject and put it into practice, and what are their conclusions, advice, and lessons learned.

Continuous Improvement

We have a closer look at the steps that follow after the digitalization.

Vision of the Future and Changing the Way We Do Science

With continuous improvements in mind, we conclude the book with insightful expert comments on the subject of the future of science. Many of the described technologies are already becoming important, and here we identify those that might shape the next 5–10 years and change the way we do science.

As you read this book, you will gain holistic knowledge on digital transformation of the laboratory. Tracking, analyzing, and leveraging the value of data you are collecting, by implementing tools that can empower the people in your lab, are the main points of this journey.

Using the knowledge, you will be able to start defining what exactly you want to achieve. Once you clarify your main goals, you will be able to go all the way back through the processes in your lab and see which need to be digitalized.

That is when you will get the real incentive to do it.

You will understand whether you are trying to just use technology as a convenience to support the system you already have, or are you ready to think about using the better technology to change and improve the system.

You will realize what kind of decisions you need to make throughout the cycle.

Selecting the right digital solutions is quite a challenge. It is important to think how the potential solutions will fit into your existing architecture. An investment of time, energy, and budget is always involved, especially if the solutions are not integrated properly or your team is not in sync.

The knowledge you will gain will enable you to measure and evaluate the impact of digitalization. How will the use of particular tools improve specific parts of your processes to reach your goals within the given time frames?

Keeping the razor-sharp focus and determination is the most potent driver of digitalization.

All solutions work, but the execution is crucial. You will learn how to take an agile approach, define the value for your team, start small, and scale up efficiently and successfully.

This book is a result of collaboration between different authors – researchers, business owners, consultants, managers, and professors who wrote about their vast experience and provided valuable perspective on the subject of digital transformation. Because every lab is different, and there are as many use cases as there are labs, our aim was to introduce you to a digital mindset that will enable you to find the best solution for your lab.

This book guides you through the aspects of taking your time to understand the basics of technology, adapt the digital mindset, include your team and address their concerns and challenges, read how other labs started to pave their digital way, and stay inspired along the way.

Let's dive in.

SciNote LLC, USA *Tea Pavlek*

Part I

Inspiration

We start this book with an inspiring overview of lab evolution, new technologies, and new science being done. It will give you a complete overview of the subject of laboratories of the future and, hopefully, add to the vision and purpose of your own career in science and technology.

1

The Next Big Developments – The Lab of the Future

Richard Shute and Nick Lynch

Curlew Research, Woburn Sands, UK

1.1 Introduction

Steve Jobs once said that "the biggest innovations of the 21st century will be at the intersection of biology and technology"; in this (r)evolution, the lab will most definitely play a key role.

When speculating on the future digital transformation of the life sciences R&D, one must consider how the whole lab environment and the science that goes on in that lab will inevitably evolve and change [1, 2]. It is unlikely that an R&D lab in 2030, and certainly in 2040, will look and feel like a comparable lab from 2020. So, what are the likely new big technologies and processes and ways of working that will make that lab of the future (LotF) so different? This section endeavors to introduce some of the new developments in technology and in science that we think will change and influence the life science lab environment over the upcoming decade.

1.2 Discussion

Before going into the new technology and science in detail, it is important to recognize that this lab evolution will be driven not just by new technologies and new science. In our view, there are four additional broader, yet fundamental and complementary attributes that influence how a lab environment changes over time. They are:

1. People and culture considerations
2. Process developments and optimization
3. Data management improvements
4. Lab environment and design

When we add the fifth major driver of change – new technology (including new science) – it becomes clear that digital transformation is a complex, multivariate concept (Figure 1.1).

Digital Transformation of the Laboratory: A Practical Guide to the Connected Lab, First Edition.
Edited by Klemen Zupancic, Tea Pavlek and Jana Erjavec.

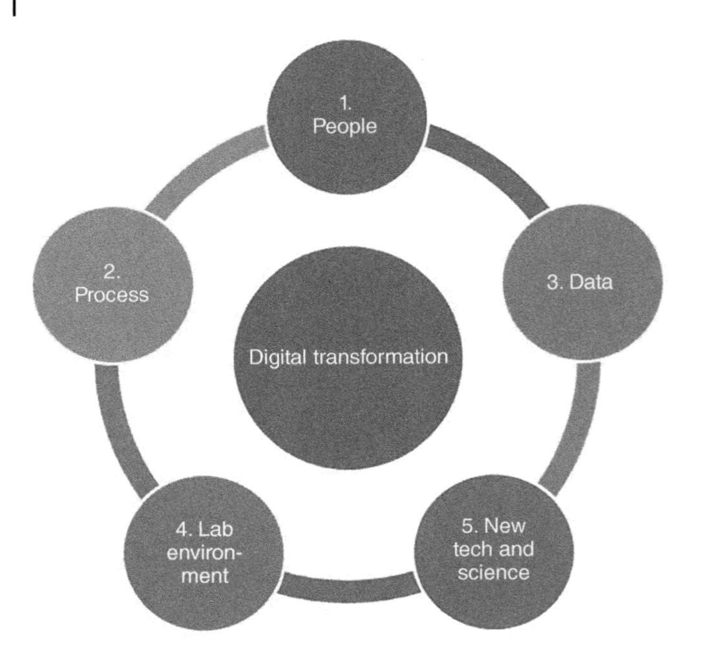

Figure 1.1 Complex, multivariate concept of lab transformation.

In this section, we discuss how each of these high-level attributes will influence the changing lab and the expectations of the users. For all five areas, we include what we think are some of the most important aspects, which we believe will have the most impact on the "LotF."

1.2.1 People/Culture

The LotF and the people who work in it will undoubtedly be operating in an R&D world where there is an even greater emphasis on global working and cross-organization collaboration. Modern science is also becoming more social [3], and the most productive and successful researchers will be familiar with the substance and the methods of each other's work so breaking down even more the barriers to collaboration. These collaborative approaches will foster and encourage individuals' capacity to adopt new research methods as they become available; we saw this with the fast uptake of clustered regularly interspaced short palindromic repeat (CRISPR) technology [4]. "Open science" [5] will grow evermore important to drive scientific discovery. This will be enabled through the increased use of new cryptographic Distributed Ledger Technology (DLT) [6], which will massively reduce the risk of IP being compromised [7]. The LotF will also enable more open, productive, collaborative working through vastly improved communication technology (5G moving to 6G) [8]. The people working in these labs will have a much more open attitude, culture, and mindset, given the influence of technology such as smartphones on their personal lives.

Robotics and automation will be ubiquitous, but with more automated assistance, the density of people in the lab will likely drop, allowing scientists to focus on key aspects and complex parts of the experiments. As a consequence, issues around safety and "lone working" will grow, and a focus on the interaction points which scientists have with automation will develop to ensure they are properly protected. For the few remaining lab technicians, not only will safe working become of increased importance, but the need for organizations to deliver a better "user experience" (UX) in their labs will become key to help them both attract the smaller numbers of more expert technicians and also retain them. The lab technician's UX will be massively boosted by many of the new technologies already starting to appear in the more future-looking labs, e.g. voice recognition, augmented reality (AR), immersive lab experience, a more intelligent lab environment, and others (see later sections).

1.2.2 Process

The lab processes, or "how" science gets done in the LotF, will be dominated by robotics and automation. But there will be another strong driver which will force lab processes and mindsets to be different in 5–10 years time: sustainability. Experiments will have to be designed to minimize the excessive use of "noxious" materials (e.g. chemical and biological) throughout the process and in the cleanup once the experiment is complete. Similarly, the use of "bad-for-the-planet" plastics (e.g. 96/384/1536-well plates) will diminish. New processes and techniques will have to be conceived to circumvent what are standard ways of working in the lab of 2020. In support of the sustainability driver, miniaturization of lab processes will grow hugely in importance, especially in research, diagnostic, and testing labs. The current so-called lab on a chip movement has many examples of process miniaturization [9]. Laboratories and plants that are focused on manufacturing will continue to work at scale, but the ongoing search for more environmentally conscious methods will continue, including climate-friendly solvents, reagents, and the use of catalysts will grow evermore important [10]. There will also be a greater focus on better plant design. For example, 3D printing [11] could allow for localization of manufacturing processes near to the point of usage.

In the previous paragraph, we refer to "research, diagnostic, and testing labs" and to manufacturing "plant." We believe there is a fundamental difference between what we are calling hypothesis- and protocol-driven labs, and this is an important consideration when thinking about the LotF. The former are seen in pure research/discovery and academia. The experiments being undertaken in these labs may be the first of their kind and will evolve as the hypothesis evolves. Such labs will embrace high throughput and miniaturization. Protocol-driven labs, where pure research is not the main focus, include facilities such as manufacturing, diagnostic, analytical, or gene-testing labs. These tend to have a lower throughput, though their levels of productivity are growing as automation and higher quality processes enable ever higher throughput. In these labs, reproducibility combined with robust reliability is key. Examples in this latter area include the genomic screening and testing labs [12, 13], which have been growing massively in the

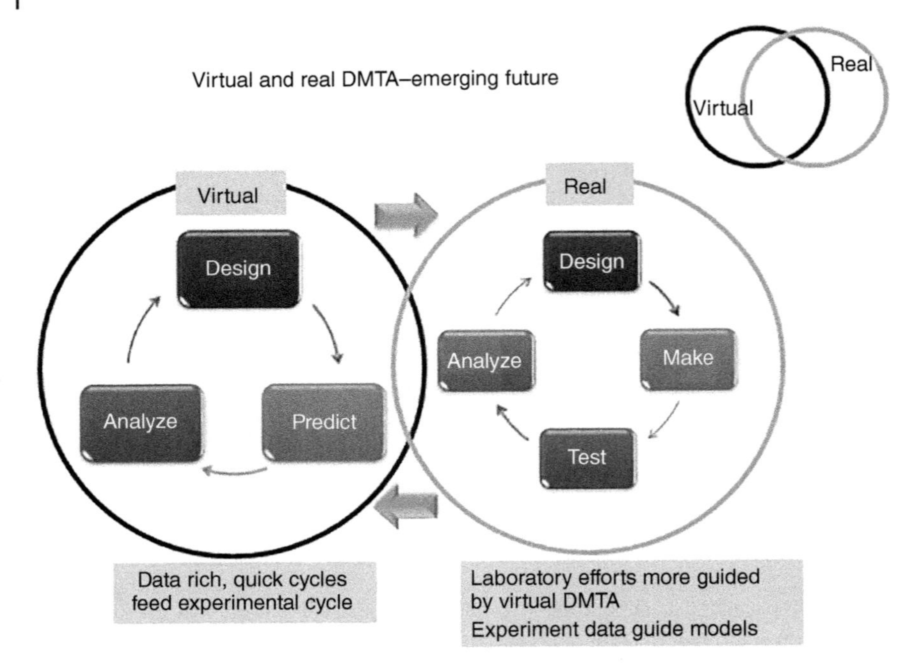

Figure 1.2 Virtual and real design-make-test-analyze (DMTA) concept.

past few years. For these labs the already high levels of automation will continue to grow.

In the hypothesis-driven lab [14] with the strong driver of sustainability combined with the growth of ever higher quality artificial intelligence (AI) and informatics algorithms, there will be more in silico, virtual "design-make-test-analyze" (vDMTA) and less, tangible Make and Test (see Figure 1.2). Fewer "real" materials will actually be made and tested, and those that are will be produced on a much smaller scale.

Finally, as labs get more sophisticated – with their high levels of automation, robotics, miniaturization, and data production (but with fewer staff) – combined with the need for those facilities to be both safe and sustainable, the concept of "laboratory as a service" (LaaS) will grow [15]. The LotF will not be a static, self-contained, and single scientific area facility. It will be a blank canvas, as it were, in a large warehouse-like facility or cargo container [16] which can be loaded up on demand with the necessary equipment, automation, and robotics to do a contracted piece of lab work. That piece of work might be a chemical synthesis or a cell-based pharmacological assay one day, and an ex vivo safety screen in the same area the next day. The key will be use of a modular design supported by fully connected devices.

1.2.3 Lab Environment and Design

The lab environment, its design, usability, and sustainability are mentioned previously in this section and elsewhere in the book, but it is fair to say that all labs will face the pressure [17, 18] to design sustainable spaces [19] that can keep up with all the emerging technical trends as well as the usability and design features needed to support a new generation of scientists. These drivers will combine to influence how the LotF evolves and experiments are performed. Research institutions are already creating more "open" labs areas to support interdisciplinary teamwork, collaborative working, and joint problem solving, rather than the previous "siloed" departmental culture. This will continue in the LotF. The growth of innovation clusters [20] and lab coworking spaces will require more consideration as to how shared automation and lab equipment can be effectively and securely used by groups, who may be working for different organizations and who will want to ensure their data and methods are stored and protected in the correct locations. Effective scheduling will be critical in the LotF to enable high productivity and to ensure that the high value of the automation assets is realized.

1.2.4 Data Management and the "Real Asset"

It is true of 2020, just as it was 50 years ago and will be in 50 years time, that the primary output of R&D, in whatever industry, is data. The only physical items of any value are perhaps some small amounts of a few samples (and sometimes not even that) plus, historically, a lot of paper! It is therefore not surprising that the meme "data is the new oil" [21] has come to such prominence in recent times. While it may be viewed by many as hackneyed, and by many more as fundamentally flawed [22], the idea carries a lot of credence as we move toward a more data-driven global economy. One of the main flaws arising from the oil analogy is the lack of organizations being able to suitably refine data into the appropriate next piece of the value chain, compared to oil, which has a very clear refining process and value chain. Furthermore, the "Keep it in the Ground" [23, 24] sustainability momentum makes the data-oil analogy perhaps even less useful. However, within the LotF, and in a more open, collaborative global R&D world, experimental data, both raw and refined, will grow in criticality. Without doubt, data will remain a primary asset arising from the LotF.

At this point then it is worth considering how data and data management fit into the processes that drive the two fundamental lab types, which we have referred to earlier, namely (i) the hypothesis-driven, more research/discovery-driven lab and (ii) the protocol-driven, more "manufacturing"-like lab.

1.2.4.1 Data in the Hypothesis-driven, Research Lab

In a pure research, hypothesis-driven lab, whether it is in life science, chemical science, or physical science, there is a fundamental, cyclical process operating. This process underpins all of scientific discovery; we refer to it as the "hypothesis-experiment-analyze-share" ("HEAS") cycle (see Figure 1.3) or, alternatively, if one is in a discovery chemistry lab, for example a medicinal chemistry lab in biopharma, DMTA (see Figure 1.2).

The research scientists generate their idea/hypothesis and design an experiment to test it. They gather the materials they need to run that experiment, which they then

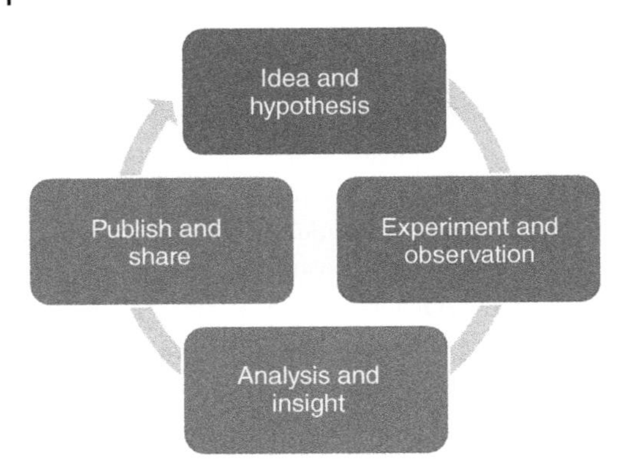

Figure 1.3 Hypothesis-experiment-analyze-share (HEAS) cycle.

perform in the lab. All the time they capture observations on what is happening. At the end they "workup" their experiment – continuing to capture observations and raw data. They analyze their "raw" data and generate results ("refined" data); these determine whether the experiment has supported their hypothesis or not. They then communicate those results, observations, and insights more widely. Ultimately, they move on to the next, follow-on hypothesis; then, it is off round the cycle they go again until they reach some sort of end point or final conclusion. All the while they are generating data: raw data off instruments and captured visual observations and refined data, which are more readily interpretable and can more easily lead to insights and conclusions.

1.2.4.2 Data in the Protocol-driven Lab

In the protocol-driven lab, whether it is in a manufacturing or sample testing domain, there is again a fundamental process which operates to drive the value chain. Unlike the "HEAS" cycle this is more of a linear process. It starts with a request and ends in a communicable result or a shippable product. This process, which we refer to as the "request-experiment-analyze-feedback" (REAF) process, is outlined in Figure 1.4.

There are many similarities, often close, between the linear REAF process and the HEAS cycle especially in the Experiment/Observe and Analyze/Report steps, but the REAF process does not start with an idea or hypothesis. REAF represents a service, which starts with a formal request, for example to run a protocol to manufacture a good or to test a sample, and ends with a product or a set of results, which can be fed back to the original customer or requester. As we noted in Section 1.2.4.1 above, it is increasingly likely that the LotF will be set up with a Laboratory as a Service (LaaS) mentality; REAF may therefore be much more broadly representative of how labs of the future might operate.

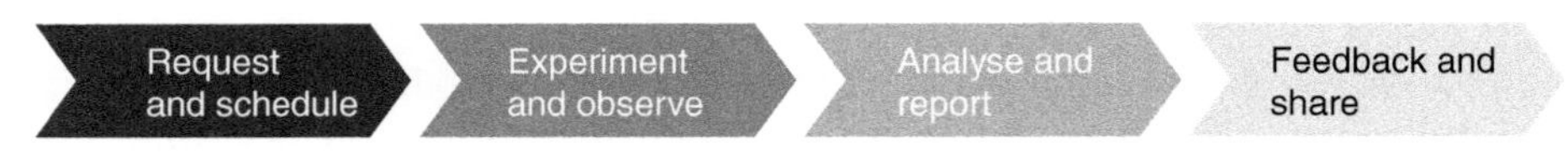

Figure 1.4 Request-experiment-analyze-feedback (REAF) process.

It is important to acknowledge that the data and information, which drive Request and Feedback, are quite different in REAF than in the corresponding sections in HEAS. With the focus of this book being on Experiment/Observe, and to a degree Analyze, we will not say anything more about Request and Feedback (from REAF) and Hypothesis and Share (from HEAS). Instead, the remainder of this section focuses on what the Experiment and Analyze data management aspects of the LotF will look like, whether that LotF is a hypothesis- or a protocol-driven lab. This is made simpler by the fact that in the Experiment/Observe and Analyze/Report steps, the data challenges in the two different lab types are, to all intents and purposes, the same. In the remainder of this section we treat them as such.

1.2.4.3 New Data Management Developments

So what new developments in data management will be prevalent in both the hypothesis- and the protocol-driven labs of 2030? In the previous two sections we asserted that these labs will be populated by fewer people; there will be more robotics and automation, and the experiment throughput will be much higher, often on more miniaturized equipment. Building on these assertions then, perhaps the most impactful developments in the data space will be:

a) The all pervasiveness of internet of things (IoT) [25, 26]. This will lead, in the LotF, to the growth of the internet of laboratory things (IoLT) environments; this will also be driven by ubiquitous 5G communications capability.
b) The widespread adoption of the findable, accessible, interoperable, and reusable (FAIR) data principles. These state that all data should be FAIR [27].
c) The growing use of improved experimental data and automation representation standards, e.g. SiLA [28] and Allotrope [29].
d) Data security and data privacy. These two areas will continue to be critical considerations for the LotF.
e) The ubiquity of "Cloud." The LotF will not be able to operate effectively without access to cloud computing.
f) Digital twin approaches. These will complement both the drive toward labs operating more as a service and the demand for remote service customers wanting to see into, and to directly control from afar what is happening in the lab. Technologies such as augmented reality (AR) will also help to enable this (see Sections 1.2.5 and 1.2.6).
g) Quantum computing [30–33]. This moves from research to production and so impacts just about everything we do in life, not just in the LotF. Arguably, quantum computing might have a bigger impact in the more computationally intensive parts of the hypothesis- and protocol-driven LotF, e.g. Idea/Hypothesis design and Analyze/Insight, but it will still disrupt the LotF massively. We say more on this in Sections 1.2.5 and 1.2.6.

The first three of these developments are all related to the drive to improve the speed and quality of the data/digital life cycle and the overall data supply chain. That digital life cycle aligns closely to the HEAS and REAF processes outlined in Figures 1.3 and 1.4 and can be summarized as follows (see Figure 1.5):

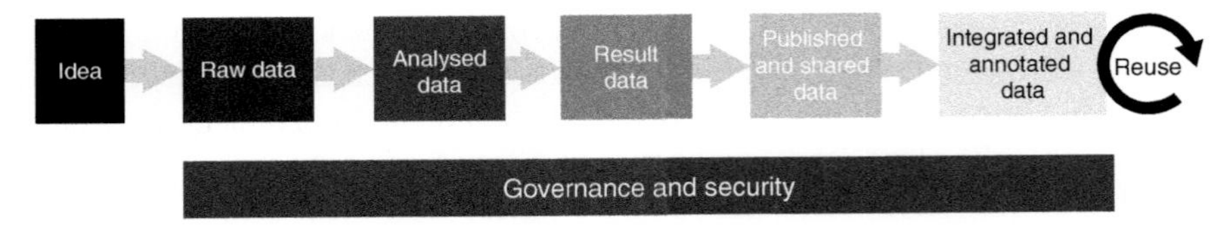

Figure 1.5 Digital data life cycle.

IoT technology [34] will allow much better connectivity between the equipment in the LotF. This will enable better, quicker, and more precise control of the lab kit, as well as more effective capturing of the raw data off the equipment. This in turn will allow the next stage in the life cycle – "Analyze Data" – to happen sooner and with more, better quality data. This improved interconnectedness in the lab will be made possible by the same 5G communication technology which will be making the devices and products in the home of 2025 more networked and more remotely controllable.

As improved instrument interconnectedness and IoLT enable more data to be captured by more instruments more effectively, the issue of how you manage the inevitable data flood to make the deluge useful comes to the fore. The biggest initiative in 2020 to maximize the benefits of the so-called big data [35] revolves around the FAIR principles. These state that "for those wishing to enhance the reusability of their data holdings," those data should be FAIR. In the LotF, the FAIR principles will need to be fully embedded in the lab culture and operating model. Implementing FAIR [36] is very much a change process rather than just introducing new technology. If fully implemented, though, FAIR will make it massively easier for the vast quantities of digital assets generated by organizations to be made much more useful. Data science as a discipline, and data scientists (a role which can be considered currently to equate to that of "informatician"), will grow enormously in importance and size/number. Organizations that are almost purely data driven will thrive, with any lab work they feel the need to do being outsourced via LaaS [37] to flexible, cost-effective LotFs that operate per the REAF process.

Supporting the growth of FAIR requires the data that is generated in these LaaS LotFs to be easily transferable back to the requester/customer in a format which the lab can generate easily, accurately, and reproducibly, and which the customer can import and interpret, again, easily, accurately, and reproducibly. This facile interchange of "interoperable" data will be enabled by the widespread adoption of data standards such as SiLA and Allotrope. We describe these new data standards in more detail in the following section.

Two additional, significant data considerations for the LotF are those of data security and data privacy, just as they are now. The more LotF services that are operated outside the "firewall" of an organization, and the more that future labs are driven by data, the more risks potentially arise from accidental or malicious activities. Making sure that those risks are kept low, through continued diligence and data security, will ensure that the LotF is able to develop and operate to its full capability. Similarly, in labs that work with human-derived samples (blood, tissues, etc.), the advent of regulations such as the General Data Protection Regulations (GDPR) [38, 39], along with the historical stringency surrounding informed consent [40] over what can happen to human samples and the data that arises from their processing, will put even more pressure on the organizations that generate and are accountable for human data to ensure these data are effectively secured. Improved adherence to the FAIR data principles, especially Findability and Accessibility, will ensure that LotFs working with human-derived materials can be responsive to data privacy requests and are not compromised.

Going hand in hand with the data explosion of the past decade has been the evolution of the now ubiquitous, key operational technology of "Cloud Computing." As explained by one of the originating organizations in this area, "cloud computing is the delivery of computing services – including servers, storage, databases, networking, software, analytics, and intelligence – over the Internet (the cloud) to offer faster innovation, flexible resources, and economies of scale." [41] In the context of LotF, assuming that the equipment in the lab is fully networked, cloud computing means that all the data generated by the lab can be quickly, fully, and securely captured and stored on remote infrastructure (servers). This book is not the place to describe cloud computing in detail, but it should be sufficient to say that the LotF will not be reliant on IT hardware close to its location (i.e. on-site) but will be highly reliant on speedy, reliable, available networks and efficient, cost-effective cloud computing.

Finally, there is a data and modeling technology, which has been present in industries outside life science for many years, which could play a growing role in the LotF which is more automated and more remote. This is the technology termed "digital twin." [42, 43] We say more on this exciting new technology in Section 1.2.5.1.

1.2.5 New Technology

In any future-looking article we can only make some best guesses as to the new technologies and science that could be important during the next 5–10 years. In this section we make some suggestions as to what new technologies we feel will impact the LotF, and what new science will be happening in those future labs. In the first part of this section, we focus on new technologies. In the second part, we suggest some scientific areas which we feel will grow in importance and hence might drive the evolution of the LotF and the technology that is adopted in that new lab environment.

New technologies will undoubtedly play a major role in driving the development of the critical components within the LotF, but their introduction and usage need to be appropriate to the type of lab being used. The role of the new technologies must be aligned to the future challenges and needs of the lab environment. These needs include, more specifically:

- Flexibility and agility of the experiment cycles, balancing between prediction (in silico) and physical (in vitro) experiments
- Improved data collection and experiment capture (e.g. "data born FAIR")

- Reproducibility of the experiment processes
- Enhancements to the scientists' UX and capabilities in the lab.

To emphasize these aspects, we focus on three broad areas in this section:

1. Lab automation integration and interoperability
2. Quantum computing and the LotF
3. Impact of AI and machine learning (ML).

1.2.5.1 Lab Automation Integration and Interoperability

Lab instrument integration and interoperability to support higher levels of lab automation have been and will continue to evolve quickly, driven by the pressure from scientists and lab managers and, above all to have better ways to manage and control their equipment [44–46]. Capabilities as diverse as chemical synthesis [47] and next-generation sequencing (NGS) [48] are seeking to better automate their workflows to improve speed and quality and to align with the growing demands of AI in support of generative and experimental design as well as decision-making [49]. An additional stimulus toward increased automation, integration, and interoperability is that of experiment reproducibility. The reproducibility crisis that exists in science today is desperately in need of resolution [50]. This is manifested not only in terms of being unable to confidently replicate externally published experiments, but also in not being able to reproduce internal experiments – those performed within individual organizations. Poor reproducibility and uncertainty over experimental data will also reduce confidence in the outputs from AI; the mantra "rubbish in, rubbish out" will thus continue to hold true! Having appropriate automation and effective data management can support this vital need for repeatability, for example of biological protocols [51]. This will be especially important to support and justify the lab as a service business model, which we have mentioned previously. It is our belief that the increased reliability and enhanced data-gathering capability offered by increased automation initiatives in the LotF will be one important way to help to address the challenge of reproducibility.

Updated automation will always be coming available as an upgrade/replacement for the existing equipment and workflows; or to enhance and augment current automation; or to scale up more manual or emerging science workflows. When considering new automation, the choices for lab managers and scientists will depend on whether it is a completely new lab environment (a "green-field site") or an existing one (a "brown-field site").

As mentioned previously, the growth of integration protocols such as IoT [52] is expanding the options for equipment and automation to be connected [53]. The vision for how different workflows can be integrated – from single measurements (e.g. balance measurements), via medium-throughput workflows (e.g. plate-based screening), to high data volume processes such as high content screening (HCS) involving images and video – has the potential to be totally reimagined. IoT could enable the interconnectivity of a huge range of lab objects and devices, such as freezers, temperature control units, and fume hoods, which previously would have been more standalone, with minimal physical connectivity. All these devices could be actively connected into expanded data streams and workflows where the measurements they take, for example, temperature, humidity, and air pressure, now become a more integral part of the experiment record. This enhanced set of data collected during experiments in the LotF will be hugely valuable during later analysis to help spot more subtle trends and potential anomalies. Furthermore, these rich datasets could play an increasing role as AI is used more and more for data analysis; small fluctuations in the lab environment do have a significant impact on experimental results and hence reproducibility. As well as this passive sensor monitoring, there is also the potential for these devices to be actively controlled remotely; this opens up options for further automation and interaction between static devices and lab robots, which have been programmed to perform tasks involving these devices. As always, it will be necessary to select appropriate automation based on the lab's needs, the benefits the new automation and workflows can provide, and hence the overall return on investment (ROI).

While the potential for these new systems with regard to improved process efficiency is clear, yet again, though, there is one vital aspect which needs to be considered carefully as part of the whole investment: the data. These LotF automation systems will be capable of generating vast volumes of data. It is critical to have a clear plan of how that data will be annotated and where it will be stored (to make it findable and accessible), in such a way to make it appropriate for use (interoperable), and aligned to the data life cycle that your research requires (reusable). A further vital consideration will also be whether there are any regulatory compliance or validation requirements.

As stated previously, a key consideration with IoT will be the security of the individual items of equipment and the overall interconnected automation [54, 55]. With such a likely explosion in the number of networked devices [56], each one could be vulnerable. Consequently, lab management will need to work closely with colleagues in IT Network and Security to mitigate any security risks. When bringing in new equipment it will be evermore important to validate the credentials of the new equipment and ensure it complies with relevant internal and external security protocols.

While the role of lab scientist and manager will clearly be majorly impacted by these new systems, also significantly affected will be the physical lab itself. Having selected which areas should have more, or more enhanced and integrated, lab automation, it is highly likely that significant physical changes to the lab itself will have to be made, either to accommodate the new systems themselves or to support enhanced networking needs.

In parallel to the lab environment undergoing significant change over the upcoming decades, there will also be new generations of scientists entering the workforce. Their expectations of what makes the LotF efficient and rewarding will be different from previous generations. The UX [57] for these new scientists should be a key consideration when implementing some of the changes mentioned in this book. For example, apps on mobile phones or tablets have transformed peoples' personal lives, but there has been slower development and adoption of apps for the lab. The enhanced usage of automation will very likely need to be managed through apps; they will therefore become a standard part of the LotF. One cultural caveat around the growth of lab apps should be flagged here. With apps enabling much more sophisticated control of automation operating 24/7, via mobile phones, outside "human working hours," there will need to be consideration of the new scientists' work/life balance. If handled sensitively, though, developments such as lab apps could offer much-increased efficiency and safety, as well as reducing experiment and equipment issues.

Voice-activated lab workflows are also an emerging area, just as voice assistants have become popular in the home and in office digital workflows [58]. For the laboratory environment, the current challenges being addressed are how to enrich the vocabulary of the devices with the specific language of the lab, not only basic lab terms but also domain-specific language, whether that is biology, chemistry, physics, or other scientific disciplines. As with IoT, specific pilots could not only help with the assessment of the voice-controlled device or system but also highlight possible integration issues with the rest of the workflow. A lab workflow where the scientist has to use both hands, like a pianist, is a possible use case where voice activation and recording could have benefits. The ability to receive alerts or updates while working on unfamiliar equipment would also help to support better, safer experimentation.

As with voice control, the use of AR and virtual reality (VR) in the lab has shown itself to have value in early pilots and in some production systems [59]. AR is typically deployed via smart glasses, of which there is a wide range now in production. There are a number of use cases already where AR in the lab shows promise, including the ability to support a scientist in learning a new instrument or to guide them through an unfamiliar experiment. These examples will only grow in the LotF. To take another, rather mundane example, pipetting is one of the most familiar activities in the lab. In the LotF where low throughput manual pipetting is still performed, AR overlays could support the process and reduce errors. AR devices will likely supplement and enhance what a scientist can already do and allow them to focus even more productively.

Another area of lab UX being driven by equivalents in consumer devices is how the scientist actually interacts physically with devices other than through simple keyboard and buttons. Technologies such as gesture control and multitouch interfaces will very likely play an increasing role controlling the LotF automation. As with voice activation, these input and control devices will likely evolve to support the whole lab and not just a single instrument. Nevertheless, items such as projected keyboards could have big benefits, making the lab even more digitally and technologically mature.

As mentioned before there is another technology which could play a significant role in enhancing the UX in the LotF; this is the "digital twin." [60] In brief, a digital twin is a representation in silico of a person, a process, or a thing. Its role has been evolving in recent years, such that digital twins can now be seen as virtual replicas of physical environments or objects which managers, data scientists, and business users can use to run simulations, prepare decisions, and manage operations [42, 61].

This technology has the potential to impact the LotF in two primary areas: (i) simulation and (ii) remote control.

Starting with simulation, digital twins, unlike the physical world, which shows you a picture of the present, can review the past and simulate the future. The digital twin can therefore become an environment to test out in pilot mode not only emerging technologies such as voice activation, AR, VR, and multigesture devices but also novel or redesigned workflows without the need for full-scale deployment. Indeed, with increased computational capability (provided by exascale computing and ultimately quantum computing – see Section 1.2.5.2), the processes that operate within the LotF will be simulatable to such a degree of sophistication that a person will be able to see, in silico, a high-resolution representation of the technology, experiment, or process they are looking to perform, in a simulation of the lab in which it will run. This digital twin will allow the operator to check, for example that the novel process is likely to run smoothly and deliver the output that is hoped for. While digital twin technology may be more applicable to the protocol-driven lab, it may also have applicability in the research lab as a means of exploring "what-if" scenarios prior to doing the actual physical experiment.

Turning to digital twin technology and improved remote control, massively improved computational technology combined with advances in AR and VR will allow operators, who might be located nowhere near the lab in which their experiment is being run, to don appropriate AR/VR headsets and walk into an empty space that will "feel" to them like they are right inside the lab or even right inside the experiment itself. The potential for scientists to "walk" into the active site of an enzyme and "manually" dock the molecules they have designed, or for an automation operator to "step into" the reaction vessel running the large-scale manufacturing of, say, a chemical intermediate to check that there are no clumps, or localized issues (e.g. overheating), will revolutionize how the LotF can operate, making it more likely to be more successful and, importantly, safer.

One final, obvious application of digital twin technology is where that LotF is not even on Earth. Running experiments in low or zero gravity can lead to interesting, sometimes unexpected findings [62]. This has led to numerous experiments having been performed on the NASA Space Station [63]. But having a trained astronaut who can effectively run any experiment or protocol, from organic synthesis to genetic manipulation, is asking a great deal. Digital twin technology could make the LotF in zero gravity a much more compelling proposition [64].

Returning to the area of instrument integration and interoperability, a more practical consideration is how different instruments communicate with each other, and how the data they generate is shared.

Within any lab there is and always will be a wide range of different instruments from different manufactures, likely bought over several years to support the business workflows. This "kit diversity" creates a challenge when you want to define a protocol which involves linking two or more instruments together that do not use the same control language. SiLA-2 [65] is a communication standard [66] for lab instruments, such as plate readers, liquid handling devices, and other analytical equipment, to enable interoperability. As indicated throughout this section, the ability to fully connect devices together will enable a more flexible and agile lab environment, making it possible to track, monitor, and remote control automation assets. This will further enable enhanced robotic process automation (RPA) as well as easier transition to scale up and transfer to remote parties. Specific devices connected together for one workflow will be easily repurposable for other tasks without a monolithic communication design and coding.

Data in all its forms will remain the dominant high-value output from lab experiments. As with protocols and communications, there need to be standards to support full data integration and interoperability within and between research communities. Over the years, data standards have evolved to support many aspects of the life science process whether that is for registration of new chemical entities [67], images [68], or macromolecular structures [69] or for describing the experiment data itself. Analytical instrument data (e.g. from nuclear magnetic resonance machines [NMRs], chromatographs, and mass spectrometers) are produced by a myriad of instruments, and the need to analyze and compare data from different machines and support data life cycle access in a retrievable format has driven the creation of the Allotrope data format [70] (ADF). This is a vendor-neutral format, generated initially for liquid chromatography, with plans to expand to other analytical data. These wide community-driven efforts such as those from Allotrope, SLAS, IMI [71], or the Pistoia Alliance [72] highlight the value of research communities coming together in life sciences, as happens elsewhere in industries such as financials and telecoms. Such enhanced efforts of collaboration will be needed even more in future.

In summary, the use of open standards will be critical for the success of the LotF, as the range of devices grows and science drives changes. There will need to be reliable, robust ways for the instruments, workflows, and data to be shared and accessible in order to support flexible, open-access, and cross-disciplinary collaborations, innovation, and knowledge exchange. The automation in the LotF will need to be effective across many different sizes and types of labs – from large, high-throughput labs doing screening or sequencing, to midsize labs with some automation workbenches, to the long tail of labs with a few specialist instruments. In planning for a new lab, creating a holistic vision of the design will be a key first element. That vision will include the future processes that your lab will want to tackle, as well as the potential new technologies to be deployed in the lab, e.g. IoT, AR, or voice control. Additionally, new skills will need to be acquired by those involved to help implement these changes, and an investment in staff and their training remains vital. Furthermore, in future there will likely be an ecosystem of lab environments both local and more disparate to consider; the LotF will be smarter and more efficient but not just through the adoption of a single device.

1.2.5.2 Quantum Computing and the Lab of the Future

The field of quantum computing is moving so fast that any review or update is soon superseded by the next breakthrough [73]. Consequently, this section focuses on introducing some of the concepts of quantum computing and how it might impact the LotF [74] and its workflows going forward especially those in the design (model/predict) and analyze stages.

Quantum computers differ [75] from our current classical computers in that they offer a fundamentally different way of performing calculations; they use the theories of quantum mechanics and entanglement [76, 77] and qubits (multiple simultaneous states) rather than bits (binary states – 0 and 1). This gives them the potential to solve problems and process tasks that would take even the fastest classical computer hundreds if not thousands of years to perform. Such performance will enable the simulation of highly complex systems, such as the behavior of biological molecules and other life science challenges [78]. A key concept in this area is that of the "quantum supremacy" [79] threshold, where quantum computers crossover from being an interesting and pure research-driven project to doing things that no classical computer can. Quantum supremacy is defined as a situation where a quantum computer would have to perform any calculation that, for all practical purposes, a classical computer cannot do because of the time involved. There is much discussion about whether we have reached "quantum supremacy," but it does seem clear that, for the foreseeable future, quantum computers will not be used for everyday activities but for highly specific, truly value-adding and accelerated tasks, much in the same way that exascale supercomputers work today. One further critical step will be needed to ensure that quantum computers are able to operate outside the research domain, that is to create quantum compilers [80] which can make code run efficiently on the new hardware, just as traditional compilers were needed with classical computers.

There are good parallels between quantum computers and the exascale, supercomputers, and clusters of today when thinking about the impact on life science and broader research. There are limited numbers of supercomputers at the moment due to their cost and the skills needed to fully utilize them. Consequently, researchers have to bid for time slots at the regional and national centers which house them. It is likely that the same process will happen with quantum computers, with regional/national [81–83] centers being established to support scientists who use them remotely to process their calculations and model building. Quantum cloud computing [84] and associated services will likely evolve in the same way that existing cloud compute and storage infrastructure and business models have evolved over the past decade.

We now focus on how quantum computing will more directly impact the LotF and the experiments which will run within it. The researchers involved will, at least in the early years, have to balance their "classical" and "quantum" calculation time with their physical experiment effort to help drive their insights and decision-making. Experiments will still be performed on complex systems, but they will be influenced even more by the work done in silico. There will likely be more rapid experiment cycles since the ability to perform quicker calculations will enhance the speed of progress and encourage research in areas that have until now been hard to explore, for example molecular simulations of larger biological entities.

Data will remain a key element of the workflow. Being able to send data to quantum computers and to retrieve the outputs from them quickly will help to influence the next experiment. However, with the rarity of quantum computing, careful planning of the holistic workflow of the data and compute will be needed to ensure best efficiency of location and data pipelines. There will be an even greater need to maintain the large volumes of data necessary for the calculations "close" to the compute power to avoid large delays in moving data around. The timelines for the impact of quantum computing remain uncertain as the jump from research to production ready is hard to judge. Not every LotF will have a quantum compute facility nearby, but in time they will be able to access this accelerator technology more readily to support their key processes. Unarguably, quantum computing will be a very exciting area for life science and will create a whole new era of scientific exploration.

1.2.5.3 Impact of AI and ML

Even with all the hype around it, AI and its subtypes, e.g. ML, will undoubtedly reshape the R&D sector and have a huge impact on the LotF [85, 86]. Many see AI as the competitive edge which will accelerate products to market, or improve patient outcomes and care, and drive cost efficiencies. As a consequence, there now exists a major talent war as organizations seek to attract the best candidates.

With the rise of AI within life sciences and health care, it has become obvious that a key blocker to success is not the maturity of the AI tools and techniques but access to data in sufficient volume and quality for the AI and ML methods to operate meaningfully. The phrase "no data, no AI/ML" is a signature of the current challenge, with much of the accessible data having been created without due care and attention to reproducibility and the FAIR principles, which are only now driving business improvement in data collection and annotation. Depending on the AI/ML model being developed, having access to a broad cohort of data from across the particular domain will be critical to ensure the necessary diversity, edge cases, and breadth. It is this which will make the analyses successful and be broadly applicable.

The LotF will be both the source of new data to drive new insights from the AI predictive workflows and a beneficiary of the AI outputs which can augment the scientists' work. As mentioned earlier, voice activation, AR, and other assisted technologies all use elements of AI to support the user, whether as chatbots or more sophisticated experiment assistants for the scientist. For example, an automated AI assistant needs to be trained on data to enhance its capability. In time it learns from the human interactions, and this helps to improve its responses and output. Even without AI though, the drive for higher quality and more abundant data remains critical.

The role of AI in generating new ideas fits perhaps most cleanly in the in silico predictive, "design" workflows that have always existed in science. However, AI has the potential to produce new ideas not previously explored or considered by humans. For the lab scientist, the role of AI will be multivariate, from supporting the initial idea to be explored, through experimental design and execution, through to finally how the data is captured and the results analyzed. AI will augment the scientists' capability during their time in the lab as well as provide new insights to guide the best outcome from the experiment. For some, the likely benefit will be decreased experiment cycle times, allowing a better outcome in a fixed time period, for others it will be quicker decision-making through linking the virtual with the physical. Complementary to this, using AI and robotics appropriately in the LotF will allow scientists to focus on the practical things that these technologies do not support well. This will make the fewer humans in the LotF more efficient and productive. Other "softer," less technical factors will also become increasingly important going forward; these include the broader ethics of AI and the possible regulatory implications of using AI for R&D decisions. The clamor to solve these issues will become louder as the potential and positive impact of AI on life science and health care is demonstrated, validated through application of the AI models to real decision-making.

1.2.6 New Science

Any attempt to predict what will be the big new, "hottest" areas of science is fraught with risk. When one overlays, for the purpose of this book, how those "hot new science areas" might impact the experiments and activities going on in the lab of the future, combined with how that LotF might look physically, then the chances of this section being at best, a bit "off" and at worst plain wrong increase rather dramatically! Nevertheless, even with this "caveat emptor," we feel that in this forward-looking section, it is important to call out a few of the new scientific areas [87] which we personally feel are worth watching for how they might impact the LotF. In keeping with the broad scope of this book we have concentrated more on the likely scientific developments in (i) the health care and (ii) the life sciences domains, but we have also picked out a small number of examples in (iii) other scientific areas.

1.2.6.1 New Science in Health Care

The biggest drive recently in health care, for both diagnosis and treatment, has been the move away from more population-based approaches toward a much more personalized focus. This has been made possible by the huge advancements in gene and genome-based technologies. Advances in gene and whole-genome sequencing will continue to assist better diagnosis, with sequencing times and costs reducing dramatically, and accuracy and quality rising significantly. These advances will make the protocol-driven labs more prevalent, more efficient, and more cost-effective. The development of better treatments based on gene expression manipulation and gene editing (e.g. CRISPR) [4]) as well as pure gene therapy [88] will continue apace. Diseases that will benefit from such developments will include many inheritable conditions such as Huntington's chorea and cystic fibrosis, as well as many cancers.

On the whole cell front, improvements in chimeric antigen receptor T-cell (CAR-T) [89] treatments to "supercharge" a patient's own immune system will mushroom. Individual CAR-T therapies to fight cancers more widely, not just leukemia and lymphoma but also more difficult-to-treat infections (e.g. tuberculosis [90] and some viruses), will become more widespread and cost-effective [91].

Other, more traditionally based approaches to the treatment of diseases, such as vaccines to combat certain viruses (e.g. influenza and novel corona viruses) and some cancers (e.g. cervical cancer), as well as stem cell therapy [92] will continue to thrive. New, more effective vaccine approaches to a greater range of cancers and novel viruses will be developed more quickly than before. Better understanding of some long-standing diseases, for example in the cardiovascular arena, will demonstrate infectious components [93], and these too will become susceptible to vaccine approaches.

Building on the infectious agent theme, research into novel approaches to treat bacterial and viral infections will continue, although probably mostly in academic and charitable trust-funded labs. Approaches such as phage-based treatment for bacterial infections will become more of a focus as traditional small molecule-based strategies are met with evermore intractable and resistant bacteria [94]. Supercharged immunological approaches to bacterial infection will also be a focus for research.

The growth of such novel therapeutics as CAR-T, alongside other whole cell-based approaches and non-small molecule agents, will complement the ever-expanding set of large-molecule therapies. The use of these so-called biopharmaceuticals or biologics has become more widespread in the past decade and will continue to grow. Similarly, the research, development, and manufacture of antibodies [95], modified RNA [96], peptides [97], conjugates, proteolysis targeting chimera agents (PROTACS) [98], antisense oligonucleotides, and other therapeutic macromolecules will continue to expand rapidly. While in vitro activity of such agents can often be demonstrated relatively clearly, they present a major challenge when it comes to in vivo efficacy. The development of novel formulation and drug delivery systems to enable effective administration of these twenty-first century therapeutics will become a major area of scientific investigation.

Finally, an area of growing interest, which could be considered the antithesis of antimicrobial research, is that of the microbiome [99]. There is increasing recognition that the commensal bacteria and other microbes which live symbiotically in and on our bodies (mainly mucous membranes in, for example the gut and also on the skin) can play a major role in our acquisition, presentation, and the severity of certain diseases (e.g. irritable bowel syndrome, Crohn's disease, and psoriasis). Research into an individual's microbiome and treatments based on "normalization" of a person's inherent flora will grow and become more mainstream over the next few decades [100].

1.2.6.2 New Science in the Life Sciences Domain

As discussed in several of the earlier parts of this section, there is one critical, global driver which will dominate new science and how it is performed in the LotF; that key driver is climate change and the supporting concept of "sustainability". There will be new research looking specifically into climate change and sustainability as areas of interest in themselves, but the need for the LotF, both the hypothesis- and the protocol-driven lab, to be more sustainable, less dependent on oil and oil-based products, and yet be more efficient, will become paramount in the decades to come. Labs that do "chemistry" will be a primary focus for these developments, but biology-focused labs will not be immune. The pressure to be more environmentally friendly, using fewer reagents and disposable materials, will lead to new research to discover, for example effective replacements for all the lab plastics currently used; greener chemistry (use of less-toxic solvents and reagents); greater use of catalysts; and more use of biological systems to perform complex chemical transformations. None of these examples are exactly new, but their importance and greater use in the LotF will be significant.

Just as the next generation of scientists is exquisitely conscious of the environment, so too is it particularly focused on animal welfare. The ever-growing drive toward minimization of the use of animals in research and product testing, while it can never in truth be completely eliminated, will continue to accelerate. Initiatives such as the "3Rs" [101] looking to replace, reduce, and refine the use of animals in the lab will gain more traction [102]. In vitro approaches to meet the goals of the 3Rs will include developments such as organ-on-a-chip [103] and the increasing use of stem cells. These new methods will become widespread in the LotF.

Finally, there is one lab technique, which has been a mainstay of the lab for hundreds of years, yet is still undergoing significant evolution and is likely to feature significantly in the LotF: microscopy. Advances in traditional imaging revolutionized life sciences over a decade ago, but current developments in microscopy are likely to transform utterly how in the future we perceive "things" both at the molecular and macromolecular levels. There are two specific examples, which we feel are worth mentioning here: firstly, the scanning tunneling microscope (STM) [104] and secondly, the cryo-electron microscope (cryo-EM) [105]. STM and other comparable new microscopy techniques [106] have the potential to take to an even higher level our ability to study cells, solid-state materials, and many other surfaces. STM has clear potential applications in biology, chemistry, surface science, and solid-state physics [107]. The STM, which operates through the principles of quantum tunneling, utilizes the wavelike properties of electrons, allowing them to "tunnel" beyond the surface of a solid into regions of space that are normally forbidden under the rules of classical physics. While the use of STM has been focused mainly on physicochemical and solid-state challenges, increasingly scientists are looking at STM as a means to see more deeply into chemical and biochemical systems, right down to the atomic level [108]. Cryo-EM is the electron microscopic imaging of rapidly frozen molecules and crystals in solution. It demonstrates its main benefits at the macromolecular level, enabling scientists to see the fine structures of proteins, nucleic acids, and other biomolecules, and even to study how they move and change as they perform their functions, but without having to use the intense electron beams and high vacuum conditions used in traditional electron microscopy [109].

1.2.6.3 Other Important New Science Areas

We have asserted throughout this section that the driver of climate change and the push to greater sustainability will dramatically affect how the LotF will look, and what will take place within it. In a final piece of speculation on what new science will be taking place in these future labs, we suggest two research areas, which we believe will occupy a great deal of time and effort in labs over the next 10–20 years. These two areas are carbon (actually carbon dioxide) fixing and sequestration [110], and R&D around new battery technology, particularly new technologies which avoid the use of heavy metals [111]. The scientists working in or near the LotFs will do a great disservice to their descendants and to the planet if they do not research these critical areas.

1.3 Thoughts on LotF Implementation

The lab environment is changing – this is certain. New and existing science demands combined with critical issues of data management and reproducibility will require a strategic direction to be set and then deployed. It will be important for lab managers to identify what they want to achieve by employing the new approaches of AI, quantum computing, and advanced automation technology. Business ambition and needs, and the assessment of the maturity of organizations beyond the lab environment in the context of initiatives such as FAIR data, will need to be investigated as a matter of urgency to help drive lab of the future decision-making. With such a pace of change it will be important to "think big" as well as be practical during implementation. Thinking in expansive terms, organizations must consider all the opportunities on offer within the key areas of technology, data, people, and process to highlight possible future visions and ways of working. They should use scenario planning to explore, influence, plan for, and manage the future. These scenarios will perhaps be most effective when they are personalized to the organization, function, lab, or team's future, rather than to a generic vision. The benefit of running pilots prior to fuller implementation in the LotF cannot be overstated. Small LotF pilots will allow experimentation across the broad themes. These will reveal what works and what needs adjustment based on the key lab environments. The successful use cases can result in new designs, collaborations, future partnering with technology groups, and new predictive models to support experiments in a timely manner. Moving beyond these smaller pilots and the learnings from them will help catalyze organizational change to support a lab environment that can adapt to new science and get the most from data, digital technology, and AI-driven transformations. All these changes will present new business opportunities, the chance for new relationships with vendors, and the need for new business partners. They will also present opportunities for all lab colleagues to take part in the transformation and to take on new roles and skills to support the implementation and future impact.

1.4 Conclusion

In this section we have endeavored to show how the LotF will potentially evolve, using five main areas as a focus for those possible changes and developments: (i) the people and organizational culture aspects; (ii) the process components; (iii) the LotF environment and design; (iv) the data management challenges; and (v) the new technologies and the new science which will take place in those LotFs. In this concluding section, we would like to pick out just a few key messages from each of these areas to highlight the promise and the challenge that "LotF" presents.

From a people and culture perspective, the point we would stress most is the importance of considering the future lab from the perspectives of the different roles working in, around, or in association with the LotF: the scientists – not just the practical hands-on chemists, biologists, biochemists, physicists, etc. but also the new breed of data scientists and engineers – the lab managers and building managers, the technicians and equipment operators, and all the other staff who will make the LotF an exciting, stimulating, and challenging place to work. The LotF will be more "open," collaborative, and more automated, if rather more sparse of people. Critical to the LotF's success will be the "UX" of all the people associated with it.

The processes in the LotF will be dominated by flexible automation and robotics, whether that lab is a hypothesis-driven, research lab, or a manufacturing, testing lab operating more in "LaaS" mode. More effective in silico modeling of the lab processes will make the LotF a safer, more productive place to work.

The lab environment, as well as being designed around large amounts of automation and robotics, flexibly configured, and interconnected, will more often than not be remote from the "customers" of the work actually being done. Good data and network interconnectedness of the LotF will be absolutely critical if it is to operate effectively and securely. The LotF will also be a markedly more sustainable and greener environment.

The data generated by the LotF, whether it is the "raw" data coming off the instruments or the "refined" result data derived from the raw data, will continue to be key to the LotF; if anything, the criticality and value of the digital assets generated by LotFs will become even more important in the future. Data-focused technologies and standards such as IoT, FAIR, SiLA, and Allotrope will ensure that the high-value digital assets are well managed and secured. The increasing focus on data privacy, security, and protection will put heavy pressure on LotFs with regard to good governance and compliance.

Finally, when considering new technologies such as AI/ML and quantum computing, and new science such as CRISPR and CAR-T, we feel we cannot overstress that science, technology, research, and development never cease to evolve. New discoveries are being made constantly, and these will without doubt have an impact on the LotF in ways we cannot predict now, in 2021. We can state quite confidently that there will be some technologies or scientific discoveries we have not mentioned here, which will affect significantly what happens in the labs of the future. We have highlighted those we feel now are important to help guide and stimulate you, the reader as you try to understand where and how the LotF is likely to develop. There will be others. In fact during 2020 a number of the themes and directions we have highlighted in this chapter have come to pass as the world has grappled with the momentous events surrounding the SARS-CoV-2 (COVID-19) pandemic. The pace of scientific and medical response to understanding the virus and its treatment has been unparalleled. Global, open and collaborative sharing of data and information on the virus itself, on the epidemiology of the disease, on its acute treatment through, for example, accelerated drug repurposing and the development of an effective vaccines, has allowed enormous progress to be made towards helping the control of the virus [112]. New technology has also enabled safer lab working to cope with COVID-19 restrictions (e.g. equipment booking, lab capacity planning, remote access to instruments supporting home working, and tracking of contacts). Technology and automation have supported the faster establishment of new test facilities, but there have been differences in approach between larger and smaller local labs, and between different countries, e.g. between Germany and UK (in the UK these are known as "Lighthouse Labs" [113]). And new science has played a huge role in the development of the many potential vaccines currently being progressed and trialled in labs and clinics across the world [114]. Nevertheless, despite this explosion of cross-border, cross-research group and company collaboration, there have still been challenges around speed of data sharing, data accuracy and trust in the information being disseminated widely, particularly as that sharing has often happened without or before robust peer review [115]. This tells us that there is still a long road ahead on the LotF journey. Most assuredly though, the lab of the future in say, 2030, will be very different from the lab of 2020; but it will be a fascinating, exhilarating and safer place, not only to work, but also to have your work done, and to do new science.

References

1 Deloitte Tackling digital transformation. (2019). https://www2.deloitte.com/us/en/insights/industry/life-sciences/biopharma-company-of-the-future.html (accessed 1 February 2020).

2 Shandler, M. (2018). Life science's lab informatics digital criteria to separate vendor leaders from laggards. Gartner G00336151. https://www.gartner.com/en/documents/3895920/life-science-s-lab-informatics-digital-criteria-to-separ (accessed 1 February 2020).

3 Open Science Massively Open Online Community (MOOC) https://opensciencemooc.eu/ (accessed 1 February 2020).

4 Vidyasagar, A. (2018). What is CRISPR? https://www.livescience.com/58790-crispr-explained.html (accessed 1 February 2020).

5 Open Science https://openscience.com/ (accessed 1 February 2020).

6 Tapscott, D. and Tapscott, A. (2016). *Blockchain Revolution*. New York. ISBN: 978-0-241-23785-4: Penguin Random House.

7 Shute, R.E. (2017). Blockchain technology in drug discovery: use cases in R&D. *Drug Discovery World* 18 (October Issue): 52–57. https://www.ddw-online.com/informatics/p320746-blockchain-technology-in-drug-discovery:-use-cases-in-r&d.html.

8 Gawas, A.U. (2015). An overview on evolution of mobile wireless communication networks: 1G-6G. *International Journal on Recent and Innovation Trends in Computing and Communication* 3: 3130–3133. http://www.ijritcc.org.

9 Chovan, T. and Guttman, A. (2002). Microfabricated devices in biotechnology and biochemical processing. *Trends in Biotechnology* 20 (3): 116–122. https://doi.org/10.1016/s0167-7799(02)01905-4.

10 Zimmerman, J.B., Anastas, P.T., Erythropel, H.C., and Leitner, W. (2020). Designing for a green chemistry future. *Science* 367 (6476): 397–400. https://doi.org/10.1126/science.aay3060.

11 (i) Notman, N. (2018). Seeing drugs in 3D. *Chemistry World* (April Issue) https://www.chemistryworld.com/features (accessed 1 February 2020). (ii) Chapman, K. (2020). 3D printing the future. *Chemistry World* (February Issue). https://www.chemistryworld.com/features/3d-printing-in-pharma/3008804.article (accessed 1 February 2020).

12 23andMe https://www.23andme.com/ (accessed 1 February 2020).

13 Ancestry https://www.ancestry.com/ (accessed 1 February 2020).

14 Plowright, A., Johnstone, C., Kihlberg, J. et al. (2011). Hypothesis driven drug design: improving quality and effectiveness of the design-make-test-analyse cycle. *Drug Discovery Today* 17: 56–62. https://doi.org/10.1016/j.drudis.2011.09.012.

15 Tawfik, M., Salzmann, C., Gillet, D., et al. (2014). Laboratory as a service (LaaS): a model for developing and implementing remote laboratories as modular components. *11th International Conference on Remote Engineering and Virtual Instrumentation*. IEEE. https://doi.org/10.1109/REV.2014.6784238.

16 Data Center Container (2019). https://www.techopedia.com/definition/2104/data-center-container (accessed 1 February 2020).

17 Francis Crick Institute https://www.crick.ac.uk/about-us/our-vision (accessed 1 February 2020).

18 Hok Architects Francis crick lab design. https://www.hok.com/projects/view/the-francis-crick-institute/ (accessed 1 February 2020).

19 Crow, J.M. (2020). Sustainable lab buildings. *Chemistry World* 17 (3): 24–29.

20 Steele, J. (2019). https://www.forbes.com/sites/jeffsteele/2019/08/12/the-future-of-life-science-and-tech-innovation-is-in-clusters/ (accessed 1 February 2020).

21 Economist The world's most valuable resource is no longer oil but data (2017). https://www.economist.com/leaders/2017/05/06/the-worlds-most-valuable-resource-is-no-longer-oil-but-data (accessed 1 February 2020).

22 (i) Marr, B. Here's why data is not the new oil (2018). https://www.forbes.com/sites/bernardmarr/2018/03/05/heres-why-data-is-not-the-new-oil. (ii) van Zeeland, J. Data is not the new oil. https://towardsdatascience.com/data-is-not-the-new-oil-721f5109851b (accessed 1 February 2020).

23 The Guardian Keep it in the ground (2019). https://www.theguardian.com/environment/series/keep-it-in-the-ground (accessed 1 February 2020).

24 Extinction Rebellion https://rebellion.earth/ (accessed 1 February 2020).

25 Weiser, M. (1991). The computer for the 21st century. *Scientific American* 265 (3): 94–104.

26 Farooq, M.U. (2015). A review on internet of things (IoT). *International Journal of Computer Applications* 113 (1): 1–7. https://doi.org/10.5120/19787-1571.

27 Wilkinson, M.D., Dumontier, M., Aalbersberg, I. et al. (2016). The FAIR guiding principles for scientific data management and stewardship. *Scientific Data* 3: 160018. https://doi.org/10.1038/sdata.2016.18.

28 SiLA Consortium https://sila-standard.com/ (accessed 1 February 2020).

29 Oberkampf H, Krieg H, Senger C, et al. (2018). Allotrope data format – semantic data management in life sciences. https://swat4hcls.figshare.com/articles/20_Allotrope_Data_Format_Semantic_Data_Management_in_Life_Sciences_pdf/7346489/files/13574621.pdf (accessed 1 February 2020).

30 Feynman, R.P. (1999). Simulating physics with computers. *International Journal of Theoretical Physics* 21 (6/7): 467–488.

31 Katwala, A. (2020). Quantum computers will change the world (if they work). https://www.wired.co.uk/article/quantum-computing-explained (accessed 1 February 2020).

32 Gershon, T. (2019). Quantum computing expert explains one concept in 5 levels of difficulty | WIRED. https://www.youtube.com/watch?v=OWJCfOvochA (accessed 1 February 2020).

33 Mohseni, M., Read, P., Neven, H. et al. (2017). Commercialize quantum technologies in five years. *Nature* 543: 171–174. https://doi.org/10.1038/543171a.

34 Perkel, J. (2017). The internet of things comes to the lab. *Nature* 542: 125–126. https://doi.org/10.1038/542125a.

35 Wikipedia Big data. https://en.wikipedia.org/wiki/Big_data (accessed 1 February 2020).

36 Jacobsen, A., Azevedo, R., Juty, N. et al. (2020). FAIR principles: interpretations and implementation considerations. *Data Intelligence* 2: 10–29. https://doi.org/10.1162/dint_r_000.

37 (2018). Laboratory automation – robots for life scientists. https://www.nanalyze.com/2018/04/laboratory-automation-robots-life-scientists/ (accessed 1 February 2020).

38 General data protection regulation. https://gdpr-info.eu/ (accessed 1 February 2020).

39 Regulation (EU) 2016/679 of the European Parliament and of the Council. https://eur-lex.europa.eu/eli/reg/2016/679/oj (accessed 1 February 2020).

40 Informed consent. https://www.emedicinehealth.com/informed_consent/article_em.htm (accessed 1 February 2020).

41 What is cloud computing? https://azure.microsoft.com/en-us/overview/what-is-cloud-computing/ (accessed 1 February 2020).

42 Hartmann, D. and van der Auweraer, H. (2020). Digital twins. *Arxiv*. [Preprint] https://arxiv.org/pdf/2001.09747 (accessed 1 February 2020).

43 Rasheed, A., San, O., and Kvamsdal, T. (2020). Digital twin: values, challenges and enablers from a modeling perspective. *IEEE Access* 8: 21980–22012. https://doi.org/10.1109/ACCESS.2020.2970143.

44 SLAS https://slas.org/ (accessed 1 February 2020).

45 ELRIG https://elrig.org/ (accessed 1 February 2020).

46 MIT Computer Science & Artificial Intelligence Lab https://www.csail.mit.edu/ (Accessed 1 February 2020).

47 Sanderson, K. (2019). Automation: chemistry shoots for the moon. *Nature* 568: 577–579. https://doi.org/10.1038/d41586-019-01246-y.
48 Buermans, H.P.J. and den Dunnen, J.T. (2014). Next generation sequencing technology: advances and applications. *Biochimica et Biophysica Acta* 1842 (10): 1932–1941. https://doi.org/10.1016/j.bbadis.2014.06.015.
49 Empel, C. and Koenigs, R. (2019). Artificial-intelligence-driven organic synthesis—en route towards autonomous synthesis? *Angewandte Chemie International Edition* 58 (48): 17114–17116. https://doi.org/10.1002/anie.201911062.
50 Baker, M. (2016). 1,500 scientists lift the lid on reproducibility: survey sheds light on the 'crisis' rocking research. *Nature* 533 (7604): 452–454. https://www.nature.com/news/1-500-scientists-lift-the-lid-on-reproducibility-1.19970 (accessed 1 February 2020).
51 Protocols.IO https://www.protocols.io/ (accessed 1 February 2020).
52 IoT Lab https://www.iotlab.eu/ (Accessed 1 February 2020).
53 Olena, A. Bringing the internet of things into the lab. https://www.the-scientist.com/bio-business/bringing-the-internet-of-things-into-the-lab-64265 (accessed 1 February 2020).
54 Dehghantanha, A. and Choo, K. (2019). *Handbook of Big Data and IoT Security*. Cham: Springer https://doi.org/10.1007/978-3-030-10543-3.
55 Palmer, E. (2018) Merck has hardened its defenses against cyberattacks like the one last year that cost it nearly $1B. https://www.fiercepharma.com/manufacturing/merck-has-hardened-its-defenses-against-cyber-attacks-like-one-last-year-cost-it (accessed 1 February 2020).
56 Lazarev, K. (2016). Internet of things for personal healthcare. Bachelors thesis. https://www.theseus.fi/bitstream/handle/10024/119325/thesis_Kirill_Lazarev.pdf?sequence=1 (accessed 1 February 2020).
57 User Experience for Life Science https://uxls.org/ (accessed 1 February 2020).
58 Gartner predicts 25 percent of digital workers will use virtual employee assistants daily by 2021. https://www.gartner.com/en/newsroom/press-releases/2019-01-09-gartner-predicts-25-percent-of-digital-workers-will-u (accessed 1 February 2020).
59 Fraunhofer https://www.fit.fraunhofer.de/de/fb/cscw.html (accessed 1 February 2020).
60 Tao, F. and Qi, Q. (2019). Make more digital twins. *Nature* 573: 490–491. https://doi.org/10.1038/d41586-019-02849-1.
61 Fuller, A., Fan, Z., Day, C., and Barlowar, C. (2020). Digital twin: enabling technology, challenges and open research. *Arxiv*. [Preprint] https://arxiv.org/abs/1911.01276. DOI: 10.1109/ACCESS.2020.2998358.
62 Borfitz, D. (2019). Space is the new Frontier for life sciences research. https://www.bio-itworld.com/2019/09/16/space-is-the-new-frontier-for-life-sciences-research.aspx (accessed 1 February 2020).
63 Castro-Wallace, S., Chiu, C.Y., Federman, S. et al. (2017). Nanopore DNA sequencing and genome assembly on the international space station. *Scientific Reports* 7: 18022. https://doi.org/10.1038/s41598-017-18364-0.

64 Karouia, F., Peyvan, K., and Pohorille, A. (2017). Toward biotechnology in space: high-throughput instruments for in situ biological research beyond earth. *Biotechnology Advances* 35 (7): 905–932. https://doi.org/10.1016/j.biotechadv.2017.04.003.

65 SiLA Standard https://sila-standard.com (accessed 1 February 2020).

66 SiLA 2 https://gitlab.com/SiLA2 (accessed 1 February 2020).

67 InCHi Trust https://www.inchi-trust.org/ (accessed 1 February 2020).

68 DICOM Standard https://www.dicomstandard.org/ (accessed 1 February 2020).

69 Pistoia Alliance HELM Standard https://www.pistoiaalliance.org/helm-project/ (accessed 1 February 2020).

70 Allotrope Foundation https://www.allotrope.org/solution (accessed 1 February 2020).

71 IMI Innovative Medicines Initiative https://www.imi.europa.eu/ (accessed 1 February 2020).

72 Pistoia Alliance http://pistoiaalliance.org (accessed 1 February 2020).

73 Brooks, M. (2019). Beyond Quantum Supremacy. *Nature* 574 (7776): 19–21. Available from: https://doi.org/10.1038/d41586-019-02936-3.

74 Cao, Y., Romero, J., Olson, J.P. et al. (2019). Quantum chemistry in the age of quantum computing. *Chemical Reviews* 119 (19): 10856–10915. https://doi.org/10.1021/acs.chemrev.8b00803.

75 First image of Einstein's 'spooky' particle entanglement. https://www.bbc.co.uk/news/uk-scotland-glasgow-west-48971538 (accessed 1 February 2020).

76 Al-Khalili, J. BBC four Einsteins nightmare https://www.bbc.co.uk/programmes/b04tr9x9 (Accessed 1 February 2020).

77 Quantum Riddle BBC four (2019). https://doi.org/10.1038/s41598-017-18364-0 (accessed 1 February 2020).

78 Quantum computers flip the script on spin chemistry (2020). https://www.ibm.com/blogs/research/2020/02/quantum-spin-chemistry/ (accessed 1 February 2020).

79 Kevin Hartnett. Quantum supremacy is coming: here's what you should know. https://www.quantamagazine.org/quantum-supremacy-is-coming-heres-what-you-should-know-20190718/ (accessed 1 February 2020).

80 Chong, F., Franklin, D., and Martonosi, M. (2017). Programming languages and compiler design for realistic quantum hardware. *Nature* 549: 180–187. https://doi.org/10.1038/nature23459.

81 Edinburgh EPCC https://www.epcc.ed.ac.uk/facilities/archer (accessed 1 February 2020).

82 Argonne National Lab https://www.anl.gov/article/supercomputing-powerhouse (accessed 1 February 2020).

83 China Super Computing https://en.wikipedia.org/wiki/Supercomputing_in_China (accessed 1 February 2020).

84 Amazon is now offering quantum computing as a service (2019). https://www.theverge.com/2019/12/2/20992602/amazon-is-now-offering-quantum-computing-as-a-service (accessed 1 February 2020).

85 Schneider, P., Walters, W.P., Plowright, A.T. et al. (2019). Rethinking drug design in the artificial intelligence era. *Nature Reviews. Drug Discovery* 19: 353–364. https://doi.org/10.1038/s41573-019-0050-3.

86 Mak, K. and Pichika, M. (2019). Artificial intelligence in drug development: present status and future prospects. *Drug Discovery Today* 24 (3): 773–780. https://doi.org/10.1016/j.drudis.2018.11.014.

87 For a set of other potentially "hot" scientific areas as picked out in 2017. https://www.timeshighereducation.com/features/what-are-the-hot-research-areas-that-might-spark-the-next-big-bang (accessed 1 February 2020).

88 FDA https://www.fda.gov/consumers/consumer-updates/what-gene-therapy-how-does-it-work (accessed 1 February 2020).

89 National Cancer Institute https://www.cancer.gov/about-cancer/treatment/research/car-t-cells (accessed 1 February 2020).

90 Parida, S.K., Madansein, R., Singh, N. et al. (2015). Cellular therapy in tuberculosis. *International Journal of Infectious Diseases* 32: 32–38. https://doi.org/10.1016/j.ijid.2015.01.016.

91 Maldini, C.R., Ellis, G., and Riley, J.L. (2018). CAR-T cells for infection, autoimmunity and allotransplantation. *Nature Reviews. Immunology* 18: 605–616. https://doi.org/10.1038/s41577-018-0042-2.

92 Stem cells: what they are and what they do. https://www.mayoclinic.org/tests-procedures/bone-marrow-transplant/in-depth/stem-cells/art-20048117 (accessed 1 February 2020).

93 Bui, F., Almeida-da-Silva, C.L.C., Huynh, B. et al. (2019). Association between periodontal pathogens and systemic disease. *Biomedical Journal* 42 (1): 27–35. https://doi.org/10.1016/j.bj.2018.12.001.

94 Kakasis, A. and Panitsa, G. (2019). Bacteriophage therapy as an alternative treatment for human infections. A comprehensive review. *International Journal of Antimicrobial Agents* 53 (1): 16–21. https://doi.org/10.1016/j.ijantimicag.2018.09.004.

95 Lu, R., Hwang, Y.-C., Liu, I.-J. et al. (2020). Development of therapeutic antibodies for the treatment of diseases. *Journal of Biomedical Science* 27: 1. https://doi.org/10.1186/s12929-019-0592-z.

96 Bajan, S. and Hutvagner, G. (2020). RNA-based therapeutics: from antisense oligonucleotides to miRNAs. *Cells* 9: 137. https://doi.org/10.3390/cells9010137.

97 Fosgerau, K. and Hoffmann, T. (2015). Peptide therapeutics: current status and future directions. *Drug Discovery Today* 20 (1): 122–128; https://doi.org/10.1016/j.drudis.2014.10.003.

98 Burslem, G.M. and Crews, C.M. (2020). Proteolysis-targeting chimeras as therapeutics and tools for biological discovery. *Cell* 181: 1. https://doi.org/10.1016/j.cell.2019.11.031.

99 Ursell, L.K., Metcalf, J.L., Parfrey, L.W., and Knight, R. (2012). Defining the human microbiome. *Nutrition Reviews* 70 (Suppl 1): S38–S44. https://doi.org/10.1111/j.1753-4887.2012.00493.x.

100 Eloe-Fadrosh, E.A. and Rasko, D.A. (2013). The human microbiome: from symbiosis to pathogenesis. *Annual Review of Medicine* 64: 145–163. https://doi.org/10.1146/annurev-med-010312-133513.

101 Russell, W.M.S. and Burch, R.L. (1959). *The Principles of Humane Experimental Technique*. London. ISBN 0900767782 [1]: Methuen.

102 (i) NC3Rs https://www.nc3rs.org.uk/. (ii) European Union: Directive 2010/63/EU. https://eur-lex.europa.eu/legal-content/EN/TXT/?uri=celex%3A32010L0063 (accessed 1 February 2020).

103 Wenner Moyer, M. (2011). Organs-on-a-chip for faster drug development. Scientific American. https://www.scientificamerican.com/article/organs-on-a-chip/ (accessed 1 February 2020).

104 Voigtländer, B. (2015). *Scanning Probe Microscopy*. NanoScience and Technology. London, UK: Springer-Verlag. https://doi.org/10.1007/978-3-662-45240-0.

105 Milne, J.L., Borgnia, M.J., Bartesaghi, A. et al. (2012). Cryo-electron microscopy–a primer for the non-microscopist. *The FEBS Journal* 280 (1): 28–45. https://doi.org/10.1111/febs.12078.

106 Gao, L., Zhao, H., Li, T. et al. (2018). Atomic force microscopy based tip-enhanced Raman spectroscopy in biology. *International Journal of Molecular Sciences* 19: 1193. https://doi.org/10.3390/ijms19041193.

107 Debata, S., Das, T.R., Madhuri, R., and Sharma, P.K. (2018). Materials characterization using scanning tunneling microscopy: from fundamentals to advanced applications. In: *Handbook of Materials Characterization* (ed. S. Sharma), 217–261. Cham: Springer https://doi.org/10.1007/978-3-319-92955-2_6.

108 Michel, B. (1991). Highlights in condensed matter physics and future prospects. In: *STM in Biology*. NATO ASI Series (Series B: Physics), vol. 285 (ed. L. Esaki), 549–572. Boston, MA: Springer https://doi.org/10.1007/978-1-4899-3686-8_26.

109 Broadwith, P. (2017). Explainer: what is cryo-electron microscopy? *Chemistry World*. https://www.chemistryworld.com/news/explainer-what-is-cryo-electron-microscopy/3008091.article (accessed 1 February 2020).

110 Aminu, M.D., Nabavi, S.A., Rochelle, C.A., and Manovic, V. (2017). A review of developments in carbon dioxide storage. *Applied Energy* 208: 1389–1419. https://doi.org/10.1016/j.apenergy.2017.09.015.

111 Heiska, J., Nisula, M., and Karppinen, M. (2019). Organic electrode materials with solid-state battery technology. *Journal of Materials Chemistry A* 7: 18735–18758. https://doi.org/10.1039/C9TA04328D.

112 (i) Osuchowski, Marcin, F., Aletti, Federico, Cavaillon, Jean-Marc, Flohé, Stefanie B., Giamarellos-Bourboulis, Evangelos J., Huber-Lang, Markus, Relja, Borna, Skirecki, Tomasz, Szabó, Andrea, and Maegele, Marc (2020). SARS-CoV-2/COVID-19: evolving reality, global response, knowledge gaps, and opportunities. *SHOCK* 54 (4): 416–437. https://doi:10.1097/SHK.0000000000001565. (ii) https://search.bvsalud.org/global-literature-on-novel-coronavirus-2019-ncov/ (accessed 16 November 2020). (iii) Lisheng, Wang, Yiru, Wang, Dawei, Ye, and Qingquan, Liu (2020). Review of the 2019 novel coronavirus (SARS-CoV-2) based on current evidence. *International Journal of*

Antimicrobial Agents 55 (6): 105948. https://doi.org/10.1016/j.ijantimicag.2020.105948 (accessed 16 November 2020).

113 (i) UK Health Secretary launches biggest diagnostic lab network in British history to test for coronavirus (2020). https://www.gov.uk/government/news/health-secretary-launches-biggest-diagnostic-lab-network-in-british-history-to-test-for-coronavirus (accessed 16 November 2020). (ii) Germany's 'bottom-up' testing keeps coronavirus at bay. https://www.ft.com/content/0a7bc361-6fcc-406d-89a0-96c684912e46 (accessed 16 November 2020).

114 Archana Koirala, Ye Jin Joo, Ameneh Khatami, Clayton Chiu, and Philip N. Britton (2020). Vaccines for COVID-19: the current state of play. *Paediatric Respiratory Reviews* 35: 43–49. https://doi.org/10.1016/j.prrv.2020.06.010 (accessed 16 November 2020).

115 Christiaens, Stan (2020). The importance of data accuracy in the fight against Covid-19. https://www.computerweekly.com/opinion/The-importance-of-data-accuracy-in-the-fight-against-Covid-19 (accessed 16 November 2020).

Part II

Knowledge Base

The inspiration provided at the beginning of this book covered pillars of scientific evolution towards the laboratories of the future. We hope it gave you a vision of the future of science.

To enable you to develop a good understanding of the advances we will talk about further on, the following *Knowledge Base* section will focus on crucial terms to understand.

It will give you the solid basis of knowledge that you will be able to apply further on as your lab grows and evolves.

2

Crucial Software-related Terms to Understand

Luka Murn

BioSistemika d.o.o., Koprska ulica 98, 1000 Ljubljana, Slovenia

2.1 Digital Revolution

Ever since the 1950s, the world has been undergoing a **digital revolution** that still lasts to this day.

2.2 Computers

Computers are everywhere around us. Often people associate computers with PCs (personal computers), our smartphones, or our smartwatches. But computers are also located in cars, traffic light controllers, airplanes, televisions, digital cameras, microwave ovens, dishwashers, thermostats, and elsewhere.

With regard to the life science industry, most laboratory instruments that have some modes of execution and are driven by a controlling mechanism of some sort also contain a computer – often called an **embedded system** in such a case. In fact, according to one study, 98% of all manufactured microprocessors are used in such embedded systems [1].

A computer is, very simply put, a machine that can respond to a specific set of instructions (called a computer **program**) and executes them by carrying out a sequence of arithmetic or logical operations.

Historically, development of digital computers really took off during World War II and subsequent years. While the predecessors of modern computers – analog computers – were very bulky devices, often filling space of entire rooms, the transistor-based computers could fit the same processing capabilities onto a small slice of silicon.

In 1936, Alan Turing – an English mathematician who was, among other achievements, pivotal in cracking the Nazi code messages during World War II – proposed a mathematical concept of a simple machine that would read instructions and – in essence – process data, both from the same medium (in his case, an infinite memory

Digital Transformation of the Laboratory: A Practical Guide to the Connected Lab, First Edition.
Edited by Klemen Zupancic, Tea Pavlek and Jana Erjavec.

tape). In computer science, such a machine is nowadays referred to as a universal Turing machine [2]. With these ideas, Turing made very important advances in the field of computability – which mathematical problems can be solved using computers, and which not. But the notion that both the program and the data reside on the same medium led to another key component of modern computer design: **programmability**.

Programmability is the key component of modern computers. It allows computers to perform different instruction sets every time due to the fact that the instructions are simply stored on computer memory, and they can be changed by the very same computer. In essence, such a computer is constantly modifying its own instruction set.

Based on Turing's concepts, another brilliant scientist of the twentieth century, John von Neumann, is credited with proposing an architecture of a **stored-program computer**, which is a computer that can read different instruction sets (called **programs**) stored in electronic memory and execute them to process data on the very same electronic memory. This architecture is often referred to as **Von Neumann architecture**, and the vast majority of all modern computers are, in core concepts, still based on this architecture. This means that any time you pick up your mobile phone or your tablet, that device is based on Von Neumann architecture (Figure 2.1).

The **central processing unit (CPU)** comprises the **control unit**, which is tasked with reading the sets of instructions from the **memory unit** and executing them as arithmetic or logical operations on the **arithmetic/logic unit**. If necessary, data can also be written back to the memory unit. The **input and output devices** are practically everything else that comprise a modern computer and are normally much larger in terms of physical size than the CPU and the memory unit. These devices include hard drives, solid-state drives (SSD) drives, keyboards, computer

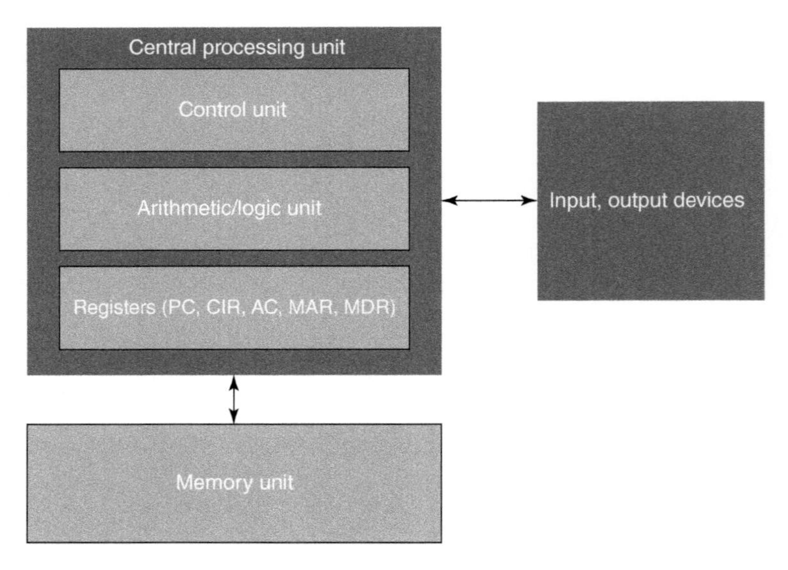

Figure 2.1 The Von Neumann architecture.

mice, display devices (e.g. monitors), touchscreens, audio speakers, microphones, USB connectors, network cards, graphic cards, and other components.

Any time a computer is turned on, it is performing arithmetic and logical operations. *Computers are often perceived as the ultimate tool that will solve all complex problems. A common misconception is that all computational problems (all problems that are solvable by performing a sequence of computational steps) can be solved with computers by dedicating enough processing power, accumulating enough data, or similar. While all complex computational problems are theoretically solvable by a computer, many of them would require tremendous computer memory or – more often – a lot of time. Even on modern supercomputers, complex problems with large data inputs can potentially require thousands of years to execute – which is a resource we simply do not have. Quantum computers – in theory – would solve this time constraint, and that is one of the reasons behind extensive research on quantum computing (see Section 1.2.5.2).*

2.2.1 Programs, Instructions, and Programming Languages

A computer **program** is **a sequence of computer instructions** that is stored on the computer memory.

Primarily, computer instructions can be grouped into at least three categories: data and memory handling, arithmetic and logic operations, and control flow operations.

Majority of modern applications and programs are written in a **high-level programming language**. As of 2019, majority of readers might have heard terms such as Java, JavaScript, C#, .NET, R, Visual Basic, Ruby, and similar. These are all examples of high-level programming languages. Such languages often use natural language elements (to be more friendly to the **programmer**, e.g. a person who writes them), automate various areas such as memory management, and offer a plethora of more complex instructions out of the box. These are usually available as part of a common standard library.

Besides the aforementioned benefits, high-level programming languages also offer another very important possibility: due to abstractions, a program written in a high-level language can be executed on different hardware, processor, and operating system types. Thus, high-level programming languages are one of the key cornerstones of cross-platform application development (see Section 2.5.5).

The most important take-away message is to understand that a ***computer program*** *is a sequence of commands that a computer needs to execute. These commands might be to perform arithmetic or logical operations; however, they might also be commands to, e.g. manipulate elements on the computer screen and wait for user input from the keyboard.*

Nowadays, instead of a program, the term application is often used. ***An application*** *is a computer program, or a group of programs, aimed at* ***fulfilling a specific task/activity*** *(e.g. word processor, spreadsheet, email client, media player, and videogame).*
A deeper dive into applications is presented in the section Applications further on in this chapter.

2.2.2 Hardware and Software

It is worth pointing out the two terms that have emerged – and are nowadays being very frequently used – with regard to computers and instruction sets.

Hardware represents everything physical about computers: CPUs, memory unit/s, and all the input and output devices; these include all the chips such as computer motherboards, various graphic, sound and network cards, and any other physical computer components.

In contrast, **software** refers to computer instructions/programs described in the following Software section of this chapter, and is a more broad definition than computer program or application, but at the end of the day, it relates to a computer program, or a set of computer programs. Throughout this book, the terms *software*, *application*, and *program* are often used interchangeably.

The origins of the words hardware and software come from the fact that hardware is "hard" to change/modify, as physical action is required, while on the other hand, software is easily modifiable due to the nature of programmable computers.

2.2.3 Operating Systems

During the digital revolution, and especially after the 1980s, computers were becoming more and more complex. The main reason for this was that their field of use was expanding: they were being used for various new purposes (e.g. automating various industry production lines, solving mathematical problems, and PCs for word processing) – in essence, they were fulfilling, and executing, a plethora of very different and complex tasks. Consequently, the computer architecture was becoming more and more complex; this was especially true for input and output devices. Display devices and graphic cards, keyboards, computer mice, audio cards, computer network interfaces, and similar were being developed at a rapid pace.

Managing all these different input and output devices, which nowadays comprise the majority of modern computers, is a very complicated task. Every such device comes with a complex array of states and behaviors that need to be captured if a computer program would want to use it.

An ***operating system (OS)*** *(sometimes simply shortened to* ***OS****) is a software (a program or a set of programs) that manages the computer's hardware resources – input and output devices and memory unit – as well as software resources (concepts of a file system, security, etc.).*

Figure 2.2 demonstrates the role of the operating system in modern computers.

There is no such thing as a "universal" operating system. Due to different nature, hardware components, and trade-offs for specific computer uses (e.g. mobile phone, PC, and tablet), there are many different operating systems on the market. For example, as of 2019, everybody knows Microsoft Windows, Apple macOS, or Linux, operating systems developed for desktop computers.

Choosing an unpopular or discontinued operating system might have a big financial impact on an organization in the future (due to the need to migrate to an alternative system), while opting for a market leader has an effect of a vendor lock-in.

Operating systems allow computer programmers to write computer programs/applications that can – in an easy way – access and manage computer's hardware and software resources without the need to implement all this management code themselves. This is normally done using a mechanism called **system calls.**

As well as management of computer hardware, operating systems also implement many other important concepts, such as (i) means for different programs to communicate with each other, (ii) multiuser environment, (iii) file systems, (iv) security features, (v) graphical user interface (UI), and similar.

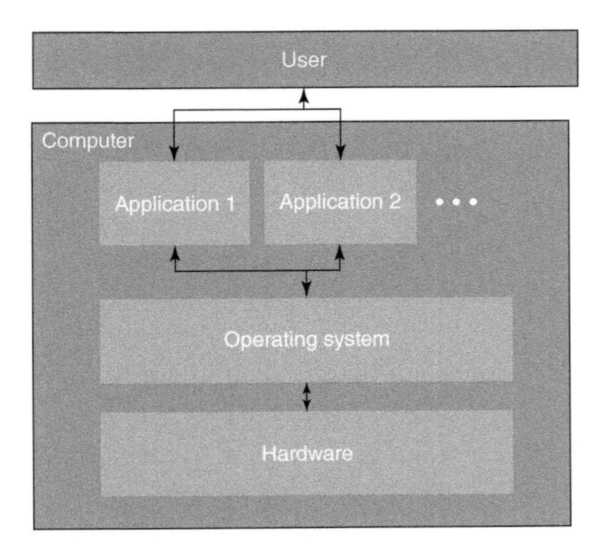

Figure 2.2 Operating system's role in modern computers.

Operating systems are very large and complex pieces of code, and that is the reason why there are either industry giants (Microsoft and Google) behind their development or they are large open-source projects (Linux).

2.2.4 Abstraction

Because of everything listed above, operating systems abstract (to a degree) hardware and software layers for programs/applications. What this means is that a computer program/application can access various different hardware and software resources (without knowing their specific behaviors) because they are abstracted behind a common interface by the operating system.

Due to multiple reasons exceeding the scope of this book, the hardware resource that is normally not abstracted (or not abstracted fully) is the microprocessor itself. As a consequence, **most modern programs/applications are written for a single specific operating system** and sometimes, the microprocessor type/ architecture.

To put it differently, an application targeting Microsoft Windows operating system will not work on a Linux operating system.

High-level programming languages allow programmers to write a program/ application that can often be compiled for different microprocessor types; some languages also offer the same flexibility when it comes to the operating system.

2.2.5 Virtualization

Virtualization is one of the key concepts of modern computer systems.

Everything cloud based that you use on a daily basis (Gmail, HubSpot, Egnyte, Trello, Slack, and Amazon, etc.) employs virtualization.

Virtualization is also used on the premises by any larger organization that has a dedicated IT department. It is a more cost-efficient solution for any server setup.

Back in the 1960s, computers were still very expensive, and enterprises soon realized that it was more cost-effective to buy a single powerful computer, and have it used by multiple users, especially back in the day when not everyone was using a computer throughout the entirety of their working day (there were no desktop computers yet at the time). Such shared, multiuser computers were called **mainframes**, and they are still extensively used, predominantly throughout larger enterprises for business-critical computations.

As computer hardware became more powerful, computer engineers were exploring how to best utilize resources of a single, powerful computer to serve a simultaneous, multiuser environment. Thus, **virtualization** was born, and it has been a pivotal approach for optimization of computer hardware resources up to this day. Vast majority of mainframes, as well as server computers, employ virtualization concepts to efficiently distribute their processing power, memory, and storage to multiple users.

Virtualization is a very extensive and active field of computer research and development. Nonetheless, these are some of the common virtualization terms briefly explained:

- *Snapshot*: a state of the entire virtual machine (that can, for example, be easily backed up).
- *VPS*: Virtual private server – a common term for web server setups (see Section 2.3 – HTTP).
- *Containers*: A type of virtualization on a level of individual processes. After 2014, an increasingly popular virtualization approach due to **Docker** technology.

2.3 Internet

It is estimated that in 2019, 53.6% of the world population was using the Internet [3]. Because of such a widespread usage, it can easily be said that the Internet as a technology has fallen under the rules of Clarke's third law: "Any sufficiently advanced technology is indistinguishable from magic." [4] The Internet is also considered as the current field where digital revolution is happening and progressing.

For everyday use, it could be argued that one does not need to understand how Internet works; however, in the field of laboratory digitalization (or digitalization in general), where Internet is taking a very pivotal role in the recent years, understanding the underlying workings of the Internet is a business-critical knowledge. More and more software vendors are moving into the direction of web applications, where their software is delivered to the customer through the Internet. It is vital to understand the benefits, and risks, associated with purchasing such a service.

But before digging into the history of how the Internet came to be what it is today, let us first describe what it is.

The Internet is a **network of networks**; it is a **global, physical network** that **connects each endpoint** (e.g. your mobile phone, PC, or laptop) to **every other endpoint** on the same network.

This network of networks is achieved by having a plethora of hubs, connected by physical cables (electrical or fiber optic) or wireless (radio waves and electromagnetic waves), all across various locations throughout the world. All these devices achieve interoperability by following the same industry-agreed protocol suites.

Any time you connect your mobile phone, or your computer, to the Internet, you can theoretically connect to any other device on the Internet and vice versa (ergo, your device is exposed to be connected to by any other device on the Internet).

To explain further – in the communication industry, a **protocol** is a set of defined rules and procedures that specify how two or more entities exchange information with each other. If industry leaders adopt a protocol, it inherently means that everybody involved in the industry must adhere to the protocol; this might be an extra complication for each individual player in the industry, but for the user base, it is a major winning point as it allows for much greater quality of the service, and degree of interoperability between various vendors.

2.3.1 World Wide Web (WWW)

In order to understand how web applications work, the World Wide Web (WWW) is an important part of the Internet that needs to be described in more detail.

By the late 1980s, the Internet allowed computers to communicate with each other over the network in a reliable way. However, above the communication/network layer, there was no standardization yet as of what data to actually transfer between the computers and how that would really work.

English scientist Tim Berners-Lee invented the WWW in 1989 while he was employed at CERN in Switzerland. **WWW** (often simply called **the Web**) originated as a conceptual idea of having a **shared information space** where documents and other digital (web) resources could be identified and accessed.

Of course, the main infrastructure where this information space would live is the Internet. The media that would be stored by this information space would be referred to as **hypermedia**, and the documents/resources would be referred to as **hypertext**.

Tim Berners-Lee, and others, developed various protocols to support the WWW, the most important of which is the HTTP (hypertext transfer protocol).

The WWW has three core concepts:

- *Each web resource (hypertext) is described using a common, standardized publishing language*: ***HTML (hypertext markup language)***.
- *Each web resource (hypertext) can be referenced and accessed by its globally unique identifier*: ***URL (uniform resource locator)***.
- *A protocol is defined to access and transfer web resources over the Internet*: ***HTTP***.

In the original WWW concept, web resources were static documents that included text (and other media, such as images, sounds, and videos) and **hyperlinks** – references to other web resources. It is important to point out that originally, WWW had no notion of web applications, and it was devised simply as a web document exchange space.

A term often used in the context of WWW is a **web page**, which – to put it simply – represents any document formatted in the HTML markup language. If a web page represents a single document, **a website** represents a collection of related web resources (web pages and other multimedia content).

These documents are usually retrieved using a **web browser** – a computer application/program that allows the user to navigate the Web's information space and access web pages.

2.3.2 Web Applications

During the 1990s and 2000s, the WWW was primarily being used as originally intended: it was an information space for exchange of static documents. Users could open their web browsers on their desktop computers and navigate to various websites that would contain static information. There was not much user interaction.

The success and the extent of use that the WWW achieved, however, was quite unprecedented and gradually, the WWW far exceeded the original idea of a mere web document information space. Gradually, programmers were starting to build more and more complex applications on the top of it.

> *The **difference between a website and a web application** might be hard to pin down, but applications are generally more complex programs that handle things such as user state and user data, authorization (e.g. user login), dynamic content, animations and rich user interactions, and so forth. They also fulfill a specific task for the user (whereas websites are merely containers of hypertext data).*

Despite its origin (a shared static document information space), by 2020 WWW has grown and emerged as the single most popular **application delivery platform** in the industry. This means that complex and rich applications are delivered to users through the WWW layer (which is built on top of the Internet layer).

The fact that **web applications** require a different set of rules, concepts, and behaviors (compared to the original concept of the Web) meant that ever since the late 1990s, very active and competitive development of various technologies on top of the WWW was taking place – with the intent to support and standardize development of web applications on the Web. All industry leaders have been involved; the technology landscape for development of web applications is finally stabilizing and slowly coming toward using standard solutions.

There are few other terms that are widely used in the context of web applications:

- *CSS (cascading style sheets)*: While HTML is used to define the content of a web page, CSS is a language that describes how the content of a web page is presented to the user (e.g. positioning of elements, colors, graphics, and fonts).
- *Responsive web design*: Represents web pages/applications that can adapt to various display devices and window/screen sizes. For example, you can choose to use a web application that does not have a mobile application for your smartphone, but its design is responsive. Therefore, your chosen web application can be used on your smartphone (or other mobile device) because it will, to a certain degree, adapt to the size of your screen automatically.
- *Single-page applications (SPAs)*: Represent a modern approach to how a web application is programmed ; they behave as any other web application, but they normally offer a better, more rich user experience (UX) than regular web applications.

- **Progressive web applications (PWAs)** – Try to combine the best of both worlds: rich UX of traditional (desktop or mobile) applications and the benefits of web applications; the most important thing about them (besides rich UX) is that they might allow for an **offline mode** (online regular web applications) – where the data is synchronized between the client and the server once the connection is established again. They might also use device **hardware access** (see Section 2.3.3) more than regular web applications would.

2.3.3 Web Applications in Comparison With Traditional Applications

Since there are quite a lot of misunderstandings, it makes sense to point out the differences between traditional applications and web applications.

To access any web application, a web browser needs to be used. A web application is delivered to the user within the web browser in the same manner as a user would be accessing a website, e.g. by visiting a specific URL. Web applications ***do not require any installation*** *to be done. This is different from traditional applications, which need to be installed on each individual computer (PC, tablet, mobile phone, etc.) from an installation media (normally, an installation file which can be downloaded from the Internet or stored on some exchangeable media such as a USB key). Web applications are simply started when the correct URL is accessed. Inherently, this means that the only traditional application that needs to be installed on the computer is the web browser.*

Normally, web applications **require constant network connection** between the client and the server. The server which holds the web application needs to be accessible by the computer on which the web browser is running; if the server is on the Internet, this means that the client computer needs to have access to the Internet; if the server is located on an **internal organization network** (e.g. intranet), client needs to be connected to it.

This might be an issue in regulated environments, where access to the Internet from within the organizational network is often not allowed due to **security and regulatory constraints**. In such a case, web applications can also be hosted on a server that is placed on the internal organizational network. In this case, however, the web application **must be installed on the internal server**, and this procedure is usually more complex than just executing an installation file. Vendors of web applications usually assume that organizations with internal networks have internal IT personnel who can perform the installation. In this case – where the server is located on the customer's premises – the setup is referred to as **on-site** or **on-the-premises installation/setup**.

The schema below (referred to in the industry as the *network topology diagram*) represents the difference between the on-site/on-the-premises setup (Figure 2.3) and the Internet/cloud setup (Figure 2.4).

Due to the fact that web applications are hosted entirely on the server, they can be maintained, and updated, at a central place; all clients (e.g. web browsers using the web application) will automatically receive an updated version of the application next time they use it. Not just the updates but also the maintenance of a web application is much easier to perform in this scenario (e.g. performing data backups and similar). This also means that if the client computer malfunctions (or gets stolen), ***data is not lost****, as it resides on the server.*

There are two important trade-offs to be considered. In the case of traditional applications, each application that is installed on a specific computer has full processing and memory capabilities of that computer. In the case of a web application, however, each additional user accessing/using the web application (this means communicating with the server via the network) at the same time **requires extra hardware resources** (computer memory and processing power) of the server. With regard to this, the term **scaling** is often used. The ability of the system to scale (there are two more technical terms, *scale-out* and *scale-up*, which exceed the scope of this book) means that the system is able to allocate (or deallocate) extra hardware

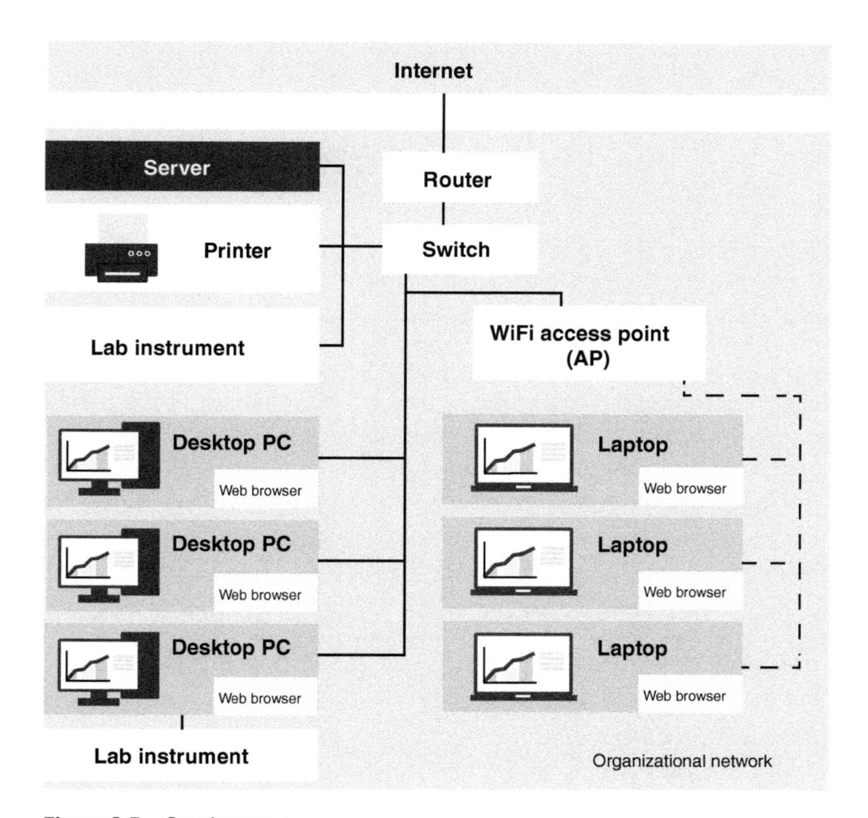

Figure 2.3 On-site setup.

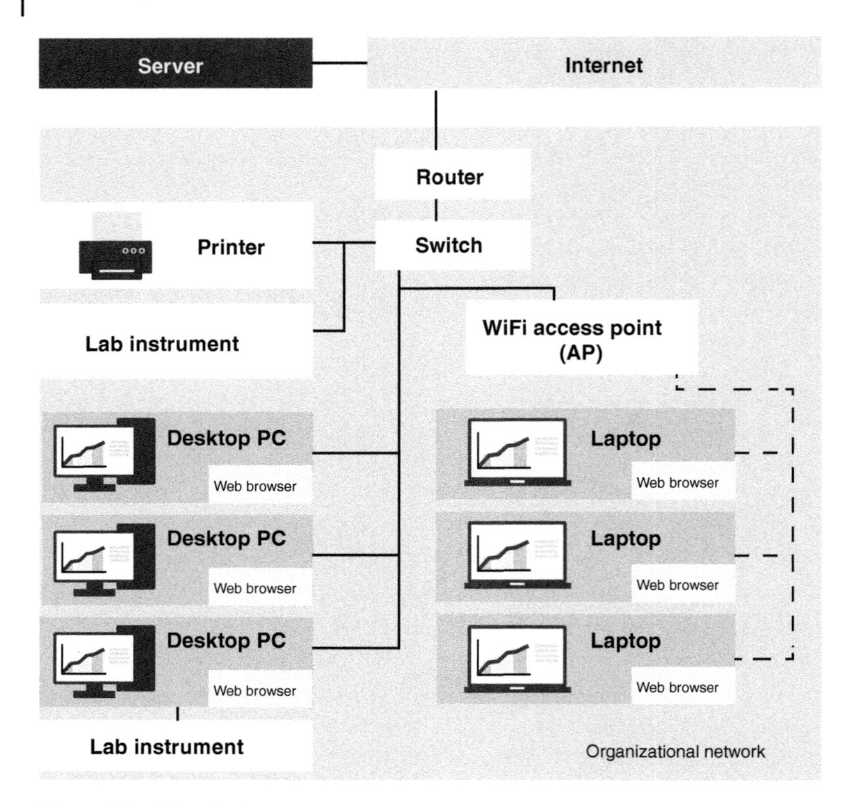

Figure 2.4 Cloud setup.

resources (to/from the server) based on the demands to ensure normal operating capabilities (and good UX).

The other observation that can be done is that in the case of web applications, (i) network and (ii) the server are two **single points of failure** of the entire system. Local computer networks can be considered very reliable; however, the Internet is still very unreliable from a pure risk assessment; if the servers are located on the Internet, it is of utmost importance to ensure stable and uninterrupted Internet connection at all times.

In the context of laboratory digitalization, another important factor is **access to hardware**. Traditional applications (that are installed on the computer) normally have direct access to the operating system functions, thus direct access to any hardware devices (including external hardware devices such as laboratory equipment) that are connected to the computer. On the other hand, a web application runs fully within the web browser. Due to security reasons – primarily associated with the Internet – web applications have a harder time accessing hardware resources of the computer, as they are executed in a much more restricted environment. This is changing, and with the popularity of the Internet of Things (IoT), industry is moving toward standardizing secure technologies to allow web applications to access hardware resources. It might, however, still take some years before the technology is really adopted and well supported by the industry leaders.

*Finally, another reason why **web applications are favored** by application vendors is because they are **much harder to distribute illegally**. A large portion of the application source code is only executed on the server (this is often called a backend), and an application vendor has full control over who has access to it. The only code that is accessible to the users is the codebase in the web browser (called frontend). Any software piracy act thus needs to either hack into the central server or reverse engineer the server-side source code, both of which are very hard to perform.*

2.4 Cloud Computing

In 2006, American technology company Amazon.com released its Elastic Compute Cloud (EC2) service. It could be said that this marked the start of modern cloud computing, as this service disrupted the current state of things on the market. Google and Microsoft soon followed in Amazon's footsteps in subsequent years, and as of 2019, cloud computing is in full swing.

From a purely technological point of view, when it was launched, **cloud computing** did not bring any disruptive, innovative technology that would fundamentally change how the Internet and web applications are used. It was leaning heavily on the virtualization concepts (see Section 2.4.2.1) already present on the market. Instead, the success of cloud computing can be attributed more to its business model and the problems it was solving.

The most simple way to explain this is to outline how the Internet landscape looked like before cloud computing, and what changed when cloud computing emerged. In the 1990s and 2000s, any organization that wanted to have a website or a web application present on the WWW had to buy server computers (usually, dedicated computers with dedicated hardware for web server needs). An organization would need to physically put these server computers on their premises, connect them to the Internet, and maintain them. Such setup is nowadays sometimes referred to as a **bare-metal server**.

*Maintaining a server computer (often referred to as having a **local installation of a web application**) to ensure that a website/web application is constantly available online on the Internet is not such a trivial task as one might think. Organization has to ensure that all hardware parts are constantly working, malfunctioned parts are replaced, power supply is guaranteed, data is backed up (possibly in a separate physical location, if something happens in the main location), and also that the* *operating system on the computer, as well as all the software, is kept up-to-date to include the latest security fixes and patches. For bigger organizations with existing internal IT teams, this is a relatively small extra overhead; for small organizations with no in-house IT expertise, this can be an exceedingly large cost.*

What cloud computing companies – which are often referred to as **cloud service providers** – offer is a possibility for organizations to host their websites or web applications on the cloud provider's physical infrastructure. Cloud providers own the **data centers** (large physical buildings where the server computers are located) and offer parts of these server computers as an **on-demand offering** to its customers (this is often referred to as public cloud, which will be explained further on in this chapter).

Main benefits of cloud computing are:

- *Maintenance of the web server computers is partially or fully (depending on the level of the service) outsourced to the cloud service providers; ensuring that servers stay up and available on the Internet is the core responsibility of the cloud service provider.*
- *Due to their extensive physical infrastructure, cloud service providers can offer much more immediate and flexible **scaling capabilities** (see Section 2.3.3) than in-house bare-metal setups.*
- *As all data is centrally stored, it allows for much easier data consolidation and aggregation. Many cloud service providers offer **business intelligence** (data analysis of business information) capabilities and tools.*

As any technology that has been very quickly widespread and popularized, cloud computing has been criticized and challenged (and still is). The industry considerations on cloud computing are represented in the further section of this chapter.

2.4.1 Classification of Cloud Services

As mentioned before, cloud computing perhaps has even more to do with the business/revenue model shift of the entire IT industry toward subscription-based models than the technology behind the cloud itself. Majority of cloud service providers nowadays (as of 2019) run a service model, where **costs are charged to the customer on a periodic (monthly/quarterly/annually) basis**, based on actual periodic usage.

It is no wonder then that the industry has classified the cloud services into multiple **"XaaS"** (*X as a service*) categories. There are more than the three outlined below, but these are still the most important categories. The schema below shows how different levels of service are offered based on the cloud service category.

2.4.1.1 IaaS (infrastructure as a service)

Infrastructure as a service offers server infrastructure to the customer. This means all the necessary server computer hardware can be allocated to the customer on-demand. The customer is responsible for ensuring that the operating systems are kept up-to-date and everything else as well.

IaaS customers are normally IT companies. In the scope of lab digitalization, organizations could consider IaaS in the following two scenarios:

- As a (possibly) cheaper alternative to on-the-premises IT infrastructure of an organization (of course, bearing in mind the cloud considerations, outlined in the Section 2.4.3).
- As an infrastructural platform if an organization is developing its own software product.

Examples of IaaS are Amazon AWS (Amazon web services), Microsoft Azure, Google Compute Engine, and DigitalOcean.

2.4.1.2 PaaS (platform as a service)

Platform as a service (PaaS) is primarily aimed at software development companies as its customers. The idea behind the PaaS is to provide everything to the customer to be able to launch and run a web application in the cloud, apart from the development of the application itself (which is the core business of a software development company).

This is probably the least interesting category in the scope of lab digitalization, unless an organization is developing its own software product.

Examples of PaaS are AWS Elastic Beanstalk, Windows Azure, Heroku, and Google App Engine.

2.4.1.3 SaaS (software as a service)

Software as a service (SaaS) is the most important cloud computing category in the lab digitalization field. A SaaS provider/vendor offers its software (which is normally a *web application*) to its customers. This can, for example, be a LIMS (laboratory information management system), an ELN (electronic laboratory notebook), or similar.

In SaaS, everything is provided to the customer by the SaaS vendor – the customer simply pays for the usage of the application (software) based on the usage of the application. Nowadays, this is usually done on a **per-user per-month basis** (you are charged for the number of supported users on a monthly basis).

SaaS vendors are often software development companies that have developed their own products (e.g. a LIMS system); it is important to note that these companies often rely on other PaaS or IaaS vendors behind the curtain for the infrastructure for their software products. For example, a company offering a cloud LIMS system to its customers might be hosting the LIMS system (a web application) on an IaaS infrastructure (e.g. AWS). This is predominately true for smaller SaaS vendors, and startup product companies, as the only option for them to have a product offered

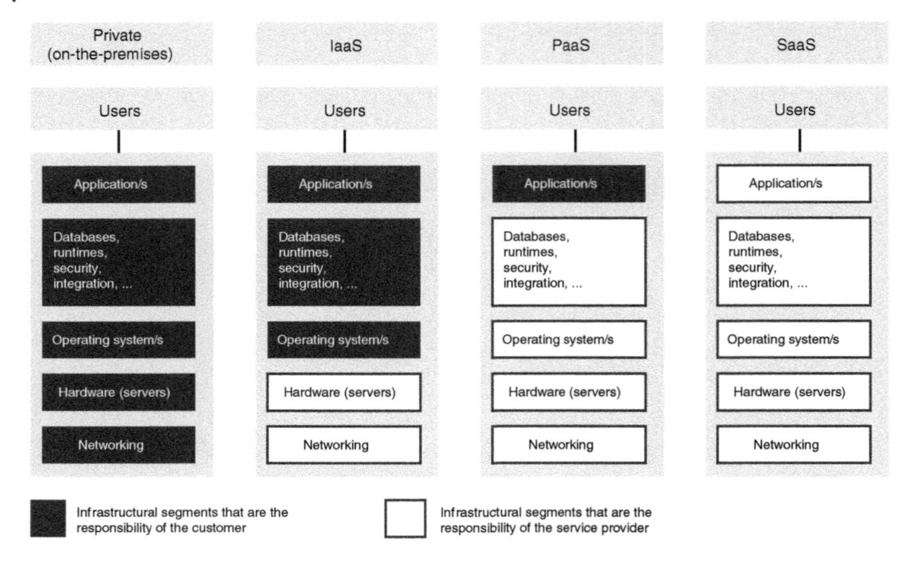

Figure 2.5 Explanation of the differences between Private, IaaS, PaaS, and SaaS.

to their customers is to rely on an IaaS or PaaS provider for infrastructure. To put it differently, they do not have necessary financial and technical capabilities to run and manage their own infrastructure.

Figure 2.5 shows the dark boxes that represent the infrastructural segments that are the responsibility of the customer, while the white boxes represent the infrastructural segments that are the responsibility of the service provider (e.g. that are outsourced).

2.4.2 Cloud Deployment Models

Depending on how the cloud infrastructure is set up and maintained, three distinct cloud deployment models have emerged in the industry.

> *Cloud service providers often use these terms when offering their services, so it is important to understand the basic differences between them.*

2.4.2.1 Public Cloud

A cloud is called **public cloud** when the cloud services are offered through an open network. In practice, this is primarily the Internet. Majority of the chapter was in truth describing a public cloud model.

The public cloud is normally hosted on the cloud service provider's infrastructure and is architectured and set up in a way to allow multiple customers to securely access the same hardware infrastructure as well as to securely communicate over a nontrusted network (the Internet).

All major IaaS providers (AWS, Microsoft Azure, and Google Compute Engine) offer public cloud capabilities.

2.4.2.2 Private Cloud

Contrary to public cloud, a **private cloud** is meant to be only used by a single organization/customer. This inherently means that the whole cloud infrastructure (servers, data centers, and network infrastructure) needs to be dedicated for a single group of organization's users; the infrastructure must also be physically separated from infrastructure used for any other organization/customer.

Private clouds can be internally deployed within an organization's network infrastructure or externally hosted; they can also be self-managed by the organization itself (organization's IT department) or managed by a third-party cloud service provider.

The big reason for the existence of private clouds is **improved data safety and security**. Private clouds offer physical data segregation as well as extra security measures that might not be available for public clouds due to their shared nature.

The trade-off, of course, is the cost. Private clouds are very capital intensive as they require significant hardware and physical space allocations. All the infrastructure also need to be maintained on an on-going basis. It could be argued that private clouds do not benefit from cost savings of the public cloud providers' business models.

2.4.2.3 Hybrid Cloud

A **hybrid cloud** is a combination of cloud setups where some segments of the cloud infrastructure are public and other segments private. As a combination of the two, hybrid clouds can represent many different configurations and setups, so it is hard to precisely classify.

Organizations often employ hybrid cloud setups to suit their business needs: one possible scenario could be if an organization would want to have their business-critical data stored in a private cloud environment; however, they want to exchange some of the less sensitive data with their customers via public cloud capabilities. An alternative scenario could be that an organization primarily uses a private cloud setup but relies on public cloud to meet temporary computing resource needs if the private cloud cannot meet them.

2.4.3 Issues and Considerations

As any technology, cloud computing has been the subject of criticism throughout its inception and up to this day. The majority of criticism relates to the fact that with cloud computing, a user loses direct control of his data as the data is physically stored on the computer that belongs to somebody else – the cloud service provider.

Due to this fact, most of the criticism revolves around data safety and security (explained further in the Chapter 4 of this book):

- As an organization does not have physical control of the computers that store its data, it needs to rely on the cloud service provider to implement measures that prevent data loss (data safety).
- Especially for public clouds, multiple organizations share the same physical infrastructure of the cloud service provider; organization needs to rely on the cloud service provider to prevent unauthorized access to its data by other service users; similarly, in ordinary scenarios, cloud service providers should not access organization's data directly either.
- Especially for public clouds, organization accesses its data (that is stored on the cloud service provider servers) via public physical infrastructure (e.g. the Internet), which means that all data is subject to being compromised while in transit.

All major cloud service providers mitigate the aforementioned risks and usually offer detailed information, use cases, and security whitepapers on the subject.

Many industries – some more so than the others (e.g. IT, healthcare, and pharmaceuticals) – operate under very strict standardization, compliancy, legal, and/or regulatory demands (e.g. Health Insurance Portability and Accountability Act [HIPAA] in healthcare and GAMP/GxP in pharmaceuticals); many of these require that the data owned by the organization needs to be physically under the organization's control as well. For such cases, cloud adoption is certainly slower than elsewhere. Nonetheless, popularity and cost savings offered by the cloud are slowly changing the perception of even the most skeptical industries, and certain changes to regulations, standards, and legislation are under way.

Choosing the right cloud service provider might seem a daunting task for a non-IT company such as a laboratory. But as with any other tool, it should be approached from a risk-based, business analysis point of view.:

- Evaluating the cloud service provider's offering based on business needs (obviously).
- Evaluating cloud service providers as a business entity.
- Review of the agreement and contracts: primarily, if (and how) they cover privacy policy, data safety, and security, and what is the exit strategy and data contingency plan (what happens to organization's data if the cloud service provider shuts down).
- Other relevant factors.

2.5 Computer Platforms

When talking about computers, the term *platform* can be misleading, as it is used quite frequently and can represent many different things depending on the context.

Definition, relevant for the contents of this chapter, is that **a computer platform** represents an environment (in most cases, combination of hardware and operating system) where software (applications) is executed.

Throughout the digital revolution, various computer platforms have emerged, some of which have become widely recognized by the general public.

2.5.1 Desktop/Laptop/PC

With the advances in semiconductor device fabrication, and the digital revolution gaining momentum, the costs of manufacturing a computer were gradually dropping. This, alongside the fact that computers became much smaller, culminated in the production of the first generation of **personal home computers** in 1977 – the three models that are widely associated with this were *Apple II* (from Apple), *TRS-80*, and *Commodore PET*. The term **desktop** computer was simply coined from the observation that these computers could fit on a desk.

The desktop computers were a big hit and commercial success throughout the 1980s, 1990s, and 2000s [5]. Many of the IT industry giants nowadays used to be early developers in the desktop computer development field (companies such as Apple, Microsoft, and IBM).

Throughout the 2000s and onward, desktop computers were being gradually replaced by battery-powered **laptops**, especially for personal and home use. The term laptop was, again, coined from the fact that it is smaller than a desktop and can be placed on a person's lap. As of 2019, desktop computers are predominantly used in enterprise environments (where things such as powerful hardware, large display, or similar are required), while for home/private use, mobile computers and laptops are having the biggest market share.

There is another term that is frequently associated with desktops and laptops – **PC**. While these terms (desktop/laptop/PC) have different origins, they all relate to **the same computer platform**. That is because the hardware components and the operating systems are predominantly the same (or similar). Throughout this book, the term desktop computer is used to refer to this computer platform.

The predominant operating system on desktop computers is *Microsoft Windows*, followed by *Apple macOS*, and *Linux* [6].

2.5.1.1 Desktop Applications

Desktop applications are, as the name suggests, programs/applications that can be executed on desktop computers. They differentiate depending on the operating system (Windows, macOS, and Linux).

As desktop applications are still very present in the field of laboratory IT solutions (as of 2019), it is important to understand certain common concepts and properties that are specific to desktop applications:

- *Local execution*: Desktop applications are executed entirely on the desktop computer and have full access to hardware – this is primarily important if the application requires high performance or advanced graphics capabilities.

- *Installation*: Desktop applications need to be fully installed on the computer (that is normally performed manually by the user); this is often done using installation wizards, package managers, or app stores.
- *Update*: Desktop applications are normally not automatically updated; while they can be programmed in a way to check and notify the user that there is a newer version, user action is required to perform an update. Update process is often similar to the installation process.
- *No cross-platform capabilities*: If the same program/application (from the user standpoint) needs to be supported on different operating systems (e.g. Windows and macOS), essentially separate applications need to be written, one for each operating system. This is often expensive, as each application needs to be maintained separately by the application vendor.
- *Access to source code*: As the entire application is installed on the computer, users have full access to the application source code (it is a bit more complicated than that, but in essence, that is true). As with any digital media, this means that applications can easily be redistributed and copied, as well as modified and extended (of course, such actions require computer programming knowledge).
- *Licensing*: Due to the fact that source code was available to users, various desktop application licensing models emerged to allow application vendors to prevent users from redistributing their applications for no cost. Below are listed few of the common licensing models for desktop applications:
  - License is tied to computer hardware configuration (e.g. processor ID and motherboard ID). Changing computer hardware components requires relicensing.
  - License is tied to a special USB key (often called **USB dongle**) that is provided by the application vendor separately.
  - License allows users to access updates and technical support; as the application is being constantly developed and updated, not having access to updates means users are left behind (this model is very similar to how most web applications are licensed as of 2019).

2.5.2 Mobile

Due to advances in small microchip manufacturing and battery technology, first hand-held computers emerged in the 1990s. One type of such computers popular at the time were **PDAs** (personal digital assistants), but soon, attention turned to cell/mobile phones. In the 2000s, companies such as BlackBerry, Nokia (Symbian platform), and Microsoft (Windows Mobile platform) were exploring the new market, but the industry really started to blossom after Apple unveiled the first iPhone model in 2007 [7]. This device is often associated as the first modern **smartphone**.

Smartphones have thus emerged as phone devices that have strong hardware capabilities, extensive mobile operating systems, and facilitate both core phone functions (voice calls and text messaging) as well as multimedia capabilities and Internet browsing.

In the 2010s, **tablet computers emerged** as a cross between a smartphone and a desktop computer. Tablet computers are larger than smartphones but lack some input/output capabilities (e.g. keyboard and mouse) that are normally used with desktop computers.

What is common to both smartphones and tablet computers is that they often share the same mobile operating system and are considered the same computer platform for the purposes of this text.

Most popular mobile operating systems are *Android* and *Apple iOS* [8].

Mobile operating systems primarily differ from desktop operating systems in the following areas:

- Support mobile microprocessors, which are architecturally different from desktop computer microprocessors.
- Allow applications to interface with smartphone functionalities such as voice calls and text messaging as well as different input/output devices (e.g. **touchscreen**).

Devices such as digital cameras, smart cards, and smartwatches are also referred to as mobile devices, but they are considered different platforms for the purposes of this text.

2.5.2.1 Mobile Applications

Generally speaking, mobile applications have the same characteristics as desktop applications (described in the Section 2.5.1.1). To differentiate them from cross-platform applications (see below), the term **native applications** is often used in the mobile platform context.

Due to various reasons, there is a large demand for the mobile applications (more so than desktop applications) to be cross-platform – ergo, that the same mobile application/program can be executed on different mobile platforms. Cross-platform applications are described in more detail in the Section 2.5.5.

On the two most popular mobile platforms, *iOS* and *Android*, mobile applications are primarily distributed to end users through the proprietary **app store** (*App Store* on iOS devices by Apple and *Google Play* store on Android devices by Google). It is crucial to acknowledge that **the company that owns the app store has full control to review and curate the applications being offered there**.

Some app stores are less strict, while others include a detailed review process before an application can be distributed. Many of the app stores (including App Store on iOS and Google Play store) are also monetized; authors who wish to distribute their applications through the app store need to pay a fee but could be paid commissions for every purchase of their application.

2.5.3 Server/Web

Contrary to mobile and desktop computers, server computers have always been considered an enterprise market and might be less known to the general public. The history behind the Internet and how server computers came to be is described in the Section 2.3.

Modern server computers run on dedicated hardware that is often more powerful and expensive than the client computers (desktop and mobile computers) connecting to them. Server computers are often using dedicated microprocessors, memory chips, and data storage devices and are normally housed within **server racks**. This holds true both for bare-metal servers and cloud server setups (see Section 2.4).

Architecturally, they are closest to desktop computer platforms. Server computers normally run for long periods without interruption and often run unattended without a display device; instead, they can be managed remotely.

Virtualization (see Section 2.2.5) is one of the key components of modern web servers that allow a single, powerful server computer to host multiple virtual servers; this allows the same physical server to simultaneously support different server types (e.g. mail server, web server, and database server) or multiple users who want to have a server computer of their own.

The two most popular server operating systems are *Linux* and *Windows Server* [9].

2.5.3.1 Web Browser

Due to the tremendous growth and popularity of web applications in the 2010s, a web browser should be considered an important environment in which applications/programs are executed. Web applications are explained thoroughly in one of the previous sections of this chapter.

2.5.4 Embedded

According to a study, up to 98% of all microprocessors are used in embedded systems [6]. An **embedded system** is a computer system (combination of hardware and software) that has a dedicated function within a mechanical or electrical system. It is *embedded* as part of a single machine/device, alongside other electronic or mechanical components. Due to this fact, embedded systems often have computing constraints and are really tailored to a single use case (e.g. they are less general-purpose as desktop/mobile computers).

Embedded systems include everything, from portable watches, audio players to car computers, traffic light controllers, toys, and avionics.

From a software point of view, embedded systems are very tied to the underlying hardware. Despite the fact that the hardware can vary tremendously throughout embedded systems, there are certain platforms that make it easy to develop embedded solutions on single-board chips, such as Raspberry Pi and Arduino.

2.5.5 Cross-platform

As the number of different computer platforms has been growing, so has the need for cross-platform applications and technologies. An application/software can be considered **cross-platform** when it can run (without significant differences) on different computer platforms.

When looking through the eyes of a software development company – especially in the consumer market – it is natural that it wants its software product to reach as big of an audience as possible. The issue arises because all individual computer platforms (and even operating systems themselves) have so many specifics and differences that a *software/application can generally be written for a single computer platform only*. A desktop application developed for Windows operating system will (normally) not work on macOS. On a mobile computer, for example, applications can work with user drag and hand gestures, as the platform assumes there is a touchscreen present;

on a desktop computer, no such assumption can be made. Such differences, however, go beyond UI and are present on all computer levels.

One solution is to have dedicated per-platform versions of the same application. Maintaining different per-platform versions of an essentially same/similar application (from a user point of view), however, is very capital intensive.

Cross-platform software offers an alternative solution. It allows software developers to write an application only once, which can then be run on different computer platforms (the slogan "*Write once, run anywhere*" is often used).

Using a cross-platform approach can certainly save costs, but it involves a trade-off; due to the necessity to abstract specifics of different computer platforms, the following need to be considered when comparing a cross-platform solution to a *native* solution:

- *Performance and access to hardware*: As there are extra layers of abstractions, cross-platform applications are even farther away from the actual hardware and thus cannot utilize hardware as efficiently as native applications.
- *UI and UX*: Different computer platforms have very different concepts and components when it comes to UI; cross-platform solutions need to support all these differences, which inherently means that the UI and interactions are usually more limited and basic than those of native applications.

There are simply too many cross-platform frameworks, programming languages, and technologies to list them all. Few (but not all) of the popular ones are *Qt*, *Java*, *.NET Core*, and *Xamarin*.

Advances in web technologies in recent years have been very pivotal in establishing ***web applications*** *as a very promising cross-platform approach. Due to the popularity of WWW and web applications, almost all computer platforms nowadays support a web browser. This inherently means that web applications can run on most computer platforms; besides that, UI technologies for web (HTML, CSS,* and *JavaScript) have become very flexible and powerful and can be used to ensure a very fluid, smooth, and rich UX across different computer platforms. The second trade-off – performance and access to hardware – however, still remains a valid concern with the web applications.*

2.6 Applications

An **application** is a computer program aimed at fulfilling a specific user task/activity. These user tasks/activities range from various things, such as word processors, spreadsheets, email clients, media players, and videogames. In the context of laboratory digitalization, an application could be a LIMS system, ELN, ERP (enterprise resource planning system), etc.

Sometimes, the term **application suite** is used to define a group of applications that solve related user tasks/activities (e.g. Microsoft Office).

From a user point of view, applications often form the basis of interaction with computers. In the context of digitalization, it is therefore crucial to understand more about how applications are made, structured, and what their behaviors are.

Throughout the rest of this chapter, the terms application and software will be used interchangeably.

2.7 Values of Software

From a business standpoint, it is necessary to acknowledge where the value of the software really comes from. As part of the digitalization process, software is normally accredited with lowering operating costs (as much as up to 90%, according to one study [10]) as well as other benefits such as more informed decision-making and incremental economic growth [11].

2.7.1 Features

The primary user-perceived value of software comes from its features. Each type of software is developed to perform a defined set of **functions** (e.g. making an online purchase). **Features** are "tools" within the software that allow users to perform the necessary functions (e.g. view online shop, fill shopping cart, and make the purchase). It is clear that features bear direct business value of the software.

2.7.2 Design

Software design/architecture, which is explored in more detail in the Section 2.10, bears significant value as well. As explained in more detail in the Section 2.9, each software product has a life cycle of its own. This inherently means that as time goes on, software continues to be updated and maintained – which entails implementing new features within the software, upgrading existing features, and fixing defects and security issues.

When software has good design/architecture, new features and updates of existing features in the software can be implemented at a predictable cost with high reliability. In a reverse situation – when software's design is bad – implementing new features in the software is always a high-risk scenario with a very unpredictable outcome (and very often, underestimated costs). To put it differently, a good software design reduces volatility in the marginal cost of features [12]. Software design is primarily the domain of the software vendor, not its users; but users will certainly be affected when a key feature that they have been requesting for a year constantly gets delayed.

2.8 Software Development

As software users, it is necessary to know the basics of how software is developed/written and the complexity of it. Companies that develop/write software (often referred to as *software vendors* throughout this book) have teams of software developers (programmers) who simultaneously work on developing/writing an application/software. Team size, seniority, and different specializations all depend on the size, budget and timeline of the project, and specifics of the individual software. Often, one (or more) software developer is tasked with being the *software architect* – having a final say in the software design/architecture. Alongside software developers, teams often include other profiles, mainly project managers, testers/QA (quality assurance), UI/UX (user interface/user experience) designers, business analysts, and domain experts (people who understand the end user of the software).

When multiple programmers work on the same application/software, it is good to understand that the more people are involved, more overhead arises, more coordination is required, and less contribution per single programmer is achieved. To put it differently, adding two more programmers to the team of four will increase salary costs by 50%, but the software will certainly not be done 50% faster (but maybe 10% or 20% faster, depending on now the work can be parallelized).

Surely every software user has heard about (and encountered) a bug. A **software bug** is a defect/error in the software that produces an incorrect or unexpected result or causes software to behave in unintended ways. From a naive standpoint, it would seem that the more effort is put into developing an application, less defects will be present. Furthermore, with enough effort, a software application with zero defects can be produced. Sadly, reality is not like that.

As a software user, it is vital to understand that the majority of software systems (applications) are **complex systems** (other complex systems include, for example, the human brain and Earth's climate). A key finding – crucial for software domain – is that all of the current research says that complex systems can and will fail [13].

This inherently means that the software that you use will contain bugs. The software that you use will also **break/crash**. There is a certain chance that the data will be lost.

Of course, there are many different levels of measures and defenses against these problems.

*A **risk-based approach** is necessary when thinking about bugs and system failures; e.g. dedicate enough resources to prevent major catastrophes with a big enough probability and dedicate enough resources to have a good response plan when a catastrophe does happen. In practice, this means that – for example – a visual bug on settings (which are rarely changed by the users) is much more tolerable and acceptable than an invalid calculation of some business-critical data, which would need to be fixed immediately. Of course, these things are mainly within the domain of the software vendor.*

2.9 Software Product Lifecycle

Most software/applications can be grouped into one of the two categories: **internal software** (used internally by an organization) or **external software**, which normally describes a software application with the market outside the software vendor as its target audience.

When viewing software from a marketing perspective, every piece of software can be considered a product (including internal software). As any other product, regardless of the industry, **software products have a life cycle of their own**.

When the product is in the *introduction* and *growth* phases, it is gaining user adoption and fitting to the user needs. In a software product, this mostly means developing correct features within the software with the aim to solve a specific user problem. During these phases, only a smaller portion of the potential user base (innovators and early adopters) is using the software. Also during these two phases, *product pivoting* – which is a significant change of the product (and its features) – sometimes occurs. Once a software product reaches *maturity* phase, it is adopted by a larger portion of the market. At some point, the market becomes saturated, other competitive technologies and products emerge, and products enter the *decline* and *withdrawal* phase.

From a software product perspective, it is crucial to understand that developing and implementing the software is almost always *a continuous process*. It does not end when the *first version* of the software is released. Almost always, software then enters a **maintenance phase**, when software is still actively being worked on and developed further.

The work in the maintenance phase includes (but is not limited to) the following actions:

- Software issues and defects (bugs) are being resolved.
- Software is being regularly updated with latest security fixes.
- New features and changes/updates to existing features are being introduced, based on the user feedback and changed requirements from users.

For a software user point, there are three key takeaway messages from this:

1. *When deciding to purchase software, it is important to consider in which lifecycle stage the software is (similarly as purchasing any other product). Often, choosing a mature product is the optimal angle of approaching this decision.*
2. *In most cases, old and unmaintained software (often called* ***legacy code****) quickly loses value* [14]. *In the IT industry especially, advances are rapid, and if a software application is not being updated to keep up with hardware and operating system changes (as well as changes of other software), it is left behind.*
3. *A software application should (almost) never be considered as a one-time cost. Almost always, it is much better to look at it from the operating costs standpoint. As of 2019, most software vendors are transitioning toward a subscription-based per-user per-month (or per-year) offerings, which make the cost very transparent to software users.*

2.10 Software Design

As every software application is developed to solve a specific user problem, so does every software application require a design of its own. Software design is, thus, a very broad and complex topic. This section focuses mainly on how the majority of general-purpose desktop, web, and mobile applications are designed.

In the most general sense, the key role of most applications is to **process data**. In order for the software to be able to process data, this data must be captured digitally. Data can come from anywhere – either directly via user input, imported from some other software, or captured directly via dedicated device (e.g. video camera and laboratory instrument), and then transformed into digital format via hardware–software integration.

During the processing, data is inherently **transformed:** aggregated, expanded, modified, reduced, etc. Finally, this (digital) data might then be provisioned back to the physical (outside) world; essentially, back to the user. Most often, this is done via display device, but it can also be done via some other piece of hardware (e.g. an audio speaker and laboratory instrument).

How this works is outlined in Figure 2.6.

The transformation of data normally follows specific, man-defined rules of a specific domain – for example, how a blood sample's DNA is analyzed and interpreted, and how the tax is calculated based on income. These rules for transformation of data are often called **business logic**. This code and logic is predominantly related to the business value of the software.

When viewing software from this angle, it entails two things: (i) **data** and (ii) **code**. Code is the actual computer program, while the data represents the state of the application and the user data.

Same as with software features, software design can and does evolve over time. The term used to describe changes to the software design and optimization of software architecture to better suit new needs is called *refactoring*.

2.10.1 Code

The code of the application normally refers to the high-level code, assembly code and the machine code (explained in Section 2.4.1.2) of the software that serves as an input to the microprocessor to be executed. Most business applications nowadays are developed in high-level programming languages.

One of the key principles, and benefits of software, is its ability to be reused. Indeed, one of the very popular phrases among software developers is called *DRY*, or "Don't repeat yourself." As a consequence of this, the software industry is booming with **libraries, frameworks,** and **components** (which are all kinds of reusable pieces of software) dedicated to performing specific functions. Some of these libraries are proprietary and need to be purchased, while many of them are licensed as open-source projects (see Section 2.14).

In practice, this means that a modern application relies heavily on using the existing off-the-shelf libraries for specific functions and applying the domain-specific

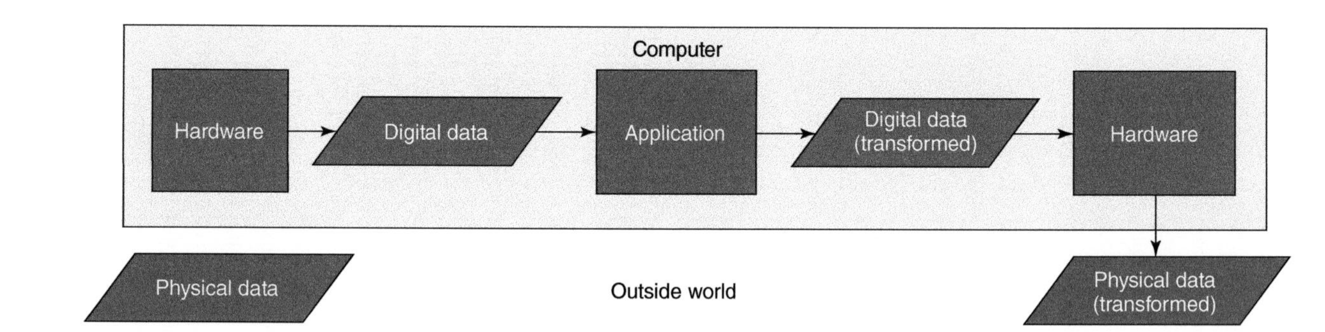

Figure 2.6 Transformation of data.

behavior onto them. For example, if an application requires user authentication, for most popular programming languages, a plethora of proprietary and open-source libraries exist. Same goes for UI components such as buttons, menus, and layouts. Indeed, developing a high-level application sometimes feels more like gluing many existing libraries and solutions together than writing any really original code. As each used software library might depend on other software libraries on its own, the number of used software libraries quickly grows high. It is not uncommon for a relatively simple application to have 100 or more such **dependencies**. These dependencies might also be *hardware dependencies*, or they might depend on the correct version and architecture of the operating system.

The problem is that all dependencies that are being developed and maintained by third parties (usually, a vast majority of them) are not under the application developer's control. Their maintainers might change a pricing model; they might shut down and stop supporting them; or they might simply change the way the library is used and force everybody to update their applications that use this library. As this is such a large issue for software developers, it has a specific name in the industry: *dependency hell*.

> *Of course, for software users, these details are irrelevant; they usually pay for the use of the software, while the dependencies are the problem for the software vendor to handle. Nonetheless, these dependencies can often drive the decisions made by the software vendor (e.g. change pricing and perform breaking changes) that, in the end, affect the end user of the software.*

This, however, is more prominent when adopting a software solution to be used on the premises infrastructure, as in this case, platform-specific dependencies directly relate to the organization's infrastructure, and they must be compatible. For example, setting up a LIMS system on the premises that was developed for Linux operating system will simply not work for an organization that has entire infrastructure running on Windows computers and servers. Such things need to be discussed, negotiated, and resolved before the software is purchased.

2.10.2 Data

The digital data that is being processed by the software is always based on modeling real-world data. This process is called **data modeling**. As with any other modeling, the right assumptions need to be made in order for the **data model** to be a good representation of the physical data world.

From a computer user standpoint, the most familiar data model is probably the computer **file system**. Most file systems assume that data is stored within files, and files are organized within a hierarchical ordering of folders. This is how people organized data before the digital revolution, and the reason for the popularity of file systems. File systems are, however, commonly not the ideal way to store the data of an individual application.

Historically, the most popular data model for digital data has been the **relational data model** that was established in 1969 [15]. This model assumes that the data is rectangular and that relations between data are well defined. For example, various spreadsheet applications (e.g. Microsoft Excel and Google Sheets) represent data in a similar way (albeit without the relations). Relational data model has been so pivotal in the industry that an accompanying language to query this relational data has emerged (and been standardized): **structured query language (SQL)**.

To a very large extent, digital data that is processed by a typical application is stored on nonvolatile data storage devices (e.g. hard disks (HDDs) and SSD) and only loaded into computer memory to speed up the processing. The system used to store this data is called a **database**. As reliable data storage and management is a very critical and complex operation, dedicated applications called **database management systems (DMSs)**, or **engines** have emerged. Some of the popular relational DMSs include *MySQL*, *PostgreSQL*, *Microsoft SQL*, and *Oracle*.

According to one study, roughly 75% of all databases in use are relational databases [16]. Alternative data models have only emerged to solve specific needs which cannot be implemented by the relational data model. Many of the alternative data models have emerged with the Internet revolution and the arrival of Big data. Some examples of nonrelational data models are *graph data model*, *time series*, *key-value store*, and *document-oriented data model*. A common term that has emerged to denote all these nonrelational data models is **NoSQL**. As of 2019, most business applications still model most of the data in a relational way and only use other data models for specific subsets of data.

From a user standpoint, sometimes there is a confusion between what is ***data*** *and what is* ***metadata*** *– which is defined as data about data. For the computer itself, this distinction is irrelevant. Both are digital data and are handled by the computer itself in the same manner.*

2.11 Software Quality

From the standpoint of a software user, **software quality** is an important attribute to consider; in essence, software quality talks about the following two topics:

- *Software functional qualities*: How the software complies to the requirements/ required specifications. In other words, this quality relates to whether the software features really implement the functions necessary to solve the user's problem.
- *Software nonfunctional qualities*: These relate more to the software design and how the software operates as needed outside the scope of the features.

The nonfunctional qualities of the software are generally less understood by software users, but they represent very important behaviors of the software system. There are many software **quality attributes** (often also referred to as "ilities," due to the suffix) for measuring nonfunctional quality. These should not be overlooked, especially when purchasing a software solution. Below are listed some of the most frequently used ones:

- *Performance efficiency*: The response of the software system to perform certain actions in a certain period; performance is a very critical attribute, as the software that is slow or latent has much less business value as well as demonstrating a very poor and frustrating UX.
- *Reliability*: The ability to continue to operate under predefined conditions; often, software systems fail when a component of the system breaks down (e.g. a database) or when the software is met with workload/traffic that was not planned.
- *Security*: Ability of the system to reduce the likelihood of malicious or accidental actions as well as the possibility of theft or loss of information. More details on software security are outlined in Chapter 4.
- *Maintainability*: The ability of the system to support changes. Maintainability of the software is heavily related to the software design/architecture and refactoring operations.
- *Interoperability*: Software system is interoperable when it is designed in such a way that it is easy to integrate with other third-party systems.
- *Scalability*: Ability of the system to handle load increases without decreasing performance or the possibility to rapidly increase the load. Scalability is especially important for web applications and SaaS, where the number of users can sometimes be hard to plan and estimate.
- *Usability*: Usability relates to how the software can be used by its end users to achieve their objectives with effectiveness, efficiency, and satisfaction in a quantified context of use. The entire domain of UX revolves around improving software usability.

Software quality management is also covered by ISO standards ISO 9126-3 and ISO 25010.

2.12 Software Integration

According to one study, it is estimated that 111 billion lines of new software code are developed each year [17]. Due to the expansion of the digital revolution, there exists an application for practically any justifiable (and solvable) problem. While it is certainly beneficial and helpful to use individual applications to solve specific user needs, computer users nowadays often use 10–20 applications on a daily basis for their work. For laboratory users, this should come as no surprise. Different applications process data during different stages of the workflow. A lot of times, however, a lot of the data transfer between applications is done completely manually by the user. Very often, applications have no means of operating with one another in automated ways. There is a great need (and large business value!) to automate and improve connections between various applications, which is only getting more necessary with the amount of software applications increasing at the current rates.

This topic is not new in the software world; software **integration**, **interoperability,** and **interfacing** are terms that have been around for decades. It is important to note that there are many ways how different software applications can interface with one another – what follows is only a selection of the most common means of how different applications and libraries can interface with one another.

2.12.1 API

By far the most frequently used term in the field of software integration is **API (application programming interface)**. The term has been in use since 1968 [18]. The problem of the API is that it is defined very broadly and can mean different things depending on the context. Nonetheless, in a simplistic sense, API relates to a part of a software component (e.g. an application and library) that represents an interface/protocol to be used when other software applications want to integrate/interface/communicate with the aforementioned component. The API is generally a part of the software component itself and is defined by the author.

Two ways of how an API can be employed to achieve software interoperability are shown in Figure 2.7.

- An application can interface with another application via its API.
  - In the domain of WWW, **REST APIs (Representational state transfer APIs)** are used heavily to achieve interoperability between different web applications.
- An application can internally use various other software libraries, which all expose an API that is used by the application to interface with them.

Due to the rapid growth of the Internet, there is quite a rich history of different technologies that have been used since the start of the twenty-first century to achieve interoperability between (i) different web applications and (ii) web applications and other platforms (e.g. mobile, desktop applications, and embedded systems).

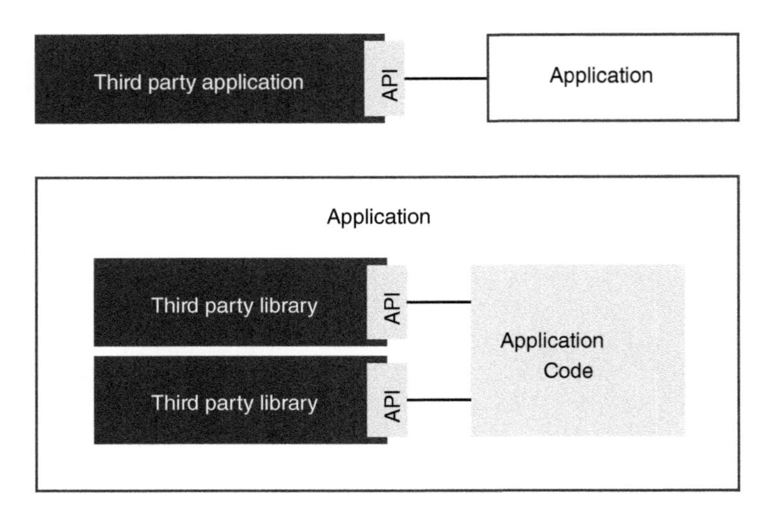

Figure 2.7 Two options to achieve software interoperability by employing API.

As of 2019, some of the popular means of interfacing are *REST APIs*, *web services*, *WebSockets*, *gRPC*, and *GraphQL*, while older, declining technologies include *SOAP* and *XML-RPC*.

2.12.2 Middleware

Another approach at interoperability is also by means of **middleware**. Middleware is a computer software application that is the centerpiece with which all applications and libraries interface. It is the "glue" that connects all the applications together. Middleware has been around for a long time, and it has been frequently used in the development of operating systems. Since the 2010s, middleware is also often considered when integrating and interfacing various web technologies.

2.12.3 Authentication and Authorization

In computer security, **authentication** means verifying a user's identity, whereas **authorization** represents the function of specifying access rights (to resources) to an individual user.

Majority of multiuser software systems must incorporate authentication and authorization mechanisms of some sort. In case of larger organizations, even more control is often required: the entire system of managing users and their access rights needs to be separated from the other software systems that the organization uses. This is such a common business requirement that there exist numerous popular protocols/standards in the industry such as *SAML*, *OAuth*, *Kerberos*, *LDAP, and Active Directory (AD).*

2.12.4 Internet of Things

IoT represents a system of embedded devices (e.g. smartwatches, home cameras, home controllers, and traffic lights) that are all interconnected via the Internet. It is important to note that from a technological standpoint, various hardware and embedded systems have been interfacing with web servers for decades; the popularity of IoT has simply surfaced in the 2000s due to the growth of the Internet and the various technological advances (e.g. Radio-frequency identification (RFID) tags, geolocation, artificial intelligence, and facial recognition) that have emerged.

2.13 Data-flow Modeling for Laboratories

In the context of laboratory digitalization, it is often useful to consider how the data flows within a laboratory or an organization. **Data-flow diagrams** represent a common technique toward modeling data flows. The key components of data-flow diagrams are as follows:

- *Data sources* – also referred to as *data producers*; the places where data originates; these can be various laboratory instruments, applications that allow users to manually input data, or third-party systems – which are, for an organization or a lab, considered an external data source.
- *Data processors* – parts of the system that process (transform, modify, reduce, extend, etc.) data.
- *Data stores* – also referred to as *data sinks, data lakes, warehouses, databases, files,* or similar; these entities hold and store data, normally for later use and for indefinite period. These can be external as well (e.g. third-party systems to which data is sent).

Drawing a data-flow diagram is helpful when assessing the current state of the art within the laboratory. Below are listed three data-flow diagrams that could represent three different laboratories in different stages of the digitalization process.

In this scenario (Figure 2.8), data is not consolidated at all; some data is still stored nondigitally in laboratory notebooks, while different digital data from different instruments (data sources) are not aggregated.

The next step in laboratory digitalization (Figure 2.9) is that all data is somehow propagated into a shared data store (e.g. ELN and LIMS system); some steps of entering the data, however, are still manual.

A fully digitalized and automated laboratory (Figure 2.10) can mean that all the laboratory instruments are automatically providing their data into dedicated data

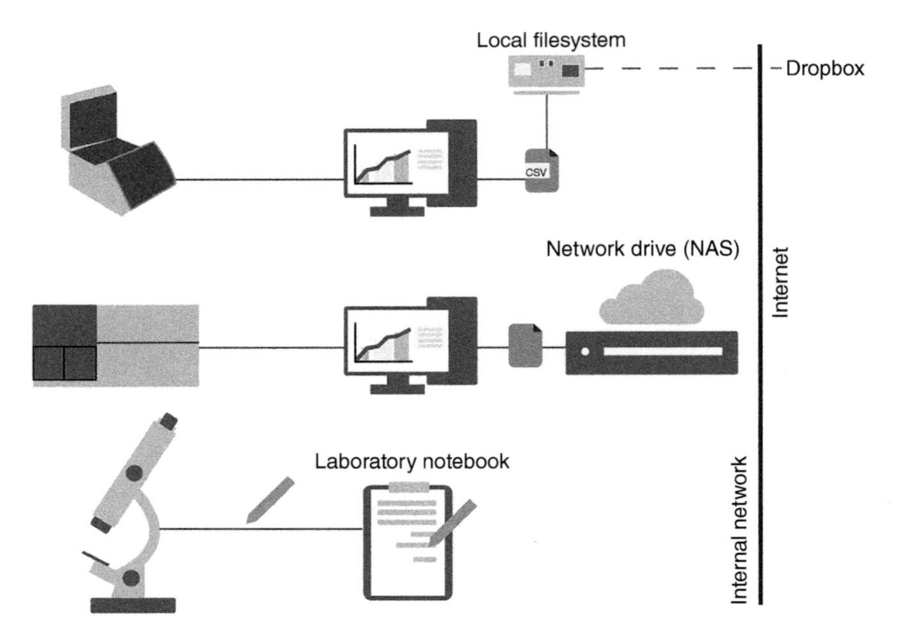

Figure 2.8 Diagram 1. Source: Courtesy of Janja Cerar.

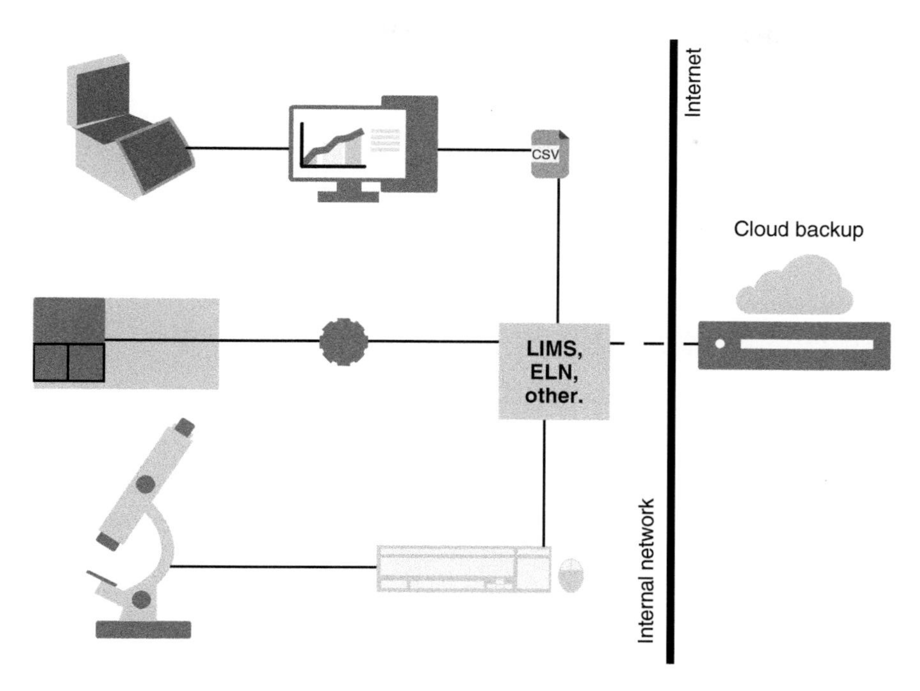

Figure 2.9 Diagram 2. Source: Courtesy of Janja Cerar.

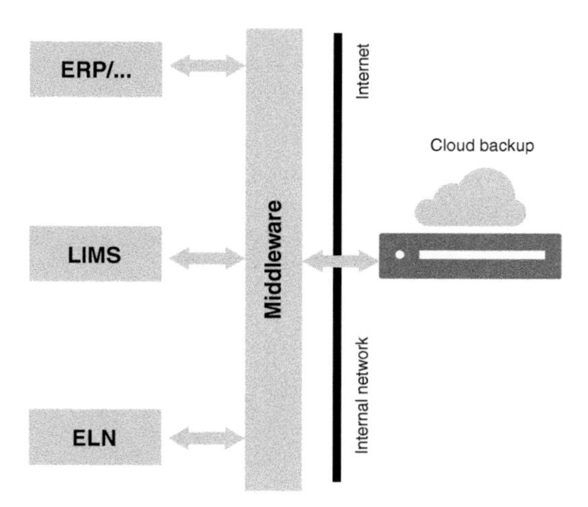

Figure 2.10 Diagram 3. Source: Courtesy of Janja Cerar.

stores; on the other hand, all the data stores as well as data processors (e.g. various user-facing applications, robotic process automation tools, and similar) can also use the middleware layer to communicate and exchange data with each other in an automated way.

2.14 Software Licensing

Similar to other digital data, software source code has all the benefits (and tribulations) that come with digital information processing. One of the big issues is that software source code can very easily be copied and redistributed. Primarily because of this, the vast majority of software source code is under copyright protection.

The choice of software license often dictates how the software pricing model works, how the software is provided and distributed to the end users, and how maintenance and updates of the software work.

Below, two of the most common software licensing models are explained.

2.14.1 Proprietary Software Licenses

The key attribute of proprietary software licenses is that the ownership of the source code copies remains with their publisher (normally, the software vendor); the publisher merely grants software users the permission to use these source code copies. These rules (and restrictions) are often specified in detail in an **end-user license agreement (EULA)**.

An example of a proprietary licensing model is how *Microsoft Windows* operating systems are licensed.

With the proprietary software licensing model, end users must agree to the software license agreement; otherwise, they are not granted access to the software itself.

2.14.2 Open Source

Open-source software licensing covers many different licenses that all allow (to a certain degree) modification and/or distribution of software source code under specified terms. This allows users to modify, customize, or simply view the software source code as well as redistribute it under the agreed terms.

Open-source licenses often require the origin of the software to be preserved (authors and/or copyright statement) or include a **copyleft** statement. Copyleft licenses require that any software that uses a copyleft-licensed software (e.g. library and application) must be licensed under the same copyleft license. For example, software developer/vendor cannot use a copyleft-licensed library (e.g. an open-source library of UI components) and then license the final software under a proprietary license. Instead, it is required that the final software will be licensed under the same degree of copyleft license. This often bears significant ramifications.

Open-source software is often available free of charge, but this is not required by the license itself. In contrast to proprietary licensing models, open-source software vendors normally make money in two ways:

- *Vendor offers a basic, community edition of the software licensed under open-source license and for free; vendor then also offers a paid, premium edition of the software (or additional software modules/plugins) that is based on the open-source version but is licensed in a more restrictive manner. The community edition often serves as a "hook" and is usually tailored to an individual user. Once the use of the software grows (e.g. the user introduces the software within his/her organization) and the need arises for more enterprise features, a premium edition must be purchased. This is often referred to as the* ***dual licensing*** *model.*
- *Vendor offers full software under open-source license and for free and offers paid support for the software.*

Contrary to proprietary software licenses, which often vary greatly from one another, several prominent open-source software license types have emerged, and vendors frequently simply reuse one of these license types for their purposes. Some of the most popular open-source license types are described below (Figure 2.11):

- *GNU general public license (GPL)* is one of the most common licenses associated with free and open-source software; it rose to prominence during the development of the Linux operating system kernel. The license is copyleft, meaning that any derivative work that uses the software licensed under GPL license must be distributed under the same or equivalent license terms as well. This copyleft statement ensures that the original software remains free, open source, and is not exploited by software companies that would not have to give anything back to the community. In the era of widespread software usage and Internet, GPL license is relatively rarely considered by software vendors as it often does not make sense from a business perspective.
- *GNU lesser general public license (LGPL)* is a less-restrictive version of the GPL license. It omits the strong copyleft clause, thus allowing any derivative work that uses/integrates the software licensed under LGPL license to be licensed under other terms (even proprietary). Most LGPL-licensed software applications are various software libraries which can be used as off-the-shelf components for building larger applications. However, whenever a developer modifies the original LGPL-licensed software (e.g. a software library), the modified software must be licensed under the same LGPL terms.
- *BSD licenses* are a family of very permissive software licenses that simply require anybody who uses a BSD-licensed software to retain/reproduce the BSD license notice (e.g. acknowledge the authors of the used software components). Some BSD licenses also include a clause that requires the user of the BSD-licensed software to acquire permission from the original authors/organizations when endorsing/promoting the derived software products (that use the aforementioned BSD-licensed software).
- *MIT license* is another very permissive software license that simply requires all copies of the licensed software to include a copy of the MIT license terms and the copyright notice unless the software is relicensed to remove this requirement.

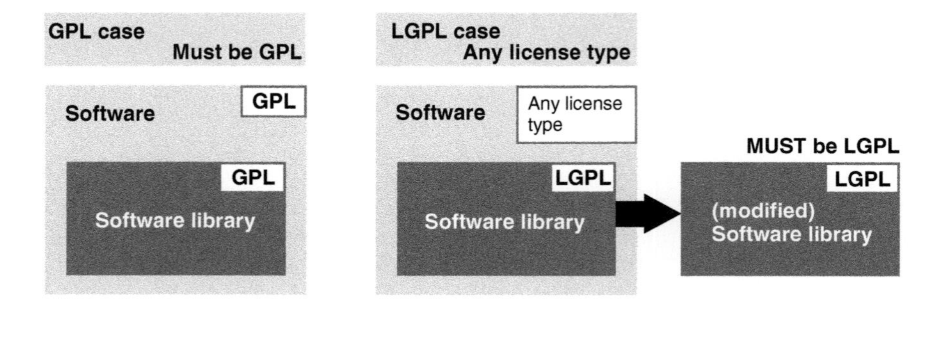

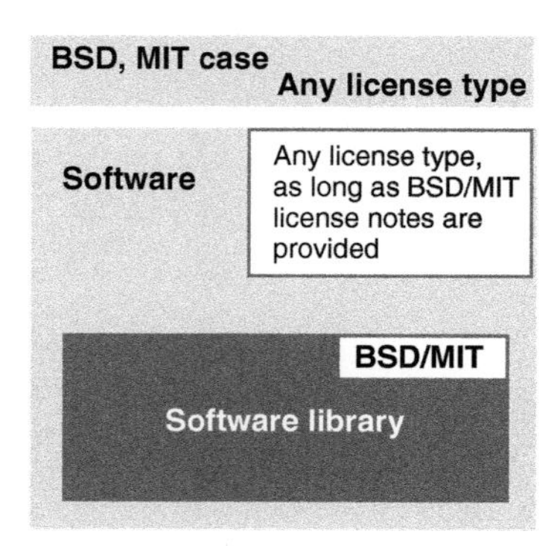

Figure 2.11 Most popular open-source license types.

MIT licensed software can therefore be used within proprietary-licensed software systems and is, as such, one of the most popular software licenses out there [19].

References

1. Barr, M. (2009). Real men program in C. *Embedded Systems Design* 22 (7): 3. http://academic.udayton.edu/scottschneider/courses/ECT358/Course%20Notes/LSN5%20-%20Embedded%20C%20Language%20Programming/Real%20men%20program%20in%20C.pdf.
2. Turing, A.M. (1936). On computable numbers, with an application to the Entscheidungsproblem. *Journal of Mathematics* 58 (345–363): 5. https://doi.org/10.1112/plms/s2-42.1.230.
3. ITU (International Telecommunication Union) (2019). *Measuring Digital Development: Facts and Figures.* Geneva: ITU https://www.itu.int/en/ITU-D/Statistics/Documents/facts/FactsFigures2019.pdf (accessed 1 February 2020).
4. Clarke, A.C. (2013). *Profiles of the Future.* UK: Hachette.
5. Ars Technica (2005). Total share: 30 years of personal computer market share figures. https://web.archive.org/web/20181007145234/https://arstechnica.com/features/2005/12/total-share/ (accessed 1 February 2020).

6 Netmarketshare (2020). Operating system market share. https://www.netmarketshare.com/operating-system-market-share.aspx?options=%7B%22filter%22%3A%7B%22%24and%22%3A%5B%7B%22deviceType%22%3A%7B%22%24in%22%3A%5B%22Desktop%2Flaptop%22%5D%7D%7D%5D%7D%2C%22dateLabel%22%3A%22Custom%22%2C%22attributes%22%3A%22share%22%2C%22group%22%3A%22platform%22%2C%22sort%22%3A%7B%22share%22%3A-1%7D%2C%22id%22%3A%22platformsDesktop%22%2C%22dateInterval%22%3A%22Monthly%22%2C%22dateStart%22%3A%222019-12%22%2C%22dateEnd%22%3A%222019-12%22%2C%22hiddenSeries%22%3A%7B%7D%2C%22segments%22%3A%22-1000%22%7D (accessed 1 February 2020)

7 Statista (2020). U.S. smartphone sales by year 2019. https://www.statista.com/statistics/191985/sales-of-smartphones-in-the-us-since-2005/ (accessed 1 February 2020).

8 StatCounter Global Stats (2020). Mobile and tablet operating system market share worldwide. https://gs.statcounter.com/os-market-share/mobile-tablet/worldwide/#monthly-201912-201912-bar (accessed 1 February 2020).

9 IDC: The premier global market intelligence company (2020). Worldwide server operating environments market shares 2017: Linux fuels market growth. http://web.archive.org/web/20190707125419/https://www.idc.com/getdoc.jsp?containerId=US44150918 (accessed 1 February 2020).

10 Markovitch, S., Willmott, P., and McKinsey-Corporate Finance Business Practise (2014). Accelerating the digitization of business processes. https://digitalstrategy.nl/wp-content/uploads/2014-J-Accelerating-the-digitization-of-business-processes.pdf (accessed 1 February 2020).

11 Parviainen, P., Tihinen, M., Kääriäinen, J., and Teppola, S. (2017). Tackling the digitalization challenge: how to benefit from digitalization in practice. *International Journal of Information Systems and Project Management* 5 (1): 63–77.

12 Rainsberger, J. and The Code Whisperer (2013). The eternal struggle between business and programmers. https://blog.thecodewhisperer.com/permalink/the-eternal-struggle-between-business-and-programmers (accessed 1 February 2020).

13 Cook, R.I. (1998). *How Complex Systems Fail*. Chicago, IL: Cognitive Technologies Laboratory University of Chicago https://web.mit.edu/2.75/resources/random/How%20Complex%20Systems%20Fail.pdf (accessed 1 February 2020).

14 Bernstein, D.S. (2015). *Beyond Legacy Code: Nine Practices to Extend the Life (And Value) of Your Software*. Pragmatic Bookshelf.

15 Codd, E.F. (2009). Derivability, redundancy and consistency of relations stored in large data banks. *ACM SIGMOD Record* 38 (1): 17–36.

16 DB-Engines (2020). Ranking per database model category. https://db-engines.com/en/ranking_categories (accessed 12 December 2019).

17 Cybersecurity Ventures (2017). Application security report 2017. https://cybersecurityventures.com/application-security-report-2017/ (accessed 1 February 2020).

18 Cotton, I.W. and Greatorex, Jr, F.S. (1968). Data structures and techniques for remote computer graphics. *In Proceedings of the 9–11 December 1968 fall joint computer conference part I 9 December 1968*, 533–544. https://doi.org/10.1145/1476589.1476661 (accessed 1 February 2020).

19 Blog – WhiteSource (2020). Open source licenses in 2020: trends and predictions. https://web.archive.org/web/20200503111426/https://resources.whitesourcesoftware.com/blog-whitesource/top-open-source-licenses-trends-and-predictions (accessed 1 February 2020).

3

Introduction to Laboratory Software Solutions and Differences Between Them

Tilen Kranjc

BioSistemika LLC, Ljubljana, Slovenia

3.1 Introduction

Computers revolutionized our workplaces, and laboratories are no exception. They changed how we record data, do experiments, and manage the laboratory. First laboratory computers came in a bundle with laboratory instruments, because they were much more advanced than the data recorders in use until then. Besides basic recording, the users could do real-time data processing as well as store methods and results for many experiments.

A few decades later, laboratory management software is starting to play a central role. Laboratory information management system (LIMS) was one of the first data management tools, mostly used in routine laboratories and the pharma industry. Today, research laboratories are starting to introduce LIMS systems and/or electronic lab notebook (ELN) to better manage the research data and projects. In order to do this, standard paper notebooks need to be replaced by PCs, tablets, and smartphones.

Laboratories also follow the industrial standards in automation. Laboratory automation is already well developed today, although the systems are expensive and only installed by laboratories that can justify their use and their cost. The core of each automation system is always the software that schedules experiments and manages the system. In this chapter, we give an overview of different types of software used in laboratories.

Figure 3.1 shows how the software for laboratory instruments and lab automation sends data through middleware to the digital data management software, such as ELN, LIMS, laboratory data management system (LDMS), or chromatography data management system (CDMS). The data from digital data management software are then fed to data analytics software. The communication between digital data management software and data analytics software can be again facilitated through middleware or specific plugins. Robotic process automation (RPA) software can be used to facilitate communication between all layers.

Digital Transformation of the Laboratory: A Practical Guide to the Connected Lab, First Edition.
Edited by Klemen Zupancic, Tea Pavlek and Jana Erjavec.

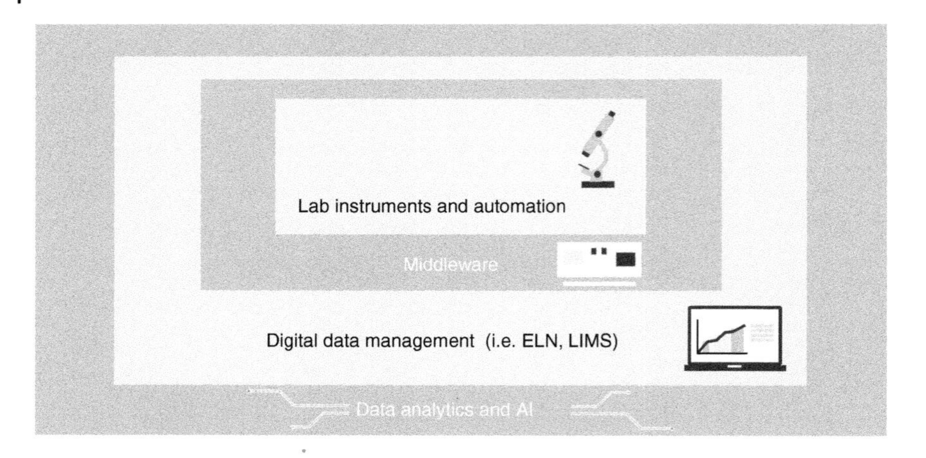

Figure 3.1 Types of software found in the laboratory.

3.2 Types of Software Used in Laboratories

3.2.1 Electronic Lab Notebook (ELN)

An ELN is an electronic replacement for the pen-and-paper notebook for recording the experimental work in a laboratory. It is more suitable for laboratories where the work is less routine. This is very typical for research and other nondiagnostic laboratories. The main benefits of using an ELN over a traditional paper notebook are the ability to search, share, and collaborate. Data is also better organized, stored more securely, and storage takes less physical space.

The first discussions about ELN are dated back to early 1993 during the American Chemical Society national conference [1]. Dr. Raymond E. Dessy from Virginia Polytechnic Institute delivered the first working example of an ELN in 1994 [2]. The first enterprise ELN solution was developed and used by Kodak in 1997 [3]. An important milestone happened in June 2000 when the Electronic Signatures Act formalized the use of digitally signed electronic records in the courts [4]. A study from 2017 reported that only 11% of study participants are actively using ELN, while 76% were strongly interested in implementing one. The main objection toward the implementation was limited budget [5].

The key feature of ELN is a digital recording of laboratory data, which changes the ways we work and interact with research data. Digital records can be searched, shared, collaborated on, attributed to real people, and retained during personnel changes. Besides this, modern ELNs have become universal tools with many extra features that help scientists to save time and be more efficient.

Data management helps users to organize and manage all data related to the experiment. The main organizational unit is typically an experiment, which includes data such as protocols, results, and reports. Data can be coauthored and collaborated on. Therefore, ELNs typically also implement user authorization control, which allows fine-tuning who has access to what content. One of the important features is also inventory management, which is typically separated from experiments and allows tracking of samples, reagents, and equipment.

Example *Sarah forks a protocol that is typically used in the research group and implements changes. She sends it to the supervisor who can directly review it. Then she saves the protocol as a new version of the original protocol. The new protocol is linked to the experiment which enables full traceability and guarantees that from this point on all users use the newest version of the protocol.*

Integrations with other software tools expand the usability of ELN. This means that ELN becomes a central point of data storage. Some common integrations are word processors and spreadsheet software as well as domain-specific software, such as statistics packages or chemical drawing software. Because all the data is stored in a central place, it is also easier to maintain the data integrity. Laboratory instruments can directly feed the data into ELN and then through other integrations with other software packages. The integrity is maintained throughout the data flow by locking of data files and versioning. In advanced settings, individual results can be inserted into reports while maintaining data integrity.

Example *A result derived from an automated analysis pipeline can be safely inserted into a report and locked for editing, while the rest of the report remains open.*

Report generation is one of the most common tasks of ELN users. ELNs can simplify with this process either through native functions or integrations with external software. The simplest implementation helps with formatting protocols or results, while more advanced use artificial intelligence (AI) tools to generate drafts of scientific papers based on researcher's data saved in the ELN, such as Manuscript Writer by SciNote, an AI-based tool that can use the information available to draft scientific papers.

Electronic signatures, access control, and audit trails are the functionalities that also provide a legal basis for the use of ELNs in industry. ELNs can be equally used in patent disputes and are therefore a credible record for the use in the court. Besides that, ELNs also comply with industry standards, such as CFR 21 Part 11, and can be therefore used in good manufacturing practice (GMP) laboratories.

Example *After Sarah finishes the experiment, she needs to confirm all steps in the protocol, upload all the data, and analyze the results. She electronically signs the experiment, and after that, the experiment is locked and cannot be changed, unless by an authorized user. Throughout the process, all changes are logged in the audit trail.*

Project management tools are less common in ELNs but can be very helpful if the organization does not use other project management tools. These functions are usually limited to the creation and assignment of tasks and progress tracking of the experiments. This is much less than a typical project management tool would do; therefore, it is mostly targeted to academic groups and smaller organizations.

Tablet PCs are present in our lives for a while now; however, their use in laboratories has been mostly discouraged until lately. Most of the chemicals used in typical biotechnology or life sciences laboratories do no harm to the hardware. Many ELNs can be used on a tablet PC, which gives the user instant access to all the protocols, data, and inventories while working at the laboratory bench. Although we are seeing more and more new technologies making their way into the labs, such as voice commands and augmented reality glasses, tablet computers seem to be the best missing link between paper and digital. They have the right balance of performance, portability, and price for the task.

The core functionality of ELNs related to digital record storage (searching, sharing, and collaboration) should simplify the management of research data. However, the feasibility is often measured by the increase in efficiency. During the implementation of ELN the team usually spends more time getting used to it and preparing the protocols and workflows. But after the initial period, the users report that they use less time than using paper notebooks (preliminary data). ELN, therefore, can become a central point of data storage and empower scientists to work smarter and more efficiently.

3.2.2 Laboratory Information Management System (LIMS)

The work processes in diagnostic and quality control (QC) laboratories are usually focused around samples. Each sample has associated metadata, such as the name of the patient or lot number and date of sampling. Then the sample gets assigned the assays to be performed on. The results of these assays are then reported back for each sample. LIMS is specifically designed for sample-centric processes and is used to manage the above process from end to end.

The core features of LIMS are directly related to the path of a sample through the laboratory. It starts by registration of samples, which is usually done by scanning a barcode. All the related information is then fetched from other databases in the organization – who ordered the analysis, required analyses, due date, etc. The samples then undergo the analyses, and the results are aggregated for each sample. The system then generates a report and provides lifecycle options – for example data entry, reviewing, signing, and delivery.

Modern LIMS systems offer much more than the core functionality. One of the most desired features is integration with laboratory instruments. This means that LIMS can send the analysis parameters to the laboratory instrument and then fetch the results after the analysis is done. Such laboratory instruments are equipped with barcode scanners and can precisely track the samples they process. Such closed integration of a device with LIMS helps to maintain the highest level of data integrity; however, many devices still cannot be integrated with LIMS (mostly "open scope" devices used in research laboratories, such as a qPCR instrument). In such a case, the laboratory technician still needs to manually transfer the data, which is a potential source of errors and data integrity risk.

Example *Many laboratory instruments allow queuing of samples, so a lab technician puts the sample into a sample hotel and the instrument automatically takes it from there, reads the barcode, gets all the required parameters from LIMS, and processes it. After the analysis, the instrument sends the data automatically to LIMS, processes according to a predefined data analysis workflow, and inserts the result into a report. The whole sequence is completely automated, and it starts when the lab technician puts the sample into the sample hotel.*

LIMS systems are designed to support audit trail logs, access control, and electronic signatures. These features are needed to comply with medical and industrial regulations, such as CFR 21 Part 11. LIMS systems can, therefore, be used in regulated environments, such as diagnostic laboratories and GMP laboratories.

A very important feature is also integration with electronic medical records (EMR), health information systems (HIS), or enterprise resource planning (ERP). EMR and HIS are often used in healthcare organizations to keep health records of patients. ERPs are more common in the industry to facilitate and automate the technical business communication between departments. Connection to customer relationship software (CRM) and invoicing software is also of importance, particularly for companies that are providing laboratory services to others, such as Contract Research Organizations (CORs).

Example *A health practitioner can order the set of laboratory tests through HIS and send the samples with barcodes to the laboratory. LIMS in the laboratory then fetches the list of tests directly from HIS. After the analysis report is generated and signed by the laboratory personnel, LIMS sends the report back to HIS, which stores it in the patient's EMR.*

Example *Drug manufacturing starts by entering a plan in ERP. This starts a sequence of events, starting with resource reservation, manufacturing of the medicines, production QC, packaging, final product QC, and product release. Throughout this process, much information is exchanged between different departments of the pharmaceutical manufacturing plant. LIMS is typically used in QC laboratories. LIMS in such a setting needs to fetch the data about the batch from the ERP, together with the analytical tests defined in the product specifications. After the tests are performed LIMS can automatically generate a Certificate of Analysis and store it in the ERP.*

LIMS can also connect to the ELN (if both support the integration) and help with regular daily activities in the laboratory. Environmental measurements, such as temperature and humidity, both from laboratories and refrigerators and freezers can be automatically fetched from sensors by LIMS at given time intervals. LIMS can also schedule system suitability tests, maintenance, and calibrations of the laboratory equipment. Some LIMS systems also have inventory management features that keep track of the stock of reagents, their location, expiry dates, storage conditions, reorder process, material safety data sheets (MSDS), and waste protocol. Users can also track the statistics about the facility, such as the number of samples processed in the given time frame, associated costs, and time spent.

3.2.3 Laboratory Execution System (LES)

Laboratories usually have a set of methods and standard operating procedures (SOP) that they need to follow to ensure the quality of the service or to follow the good practices in the industry. In a very rigid setting, they might be using a computer system, laboratory execution system (LES), that would guide and track the user through the process. The laboratory technician would therefore follow the instructions on the computer or tablet and sign off each step to make sure it was done according to the guidelines. Such a system could come in different forms, for example printed forms that are then digitized, or tablet apps that are used at the bench. There are few independent LES systems; however, more often LES became a part of the LIMS or ELN system.

3.2.4 Laboratory Data Management System (LDMS)

The most common way to store laboratory data is still a computer file system (hard drive, USB stick, CD/DVD, and network storage). While practical, it has limited functionalities for the organization, metadata storage, searching, accessibility, and access control. A solution is LDMS which provides a layer of abstraction above the file system to form a sort of a data lake. LDMS collects the raw data from laboratory equipment and stores it together with metadata (descriptors of the data).

LDMS provides organization functionalities. The data can be stored into project or product folders, organized by time/date, experiments, etc. The raw data can be also searchable, has audit trail capabilities, and access control functions. We can think of it as a document management system for laboratory raw data.

By adding advanced functionalities to LDMS it can become an ELN or LIMS. More commonly, ELN or LIMS is integrated with LDMS. The role of LDMS is to securely store the raw data and maintain the data integrity from the laboratory device to the LDMS. Then, ELN or LIMS can access raw files from LDMS and process them.

So why would anyone use LDMS if ELN does the same job but has more advanced features? One possible reason is that the organization wants to store literally all data that comes out of the laboratory instruments, which would quickly turn ELN into a mess. In practice, LDMS can replace the storage of raw files on the file system of the computer connected with the laboratory instrument and act as a central storage point for all raw data. The benefit is better security, safety, and integrity of data, as well as additional metadata that can be used for filtering. The ELN does not need to keep a copy of raw data and hence rather integrate directly with LDMS and store only the link. This would further improve data integrity and reduce the amount of storage needed.

3.2.5 Chromatography Data Management System (CDMS)

Also referred to as chromatography data system (CDS), this system is basically an LDMS specifically designed for handling chromatography systems and data. On top of LDMS data storage functions, it provides functions for managing chromatography systems and working with chromatography data.

CDMS has a client–server architecture to implement a central point of data storage as well as allow multiple users to work with the data. The central server manages the chromatography systems and allows remote control, access control, and management of instruments. Most of the work is done through the remote clients that run from the users' PCs. CDMS usually implements advanced access control and separate workspaces, so devices from R&D and production (GMP) environments can be managed by the same server.

The most important feature of CDMS is the chromatography toolkit. Users can create the whole analysis workflow, starting with chromatography method development and validation, data analysis pipeline, and reporting. CDMS also implements life cycle management for methods and results so the managers can plan and track reviews and signoffs. Methods, results, and other data elements retain the whole history of changes, and all events are logged in audit trails.

Example *A scientist in an R&D lab in a pharmaceutical company developed a chromatography method and data analysis pipeline in CDMS. After the method is developed, the CDMS asks other scientists to review it and the manager to sign the development and validation report. The method is then transferred to a GMP production laboratory where it is routinely used. The method cannot be changed. The results from the chromatograph are again reviewed and then signed by the manager.*

CDMS systems are designed to control multiple chromatography systems, which means they also provide bidirectional communication between the CDMS and each chromatography system. Beyond that, CDMS can be integrated with LIMS and therefore can work in an automated way – the lab technician puts barcoded samples on the sample tray, the system reads the barcode before the analysis, analyzes the data, and sends the report to LIMS.

CDMS systems are usually chromatography vendor specific. This means that they support the vendor's chromatography systems out of the box. Usually, they can be extended with plugins to support third-party chromatography systems or software solutions.

3.2.6 Process Analytical Technology (PAT) Software

Manufacturing of pharmaceuticals has undergone a transformation from static batch manufacturing to a more dynamic process. This requires real-time monitoring of critical parameters and process adjustments to meet the desired product quality. In such manufacturing facilities, process analytical technology (PAT) systems are used to manage and execute real-time monitoring and process adjustments.

PAT covers the whole analytical process, starting with sampling and analysis, followed by the logic that directly adapts the process parameters to meet the desired critical parameter specifications. An important part of PAT is also software that manages the process and provides some extra features, such as method development, visualizations, reporting, and overview of multiple PAT systems.

3.2.7 Automation Scheduling Software

Laboratories that perform many routine tasks often use laboratory automation to reduce the manual workload and increase repeatability of experiments. Laboratory automation is commonly used in QC laboratories in the pharma industry and some diagnostic laboratories. Laboratory automation systems are managed by automation scheduling software. Its core functionalities are all related to managing the automation work cells, planning the processes, and overview of the results. Most of the software solutions offer simple programming of automation workflows by drawing flowcharts. One of the core features is also integration with LIMS or ELN systems to store the results of the assays.

3.2.8 Laboratory Instrument Software

Most modern laboratory analytical equipment need a computer and dedicated software to operate the device. Most software solutions are designed as desktop applications or embedded systems. The trends show a rise in cloud applications in recent years, although this is isolated to a few companies that explore new possibilities that cloud applications bring.

One of the features most demanded by laboratory users is a good user experience, which is still mainly neglected by device manufacturers. The consumer market has set high standards, and laboratory users expect the same user experience from lab devices as they get it from their smartphones and home entertainment devices. Software user experience has also become one of the important factors when deciding between multiple instrument vendors.

Among the technical features, the trends also show the importance of connectivity options for devices. Connectivity features are important for direct integration of instruments with central data storage systems, such as ELN or LIMS. Device integration allows remote control of the laboratory instruments as well as a direct data feed to ELN or LIMS. This, in turn, increases efficiency and improves data integrity.

The integration of laboratory instruments with ELN or LIMS is not difficult to implement per se. The main challenge is the number of different integrations that would need to be implemented. Even if both instrument and ELN expose their APIs, there still needs to be a middleware that provides end-to-end communication between the APIs, acting as a translator. Given the number of ELN and LIMS options on the market, this would be a big chunk of work. Several standardization initiatives have been founded lately to overcome this challenge. Standardization in lab automation (SiLA) standard aims to become a remote control and data exchange communication standard; however, there are only a few implementations up until now. Analytical information markup language (AniML) is another initiative to standardize the data file format for experimental data. It got more traction, mostly in large laboratories where they are combining different data sources in analysis pipelines. The vendors are somehow reluctant to implement standards, mostly due to business reasons. The situation will likely revolve similarly as in the other industries. The laboratory instrument software will migrate to the cloud, in turn developing interfaces for programmatic access. Some vendors will support integrations/compatibility with *some* other vendors. There might also be a service such as Zapier that would act as a middleware between different applications. However, we are unlikely to see grand standardization in scientific data management.

3.2.9 Middleware and Robotic Process Automation (RPA)

Even though the device integration with ELN or LIMS is still far from reality, there are other solutions on the market that provide direct integration, although not native. Such software is called middleware, which refers to software solutions that act as bridges between different applications. They are most often used on networks but can work locally as well. In laboratories, middleware usually takes care of data exchange between instruments and data management systems, such as LIMS or ELN. Most middleware on the market today are not out-of-the-box solutions. Buyer of a middleware solution usually needs to sponsor the development of integration plugins for their devices. This is quite a significant chunk of work because laboratories usually have a number of devices from different vendors and of different ages. Developed integration plugins are usually reused and eventually, the "library" of plugins will fill up to cover the vendor landscape. A big step forward would be also more widespread use of standards, such as SiLA and AniML. Until then, middleware is a solution that works well, but the setup can be complex, and maintenance expensive.

An alternative that might help with integrations in some cases is RPA software. RPA is a piece of software that helps with automating repetitive computer tasks. RPA can fetch data from different sources, such as files, spreadsheets, databases, or even APIs of other applications, such as ELN or LIMS. The data can then be processed according to a configured pipeline. RPAs are not limited to processing data – they were designed for automating software testing, and today they have much broader use. They are heavily used to perform automated tasks in other industries, such as repetitive and scheduled marketing tasks, e.g. posting on social media.

Many laboratories perform repetitive tasks daily, which mostly involves copying data between different applications. The manual nature of work requires multiple data integrity checks during these processes, which is inefficient. RPAs are therefore being tested in laboratories to automate data copying between applications and maintain the data integrity during the process. RPAs can therefore potentially solve many challenges in laboratories. Although they are off-the-shelf products, the configuration is still rather complex. Due to a large user base, there is much more documentation available as well as help forums. Besides commercial solutions (UIPath, Blue Prism, Automation Anywhere, etc.) there are also a few open-source solutions, such as Robotic Framework Foundation, TagUI, and Taskt.

3.2.10 Data Analysis Software

Data analysis software is probably the most widely used software type. There are many solutions for data analysis, and most of them are specific for an assay, type of experiment, or type of data. For example, the largest field of data analysis software is next-generation sequencing. Even in such a narrow field, we can find different types of platforms, from command-line tools, standalone desktop apps, and cloud platforms. Some cloud platforms can even talk directly with a sequencer and can do the whole pipeline processing, so the user just needs to start the experiment. The widely popular and open-source cloud solution for processing sequencing data is Chipster, and there are also several commercial solutions.

Most scientists still use spreadsheet software solutions for data analysis, simply because they can be used for any type of data. This works fine for analysis of simple data, but when we approach big data, spreadsheets cannot handle it anymore. Sometimes this can be solved by statistical packages, such as R, SPSS, and Minitab. Some popular solutions also allow the visual design of data processing workflows, such as Orange and Knime.

3.2.11 Enterprise Resource Planning (ERP)

ERP is used for tracking business processes. Although it is not specifically made for laboratories, they often need to interact with ERP. Common use cases would be tracking orders, consumable use, inventories, invoicing, and planning order execution. Some of these features can be already implemented in LIMS. They are also tracked in ERP if they need to be tracked at a company-wide level. This is usually important in large organizations, where laboratories daily communicate with other departments, such as procurement and finance. ERPs are one of the most common software solutions on the market; therefore, there are numerous vendors offering their solutions. The market leaders are Oracle, SAP, Sage, and Microsoft.

References

1 American Chemical Society (1993). Division of chemical information. *Chemical Information Bulletin* 45 (3): 46.

2 Borman, S. (1994). Electronic laboratory notebooks may revolutionize research record keeping. *Chemical and Engineering News* 72 (21): 10–20. https://doi.org/10.1021/cen-v072n021.p010.

3 National Research Council (1999). *Impact of Advances in Computing and Communications Technologies on Chemical Science and Technology: Report of a Workshop*, vol. 15, 163. Washington, DC: The National Academies Press https://doi.org/10.17226/9591.

4 Zall, M. (2001). The nascent paperless laboratory. *Chemical Innovation* 31 (2): 14–21.

5 Kanza, S., Willoughby, C., Gibbins, N. et al. (2017). Electronic lab notebooks: can they replace paper? *Journal of Cheminformatics* 9: 31. https://doi.org/10.1186/s13321-017-0221-3.

4

Data Safety and Cybersecurity

Luka Murn

BioSistemika d.o.o., Koprska ulica 98, 1000 Ljubljana, Slovenia

4.1 Introduction

With the progress of digital revolution and global access to the Internet, digital data has become a very valuable business resource – it has even been suggested that data is the new oil. In the past, businesses relied on data primarily to support, improve, and optimize their processes. Nowadays, data processing itself is a very viable business and is tied heavily to the domination of the tertiary sector (services) across the developed countries.

As organizations and users store increasing amounts of their data in a digital form, it is necessary to consider how to ensure that this data is (i) not lost and (ii) protected from unauthorized use and access, either due to business value or, in the case of end users, data privacy and protection.

This rings very true for life sciences as well. As laboratory instruments become smarter and smarter, more digital data is generated per experiment as ever before. More so, with the advent of cloud and IoT, digital data is being transferred over computer networks at an unprecedented pace. As (digital) data is one of the key assets, both in research and diagnostics, it is a crucial responsibility of any life science company to ensure that the data safety and cybersecurity is met with adequate measures.

4.1.1 Magnetic Storage

While digital data can be stored on different types of physical media, the majority of long-lasting digital data – as of 2019 – is stored on **magnetic** storage devices, mainly **hard disk drives** and **magnetic tapes**.

Magnetic media are considered types of *nonvolatile storage*: this means that the information can be stored and retrieved even after power has been turned off and on again. While magnetic tapes have been invented before the digital revolution itself in 1928, hard disk drives have been extensively used since their invention in the 1960s. The main difference is that magnetic tapes need to be read sequentially to access data, while the benefit of hard drives is that the data can be accessed in a random-access manner (in any order).

Digital Transformation of the Laboratory: A Practical Guide to the Connected Lab, First Edition.
Edited by Klemen Zupancic, Tea Pavlek and Jana Erjavec.

While the random-access properties of hard disk drives are generally preferred to sequential access of magnetic tapes, magnetic tapes are still widely used as a data backup medium due to a lower manufacturing cost per bit of data. Hard disk drives, on the other hand, are predominantly used as a secondary data storage for general-purpose computers.

It is important to acknowledge the **lifespan of magnetic storage**. For modern magnetic tapes (such as Linear Tape-Open media), 15–30 years of archival data storage is cited by manufacturers; for hard disk drives, however, this number is often significantly lower. An extensive study has shown that most hard disk drives live longer than four years, but the failure rates of hard disks start to go up significantly after three years of use [1]. A direct consequence of this is that digital data needs to be copied often, and faulty hardware needs to be replaced as time goes on.

4.1.2 Solid-state Drives

There is also another, nonmagnetic technology that is used to store digital data since the 1990s: **solid-state drives** (SSDs). These are often used instead of hard disk drives as a secondary data storage for general-purpose computers; they offer better read times, are more resistant to physical shock, but have a more costly price per bit of data. Therefore, they are usually not used for backups.

4.2 Data Safety

Data safety revolves primarily around preventing a scenario of data loss from occurring. Any digital data that is stored on physical media must be evaluated from the data safety standpoint. In the scope of laboratory digitalization, an important consideration is that every organization which stores digital data must take necessary measures to ensure a high-enough degree of data loss prevention.

The first step for every organization is to identify where its digital data is stored. For bigger organizations, this is the responsibility of the IT department.

- Virtually every organization has various PCs, laptops, and workstations that contain data.
- Often, computers, servers, and devices located on the internal organization network also hold valuable data.
- Lastly, any data that is stored on third-party cloud providers (accessed through the Internet) must also be considered. Generally, data safety measures in such a scenario are the responsibility of the third-party cloud provider (e.g. the software-as-a-service provider – see the previous chapter: Crucial Software Terms to Understand).

4.2.1 Risks

In order to set up a mitigation strategy for ensuring data safety, risks that lead to data loss must be well understood and evaluated. Following is a list of possible risks that can result in digital data loss:

- *Software code/design flaw, crash, freeze, etc.*: As described in Section 2.8.13, there is always a chance that software systems will fail. The more software systems are used to process data, the greater the chance that one of them will fail. When a software system fails, it can corrupt, delete, or lose the data that it was processing prior to the error.
- *Privileged user administration error*: Often, IT administrators are manually performing data loss mitigation tasks such as making backups of data; as any manual process, human error can be a factor at this point; if such privileged user makes a mistake, entire sections of data can be deleted/lost.
- *Power or hardware failure*: Whenever a power shortage (e.g. loss of electricity) occurs in the computer system – or when a hardware component of the system fails, as is often the case with hard disk drives (see Section 4.1.1) – digital data can be lost. Often, data that was being processed prior to the error is corrupt, deleted, or lost.
- *Business failure (vendor bankruptcy)*: When using third-party vendors/services for processing of digital data, there is always a risk of vendor filing in for bankruptcy. It is very important to evaluate the contracts and agreements and review what the vendor guarantees in relation to the digital data in such a scenario. This subject is often called **business continuity**. The vendor should, at minimum, allow for export of all its customer's digital data within a reasonable timeframe.
- *Natural disaster, earthquake, flood, fire, war, "acts of god"*: When the devices and hardware that are used to store the digital data are physically destroyed, naturally, the digital data is lost as well.

4.2.2 Measures

There is a universally accepted mitigation measure against all risks associated with data safety: backups.

4.2.2.1 Backups

One of the key advantages of digital data (compared to, e.g. paper) is the ease of copying data. **Backup** represents a duplicate/copy of the digital data at a single point in time.

By having many backups of the digital data from different time points in the recent (and not so recent) past, data loss can be minimized. Even if entire digital data collection is corrupted or lost (due to one of the risks mentioned above), the last backup can be used to restore the state before the incident occurred; thus, the only data that was really lost is the data that has been processed in the time between the last backup time and the data loss event. Backups are sometimes also referred to as **snapshots**.

The frequency and strategy of how backups are made must be evaluated in a risk-based manner. Backups require more digital storage space, which inherently means extra costs. There are many strategies and approaches to this. For example, an organization might be performing daily backups, and storing daily backups for past 30 days, and then only a single monthly backup for months prior to that.

The backup mechanisms are usually a combination of automated and manual (human) effort. The entire process – how backups are performed and also importantly, how data is **restored** from an older backup in an event of data loss – must be validated, performed, and reviewed on a periodic basis. Most organizations perform this validation of backup and restore procedures at least on a yearly basis. These are often referred to as **disaster recovery plans**.

In case of third-party vendors and services, data backup mechanisms and disaster recovery plans are normally their responsibility and should be reviewed by the customer.

4.2.2.2 Data Replication

An extra safety measure with data loss prevention is to have data and backups replicated to different physical locations, possibly within different countries or even different continents on the globe. Normally, this is achieved by copying the data over the Internet connection to different parts of the globe. This adds an extra degree of insurance especially against natural and man-made disaster scenarios.

It is worth pointing out, however, that more and more laws and regulations are starting to be very strict as to where the digital data is physically stored and might impose certain restrictions on the location of the data. Therefore, physical locations that are considered to be used for storing data must be reviewed with the appropriate legislation.

4.3 Cybersecurity

As mentioned in the introduction of this chapter, digital data has become incredibly valuable in the modern world; data is very often related to business value; alternatively, it can be personal in nature – e.g. it could be used to uniquely identify an individual. These are just two examples of the value of data. Inherently because of this – depending on the perceived value of the data – there might exist hostile external adversaries (individuals and/or organizations) that would want to gain illegal access to the data.

Secondly, an organization's computer infrastructure (software and hardware) itself can also be considered an asset, especially for the organizations whose primary purpose is processing of data. If an adversary were to gain control over the organization's computer resources, such an act could also result in business damage; besides the unauthorized access to digital data, the business processes of the organization would inevitably be disrupted.

In order to prevent such a scenario from happening, cybersecurity measures need to be employed. **Cybersecurity deals with protecting software, hardware, and digital data from unauthorized use and/or access. Data security (a subdiscipline of cybersecurity), on the other hand, focuses solely on the protection of digital data.**

Cybersecurity is also a rapidly changing theater, as hostile agents are constantly trying to find new and innovative ways to break into hardware and software systems and gain unauthorized access to data.

Aforementioned adversaries – either individuals or organizations – are often referred to as **hackers**, or **hacker groups**, in popular culture.

4.3.1 Threat Model

Similarly as with data safety, **risk-based approach** should be taken to manage cybersecurity. An organization should start with identifying:

- all of its software and hardware resources it uses and
- all the different types and categories of digital data that it stores and processes.

Once this is identified, each software/hardware component or data type/category should be evaluated from a threat perspective: who (organization or individual) would profit (and how much) from gaining unauthorized access to it, and what is the probability of such an event happening?

Once such a risk management matrix is prepared (multiplication of risk business damage and chance of it happening is sometimes referred to as **exposure**), dedicated security measures can be put in place for high-exposure risks.

Modeling the security threats and assessing vulnerabilities are referred to as defining a **threat model**.

Primarily, malicious cyberattacks on organizations fall into one of the two categories.

4.3.1.1 Untargeted/Opportunistic Attacks

The majority of cyberattacks, especially those that concern small- and medium-sized organizations, are untargeted [2]. Such attacks normally take the form of malware (malicious software), worms and viruses that can potentially reach an organization's computer systems, and are predominantly executed over Internet connections. Methods include randomly testing organization's publicly exposed Internet interfaces, random malicious ads in websites or e-mails, and similar.

These attacks have no single organization as a target but simply want to cast as wide a net as possible and gain unauthorized access into as many vulnerable computer systems as possible. Individuals who use the Internet during their private, off-work, time are also predominantly targeted by this kind of cyberattack.

These are some of the most common intents behind nontargeted cyberattacks:

- *Botnets:* A botnet is a group of compromised computers which (normally in the background, unbeknown to its users) all run a **bot** application managed by the adversary. As all such computers have access to the Internet, adversary can perform various malicious targeted cyberattacks using a mass number of compromised computers, such as:
  - **Distributed denial of service** (**DdoS**) attacks – all computers in the botnet send Internet traffic toward a targeted website; such a website becomes overloaded and is not accessible (therefore "denial of service") to its actual users.

  - Sending vast amounts of **spam e-mail** toward e-mail service providers.
  - Use the hardware resources of compromised computers for **Bitcoin mining**.
  - Gaining access to all data that is processed on compromised computers (**spyware**).
- *"Spear phishing"*: By gaining access to an individual's computer, and therefore the individual's e-mail account, adversaries can send authentic e-mails to targeted individuals as a trusted sender, with the intent of obtaining sensitive data.
- *Ransomware*: A ransomware virus encrypts the compromised computer system's data with a key that is only known to the adversary. Adversaries then usually request a large amount of ransom to be paid (usually in one of the cryptocurrencies) for the compromised organization to reobtain their data.

Untargeted attacks are more common and widespread than targeted attacks, but they are normally associated with less risk than targeted attacks. While they can still result in a very extensive business damage for compromised organizations, they can – generally speaking – be mitigated more easily. Due to their wide angle of attack, they usually rely on well-known security vulnerabilities of computer systems and negligence in adhering to good security practices.

4.3.1.2 Targeted Attacks

Targeted cyberattacks are attacks where a malicious adversary dedicates significant resources to target a specific entity, organization, or an individual. Because of the nature of these attacks, targeted attacks are considered a much more dangerous threat than opportunistic attacks; adversaries often expend significant effort to achieve their goals. The intent of targeted attacks is also usually more serious, e.g. obtaining financial gain, national secrets, critical intellectual property, or similar. One study found that most frequently targeted computer systems are various governmental systems.

Due to the significant effort spent by the adversaries, implementing good security mechanisms against targeted attacks is hard. For high-value organizations (government, military, etc.), cybersecurity also exceeds the computer infrastructure and entails security processes and organizational structure. For example, if an adversary is prepared to kidnap an IT administrator and torture him/her to gain unauthorized access into the system, organizational measures need to be implemented to battle this.

4.3.2 Risks

Due to the fact that the IT infrastructure and the Internet are constantly changing and evolving, and that adversaries are constantly finding new ways to break into computer systems, it is impossible to list all the risks and vulnerabilities associated with cybersecurity.

As is always the case, any security system is evaluated by **its weakest link**; organizations that seek to protect themselves from targeted cyberattacks should strive to implement adequate security measures on all vulnerable layers across their computer systems and organizational structure.

A list of interfaces/pathways that malicious adversaries can potentially use to gain unauthorized access to computer systems is shown in Figure 4.1.

Often, people associate cyberattacks with adversaries (or *hackers*) *hacking* into software systems via network connections. However, more and more organizations and communities are realizing that the biggest threat to their digital data is often **social engineering** (described in Section 4.3.2.5), resulting from lack of digital security awareness and lack of proper training of employees within an organization.

4.3.2.1 Physical Access

At the end of the day, all digital data resides on physical hardware – computers, hard disk drives, and/or magnetic tapes. An organization might have its business-critical data stored within the organization premises (e.g. internal PCs and servers); larger organizations might also have backup locations on different geolocations for storing backups and archives of data.

Most services that are offered through the cloud (PaaS, IaaS, and SaaS – see Section 2.2.5) store digital data within large data centers, which host large amounts of digital data for many customers in the same physical location.

The most straightforward approach for the malicious adversary to gain access to unauthorized digital data and computer resources is to gain physical access to the computer hardware – e.g. break into the organization's premises or into a data center.

To battle this, physical access to these resources needs to be restricted; this is normally more of an organizational matter. Larger organizations (and data centers), for example, implement strict security policies and personal identification measures to ensure only personnel with enough security clearance can physically access servers/computers that contain critical data to, e.g. perform maintenance, replace hardware, and similar.

With widespread usage of mobile phones and portable computers (tablets and laptops) for business, these mobile devices often pose a security vulnerability. Salesmen, business developers, project managers, and many other professions require frequent travel; in order to support working remotely, their computers almost always store valuable digital data. This is a security risk as these computers can be stolen by an adversary (or lost).

4.3.2.2 Software Access

Computer systems rarely run in isolation but often involve complex interfacing and integrations; this inherently means that digital data travels across computer networks from one computer to another, where it is processed/stored/augmented/aggregated, or similar.

Software access represents any programmatic access through computer networks (most often, Internet, or internal organizational networks) to the organization's IT computer systems.

Any computer (hardware or software) system that is interfacing with other computer systems has at least one **network interface** (sometimes called an application

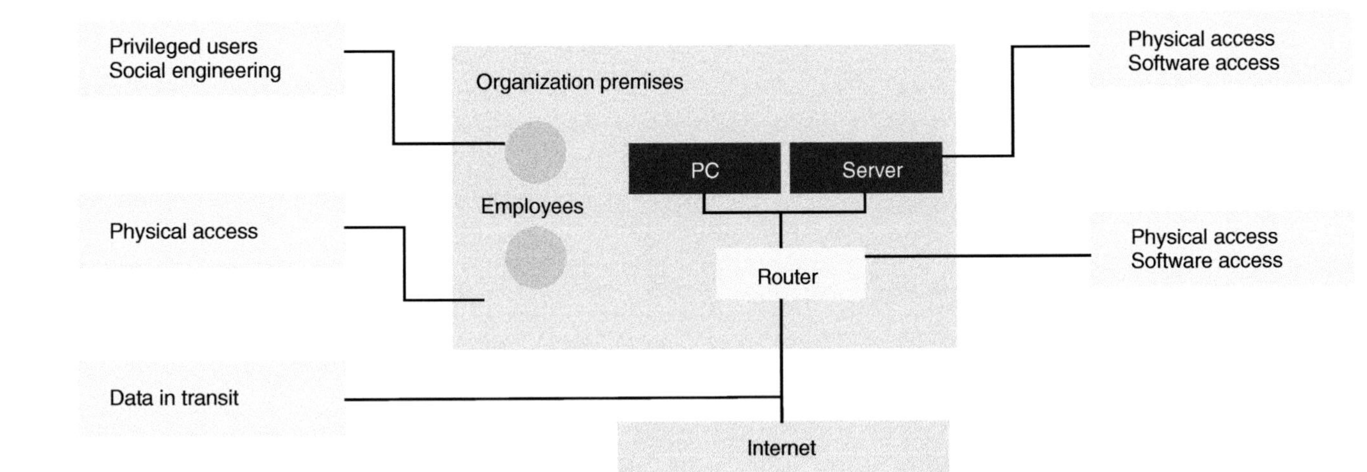

Figure 4.1 A list of interfaces/pathways that malicious adversaries can potentially use to gain unauthorized access to computer systems.

programming interface (API)) that is used as the channel for communication, and therefore, for digital data exchange. Especially when the computer system is exposed to a public network – the Internet – but also for internal networks, each such interface is the single most critical piece that needs to be configured in a very secure manner and kept updated (to include security patches) and regularly monitored.

Often, network interfaces are misconfigured, or not regularly updated and reviewed, leaving critical security vulnerabilities and **backdoors** (dedicated means of accessing remote software and hardware, often implemented by the manufacturer) that can be exploited by expert adversaries.

4.3.2.3 Privileged Users

As already mentioned in Section 4.2.1, dedicated organization personnel (normally, IT administrators/system administrators/DevOps or operations teams) need to have privileged access to computer systems. These include privileged software access (sometimes this is referred to as having **superuser access**) as well as physical access to the hardware.

It is often mandatory for privileged users to have privileged access in order to ensure normal operation of the hardware and software. These individuals, however, must be well trained in cybersecurity topics. Besides that, companies should employ organizational measures to ensure restricted physical access to hardware, as well as to monitor the privileged access to detect any unusual behavior.

When having access to computer systems with very sensitive data, privileged users are also a potential target for physical attacks, kidnaps, blackmailing, and similar. In such a case, an organization must employ dedicated security measures (e.g. not relying on privileged passwords/credentials that could be retrieved by the adversaries, requiring multiple authorization levels by multiple personnel before granting access, etc.) to battle this.

4.3.2.4 Data in Transit

As outlined in Section 4.3.2.2, modern computer systems often interface with one another; in fact, using a group of different computer systems, each specializing for its own task, is one of the key benefits of using computer systems in the first place. As a consequence, however, this means that a lot of business-critical digital data is sent over computer networks as these systems are interfacing with one another.

In the case of internal organizational computer networks, adversaries first need to gain physical access to the network (e.g. connect their computer to the UTP Ethernet cables or network hubs/switches/routers). When an organization uses wireless networks (e.g. Wi-Fi), an adversary can also attempt to break into it by simply being in the vicinity of the wireless access points. Once physically connected to the computer network, an adversary can tap into all computer network traffic. This technique is often referred to as **packet sniffing**, as the adversary can practically intercept any traffic that is being sent throughout the section of the network that was compromised.

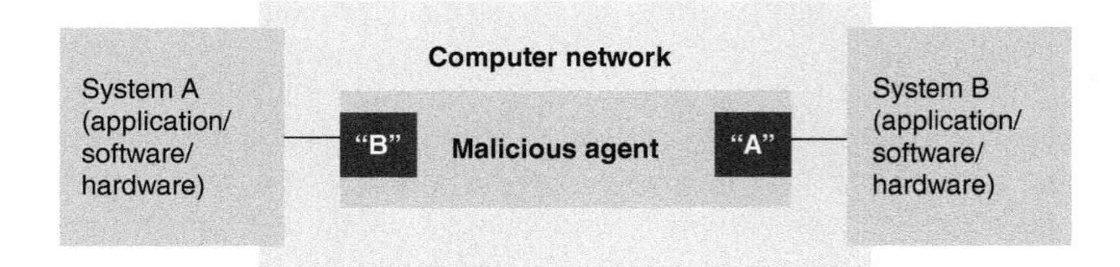

Figure 4.2 A man-in-the-middle attack.

With the shift of organizations toward using cloud service providers, it is vital to understand that the Internet poses a much bigger security threat for digital data in transit. As the Internet is a global network of networks, it is impossible for an organization to restrict access to it, and it is safe to assume that any digital data sent through the Internet can be accessed by anyone. Again, extra security measures need to be implemented to ensure that the data is only used as intended by the two computer systems that are interfacing with one another.

Especially in case of Internet connections, it is worth pointing out a type of attack that is frequently used by adversaries. A **man-in-the-middle attack** (Figure 4.2.) is when the adversary puts his/her hardware or software system between the two computer systems that should be interfacing directly – unbeknown to either of them. If an adversary succeeds, they can tap into all the traffic that is transitioned between the two computer systems.

As already mentioned in Section 4.3.2.1, salesmen, business developers, project managers, and other professions frequently travel for work; while they do, they often connect their mobile computers to various public wireless or wired networks (e.g. conference, hotel networks, and similar) in order to work remotely. Especially unprotected wireless networks (but also wired) pose a big security risk, as they might not have the necessary security measures in place, thus allowing adversaries to tap into the network traffic, as well as to perform man-in-the-middle attacks.

4.3.2.5 Social Engineering

In the context of cybersecurity, **social engineering** talks about a scenario when an adversary tries to psychologically manipulate people – normally, employees of a targeted organization – into performing actions and divulging valuable information, all with the intent of gaining unauthorized access to the organization's computer systems and its digital data.

Social engineering entails many different techniques. Perhaps the most commonly known approach is **phishing**: with this attack, adversaries provide a fake website that looks as authentic as an original website and provide a login form. Users are led to believe that this is the authentic website (e.g. online bank, cloud management web console [Amazon, Google], organization's enterprise resource planning [ERP] system, or similar) and proceed to login, thus granting the adversaries their login credentials for the real website. Such fake websites are often distributed by fake e-mails (this is often called **e-mail spoofing**).

Not really a case of phishing per se, but an approach many e-mail users might be familiar with is the famous *Nigerian prince scam*, where an adversary sends out mass e-mails to people, asking for a small amount of money loan to be transferred to their bank account, promising a refund of a much larger sum of money (which, naturally, never happens).

When phishing is performed via telephone system, it is referred to as **voice phishing** or **vishing**. An adversary could call an employee, claiming that he or she is from a, e.g. bank and ask for the credentials of the called employee.

There are a plethora of other social engineering approaches, many of which employ **impersonation** and **deceit**. Adversaries are often employing fake (or impersonating) profiles on platforms such as Facebook, Twitter, or dating services (such as Tinder) to establish trust with a specific person, normally an organization's employee. Especially dating services tend to be very hard to battle against social engineering, as humans naturally have blind spots when it comes to interpersonal relationships.

One of the more famous social engineering attacks, for example, happened to Ubiquiti Networks company in 2015. The adversaries were able to forge e-mails, pretending to be executive members of the company. The e-mails were sent to the company's financial department, requesting wire money transfers to specific bank accounts (which were under the control of the adversaries). As a result, the financial staff transferred $46.7 million dollars into the adversary-controlled bank accounts [3].

As a threat that needs to be mitigated by organizational processes and internal education of personnel, social engineering is very often overlooked and under-mitigated by organizations. It is vital to acknowledge that while setting up sound cyber defenses against hacking attacks can be implemented by a dedicated team of IT personnel, **protecting from social engineering attacks is a much harder endeavor, especially for larger organizations, as each employee (or contractor) is a vulnerability and requires diligent training;** every such individual needs to be familiarized with different psychological manipulation techniques that adversaries could use, to be able to detect such behavior in the future. While many IT personnel – and, especially, security experts – are justifiably paranoid and are skeptical of everything, other people tend to be more trusting. In truth, it is part of human nature to be helpful and kind, and that is why it is so hard to battle social engineering.

In 2003, *Harris Interactive Service Bureau* performed a survey of 500 US workers and managers who – at the time – handled sensitive customer information on a daily basis as part of their work. The data was later analyzed by the company *Vontu* [4]. Despite the fact that significant amount of time has passed since then, the survey revealed some insights that are still very relevant (even more so):

- 62% reported incidents at work that could put customer data at risk for identity theft;
- 66% say their coworkers, not hackers, pose the greatest risk to consumer privacy;
- 46% say it would be "easy" to "extremely easy" for workers to remove sensitive data from the corporate database.

All these relate to the complex and real problem that social engineering poses to organizations of all sizes.

4.3.3 Measures

In contrast to data safety, measures to battle cyberattacks are much more varied; as new angles of attacks are employed by adversaries, new measures need to be put in place to battle them. What follows is a list of commonly acknowledged security measures that address most of the risks mentioned in the previous section.

4.3.3.1 Physical Protection

In order to prevent an adversary from physically breaking into the organization premises and accessing/stealing the computers, hard disk drives, magnetic tapes, and similar, various physical security measures can be put in place such as locks, surveillance cameras, security guards, and similar.

Besides protecting computer hardware, network infrastructure itself – physical access to things such as network cables – needs to be restricted as well.

4.3.3.2 Software and Infrastructural Measures

Computer systems, as described in Section 2.8.13, are complex systems. Not only are they prone to errors and failures, even moreso, they can potentially contain security vulnerabilities. Especially computer systems that are connected to the Internet (or other public computer networks) must ensure that their software and hardware are of sound quality from a cybersecurity standpoint.

Most modern software systems use a lot of off-the-shelf software libraries and components which have been developed by a third party (see section Code). This happens partially to avoid reinventing the wheel and partially to cut costs off the development phase.

When security of the software is considered as part of the software development life cycle, such software is often referred to as being **"secure by design"**. There is a plethora of good development and coding practices that software system vendors can utilize to ensure the software code is secure.

The problem of having many software dependencies is that it is very hard to control and ensure adequate security level of third-party software code. To mitigate third-party software components, a list of all software dependencies of a software system needs to be maintained. Secondly, all third-party software components need to be regularly reviewed and monitored; when the authors of such components release updates – especially **security updates/fixes** – such updates need to be included in the main software system as well. Especially critical are the so-called **zero-day system vulnerabilities**: vulnerabilities that are unknown to the system vendor. Adversaries can exploit such vulnerabilities until the vendor learns of the vulnerability; after this time point, delivering a security fix to all existing software systems in as short time as possible is crucial for the vendor.

Besides software code, computers and computer systems must also ensure that hardware, operating systems, and network interfaces are configured in a secure way. Computers that are used by the organization can run **antivirus** software and be regularly updated with security updates. Especially on the computer network level, network interfaces need to be configured by a network security expert. **Firewall**, for example, is a software/hardware component that monitors and manages all network traffic. Often in small organizations, network interfaces are configured by junior IT personnel or software developers who lack adequate knowledge about network security. A misconfigured network interface can often be exploited by adversaries to gain unauthorized access.

Active **monitoring** of network traffic, incidents, and suspicious behavior on the network is also one of the security measures toward early detection of intrusions into computer systems.

Penetration Testing, Consulting

To ensure adequate security of a complex computer system, it is often advisable to consult a third-party security company for evaluation. The selection of the company should be done in a selective manner – a company with a good reputation and solid track record should be chosen (indeed, there have been cases of fake companies that exploited the situation).

The third-party security company can review and evaluate the security measures of a computer system from the *out-of-the-box* standpoint, outlining critical details that computer system authors might have missed.

One service that security consultant companies also offer is **penetration testing**, which is a simulated cyberattack on the computer system with the intent to find the weak security points of the system. For complex systems in use, these tests should be performed on a regular basis (e.g. each year). Penetration tests are a common process in the industry – when evaluating a cloud service provider, for example, a good measure is to check and review their penetration test reports.

4.3.3.3 Encryption

In layman terms, *encryption* of digital data is the process in which the data is transformed into a seemingly random, unstructured data that can only be transformed back into the original data (*decrypted*) by knowing a **secret key**.

In **symmetric cryptography**, the same key is used to both encrypt and decrypt the data. Because anybody who has the secret key can decrypt all the data, secret keys must be stored, managed, and distributed in a very controlled and secure way.

Asymmetric cryptography (often referred to as **public-key encryption**), on the other hand, uses two different keys for encryption and decryption of data. Public-key encryption is primarily used for establishing secure computer network connections as any client can encrypt its data (with the server's **public key**) and send it to the server, which is the sole entity that can decrypt the data with its **private key**. In such

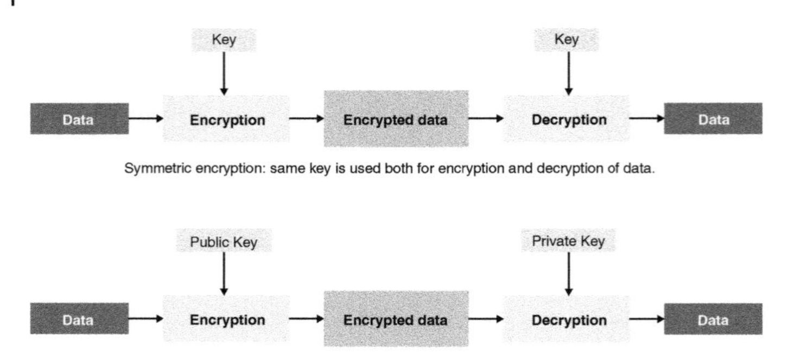

Figure 4.3 Differences between the two types of encryption.

a scheme, public keys are really public; only the private key needs to be stored very securely.

The differences between the two types of encryption are shown in the following sections (Figure 4.3).

Encryption of Data at Rest

When encryption of data at rest is used, all data that is stored on physical digital data storage devices – hard disk drives, magnetic tapes, USB sticks, and similar – is encrypted with a symmetric cryptographic algorithm. Naturally, the secret key must be stored elsewhere, not on the same storage device.

The value of having encrypted data on storage devices is that without knowing the secret key, the data cannot be decrypted (and, thus, read). This makes encryption of data at rest a good measure against adversaries that would attempt to gain physical access to the data (e.g. break into the organization premises and steal hard disk drives). It is also a great measure to ensure that mobile computer devices (laptops and tablets) that are stolen or lost cannot be compromised.

It is a common misconception that encryption of data at rest mitigates software access-based cyberattacks. When an adversary gains unauthorized access into a computer system, the secret key used to encrypt and decrypt data is normally stored within the computer memory (and, thus, accessible to the adversary).

Encryption of Data in Transit

As explained in Section 4.3.2.4, modern computer systems communicate with each other in complex ways over internal network connections, and – more and more – public networks such as the Internet. Especially on public networks – where anybody can connect to the network and monitor its traffic – encryption of data in transit is absolutely mandatory.

Due to the widespread usage of the Internet, major industry effort has been put in to ensure that the digital data propagated throughout the Internet is encrypted. Industry has adopted various standardized encryption protocols such as TLS (formerly known as SSL), HTTPS, SSH, and similar.

TLS (and similar protocols) mostly employ public-key encryption to exchange a secret key and then communicate over the network using symmetric encryption.

Any computer system that is interfacing with a public network such as the Internet should, if possible, employ the TLS protocol (TLS version 1.3) or protocols based on top of it.

Due to the extensive support and relative ease of implementation, it is advisable that computer communication on internal networks is also done in the same, TLS-based way.

Adhering to secure network interfaces that follow up-to-date security protocols is primarily the responsibility of the software vendors and IT administration that needs to configure network infrastructure in a secure way.

VPN

A **virtual private network (VPN)** extends a private computer network across a public network, such as the Internet. VPNs are primarily used by organizations to allow for their off-site employees to work remotely and access their internal network resources (e.g. files, computers, and data) from public networks via a secure channel.

An encrypted VPN can be used to ensure that digital data in transit on public networks (such as various hotel Wi-Fi networks and similar) is encrypted in transit.

4.3.3.4 Policies and Processes

To mitigate physical access attacks, limit the permissions of privileged users; as well as to battle social engineering attacks, an organization should set up strict security policies and processes in place. These policies should be enforced, reviewed, and updated on a regular basis.

This is normally more of an organizational (rather than a technical) measure. An organizational security policy normally covers factors such as rules for visitors, locking of office premises, surveillance, alarms, policy on passwords, policy on remote work, data encryption, data at rest, clean desk policy, classification and confidentiality of data, security incident response plan, and many more.

In smaller organizations, privileged users often have (almost) unrestricted control. A good mitigation practice is to implement a **role-based authorization system** within the organization, where each role has a defined set of permissions, and each organizational entity (user, department, etc.) is assigned one of the roles. A standard security practice is to follow the **principle of least privilege** – e.g. grant just the permissions necessary for an entity (user, department, etc.) to work and nothing more. Various off-the-shelf electronic authorization systems that target different organizations are available on the market.

4.3.3.5 Education

To battle the rising threat of social engineering attacks, education of all organization members (as well as external collaborators) on the matters of data safety and cybersecurity can be considered. As entire departments are switching to practically using computers for daily operations, it is really important that everybody within the company has at least basic knowledge about the threats of cyberattacks.

The education can take the form of regular internal workshops and presentations and should also be a part of the onboarding process for new organization members.

4.3.3.6 Third-party Security Review

One way to evaluate the state and level of security measures within an organization is to hire third-party security consultants for evaluation. Similarly, as outlined in Section 4.3.3.2.0, selection of the consultant company should be carefully done.

Third-party security consultants can perform a range of security services such as full IT/system security reviews, penetration tests (similarly as in Section 4.3.3.2.0 but targeted at an organization rather than a single software system), social engineering attack simulations, internal security reviews, and more.

After evaluation, the security consultant will normally present results and identified security risks and suggestions on how to mitigate them.

References

1 Beach, B. (2013). How long do disk drives last?. Backblaze Blog | Cloud Storage & Cloud Backup. https://www.backblaze.com/blog/how-long-do-disk-drives-last/ (accessed 1 February 2020).

2 Harrison V. and Pagliery, J. (2015). Nearly 1 million new malware threats released every day. CNNMoney (London). https://money.cnn.com/2015/04/14/technology/security/cyber-attack-hacks-security/ (accessed 1 February 2020).

3 NBC News (2015). Ubiquiti networks says it was victim of $47 million cyber scam. https://www.nbcnews.com/tech/security/ubiquiti-networks-says-it-was-victim-47-million-cyber-scam-n406201 (accessed 1 February 2020).

4 Testimony of Joseph Ansanelli, Chairman and CEO of Vontu, Inc. (2004). Before the United States house of representatives subcommittee on financial institutions and consumer credit: "fighting identity theft – the role of FCRA". https://web.archive.org/web/20120322110212/http://financialservices.house.gov/media/pdf/062403ja.pdf (accessed 1 February 2020).

5

FAIR Principles and Why They Matter

Keith Russell

Australian Research Data Commons, Monash University, Caulfield East, 100 Sir John Monash Drive, VIC 3145, Victoria, Australia

5.1 Introduction

When considering your lab practice and the data coming out of the lab, one of the current developments to keep in mind are the findable, accessible, interoperable, and reusable (FAIR) principles, a set of principles that has received a large amount of support across the research sector. In 2014, a group consisting of a range of stakeholders met in Leiden at a Lorentz workshop. There they launched the concept of FAIR data. This resulted in a journal article in Nature in 2016 [1] that described the 14 principles and provided more context around them.

These principles have attracted an impressive amount of attention and uptake from around the globe and have now been adopted by a range of different stakeholders in the research process as a very helpful way of thinking about data sharing. This includes research funders [2–4], publishers and learned academies [5], universities [6], research libraries [7], and research infrastructure organizations [8].

There are probably a few reasons for their popularity. First of all, they break down the four high-level areas into more detailed considerations, which is helpful in translating the concepts into a more practical implementation. However, in that translation the principles are technology agnostic and do not prefer one solution over the other, which means they can be applied quite broadly. Furthermore, they are discipline agnostic and can be applied across all disciplines. Rather than proposing to make all data open the principles allow for a much more nuanced view, taking into account the specific demands of, for example, sensitive data that cannot be made openly available. And finally, the four-letter acronym appears to resonate, who would want to do the wrong thing and make their data unfair?

5.2 What Is the Value of Making Data FAIR?

So why would you put the effort into actually making your data FAIR? There is a range of arguments for making your data FAIR, and this may differ depending on the

Digital Transformation of the Laboratory: A Practical Guide to the Connected Lab, First Edition.
Edited by Klemen Zupancic, Tea Pavlek and Jana Erjavec.

stakeholder involved. Much of this underlying value had been identified in earlier initiatives such as the push for open data and linked data, but the FAIR principles provided a neatly bundled set of principles to achieve this value [2].

For one, making data FAIR allows the verification of research findings. By publishing data and associated software, models, and algorithms alongside the journal article it can be possible to recreate the argument for the findings, and it makes it more possible to verify or reproduce the results [9].

Another argument for making data FAIR is to not spend money again to collect the data. This is especially relevant in the context of publicly funded research where taxpayer dollars have been used to create the data. A very clear example of this is the Wellcome Trust policy on sharing of data, software, and materials [10].

Having more data available for reuse will accelerate research discovery and enable us to tackle the grand challenges the world is facing. One example of this is the push to rapidly share data to combat epidemics such as the Ebola and the SARS-CoV-2 virus [11].

In some cases going back to collect the data again is not ethical or simply not feasible. One example of this is in the case of clinical trials or animal trials where it is important to minimize the exposure to potentially harmful drugs. In other cases it is not possible to recollect data, especially where it is historical data which was collected in a specific context and time. This data can be invaluable for longitudinal studies.

In the current age where data science and machine learning are booming, there is a growing capacity and interest in combining large arrays of data to find patterns, train algorithms, and develop new potential hypotheses. However, the bringing together of disparate unstandardized data sets can be a huge amount of work. A key aspect throughout the FAIR principles is the consideration how data can be more FAIR for machines. By also addressing this, machines will be able to locate data and bring it together with other related data and accelerate discovery [12].

5.3 Considerations in Creating Lab-based Data to Prepare for It to Be FAIR

As mentioned earlier, the FAIR data principles are consciously high level and not specific to a discipline or not prescriptive of a technology solution. Consequently, there has been a push to uncover what the FAIR principles mean in practice. This has resulted in reports and articles on what FAIR data means in practice [2, 8, 13], a range of FAIR assessment tools [14], a collection of 10 FAIR Research Data things exploring what FAIR means in specific disciplines or areas [15], and the FAIRsharing facility that is a curated resource on data and metadata standards, policies, and databases [16]. So, there are a growing range of assets and standards out there to build on.

When considering your data practices in the lab, there are a number of aspects you can take into account.

Assuming that the researcher will take the data away from the lab and conduct further analysis on the data and be the decision maker on how the data will be shared, there are still a number of activities you can do to make it easier to make the data FAIR at the end of the process.

1. Where relevant, keeping stable persistent copies of the data and assigning persistent identifiers for the resulting data sets.
2. Using community-agreed standard data formats.
3. Using community-agreed vocabularies and ontologies for values in the data and metadata. These vocabularies and ontologies should be accessible and should be referenced (i.e. these should be FAIR themselves) [17].
4. Providing rich provenance information alongside the data. This can, for example, include identifiers for the instruments used, data on the settings and environment in which the data was collected, and software used to process the initial observation data.
5. Where relevant, express an expectation to the lab users on how the data will be made accessible, with which license assigned to the data. This will obviously depend on the agreements between the lab and the researchers using the facility.

If as a lab manager you would like to pick up these activities, there are probably a few steps you can take:

1. Start by getting the team onboard. Provide them with some background on FAIR and announce that you will be looking into enabling the making of data FAIR.
2. Consult with a few key users of your lab and work with them to uncover best practice in their discipline. Questions that would be worth raising include: Are there existing standard data formats and approaches, vocabularies, and ontologies in use? Are there requirements from publishers, funders, or data repositories that they are expected to meet? This could include expectations around the verification of research findings, the publishing of the data in specific formats under a certain license, using a designated metadata standard, or attaching provenance information. Where are they expected to keep or publish their data? Are there relevant articles on best FAIR practice in their discipline or area of research that provide standards and approaches? For relevant existing repositories and standards, it is worth checking out Re3Data and FAIRsharing.
3. It can be useful to consult with other lab managers running similar instruments or processes on approaches they are taking in identifying instruments, data outputs, providing provenance information, providing guidance to researchers, etc.
4. If you are dependent on changes being made to the instruments or processes, discuss with providers what your needs are and whether these can be met.
5. Based on the input collected above, decide on key areas of change and check with key users of the lab whether this will fit in with their practice.
6. Run through the changes with the team and ensure they understand what they are and how researchers can benefit from them.
7. Develop or adjust supporting documentation or guides for users of the lab. This will include information on the data and metadata that is provided to them and how they can use this further on in the research process.
8. Inform researchers of the changes and how this will help them make their data more FAIR.

9. Make the changes.
10. Evaluate after a while whether the changes have been adopted. Also check with key users whether practices and standards have since changed or developed. In some disciplines, standards are now being developed in response to the FAIR principles.

It can be helpful to pick off some quick wins that will fit well with current research practice and help researchers meet funder or publisher requirements. Making data FAIR does not stop in the lab though. In the end, researchers will usually be the ones to bring all the data and metadata from the lab and subsequent analysis together and publish the end results in a FAIR form. But providing data and metadata in the right formats with guidance alongside it can make the process easier for the researcher.

5.4 The FAIR Guiding Principles Overview

To be Findable:

F1. (meta)data are assigned a globally unique and persistent identifier

F2. data are described with rich metadata (defined by R1 below)

F3. metadata clearly and explicitly include the identifier of the data it describes

F4. (meta)data are registered or indexed in a searchable resource

To be Accessible:

A1. (meta)data are retrievable by their identifier using a standardized communications protocol

A1.1 the protocol is open, free, and universally implementable

A1.2 the protocol allows for an authentication and authorization procedure, where necessary

A2. metadata are accessible, even when the data are no longer available

To be Interoperable:

I1. (meta)data use a formal, accessible, shared, and broadly applicable language for knowledge representation

I2. (meta)data use vocabularies that follow FAIR principles

I3. (meta)data include qualified references to other (meta)data

To be Reusable:

R1. meta(data) are richly described with a plurality of accurate and relevant attributes

R1.1. (meta)data are released with a clear and accessible data usage license

R1.2. (meta)data are associated with detailed provenance

R1.3. (meta)data meet domain-relevant community standards

References

1 Wilkinson, M., Dumontier, M., Aalbersberg, I. et al. (2016). The FAIR guiding principles for scientific data management and stewardship. *Scientific Data* 3: 160018. https://doi.org/10.1038/sdata.2016.18.

2 Hodson, S., Jones, S., Collins, S.. et al. (2018). Turning FAIR data into reality: interim report from the European Commission Expert Group on FAIR data. https://doi.org/10.2777/54599

3 National Institutes of Health (2019). New models of data stewardship – data commons pilot. https://commonfund.nih.gov/commons (accessed 11 April 2020).

4 National Health and Medical Research Council (2019). Management of data and information in research, A guide supporting the Australian Code for the Responsible Conduct of Research. https://www.nhmrc.gov.au/file/14359/download?token=0FwepbdZ. (accessed 11 April 2020).

5 Hanson, B. (2019). *Progress and Challenges to Moving to a FAIR-Enabled Research Culture*, vol. 21. Geophysical Research Abstracts.

6 League of European Research Universities (2020). Data summit in Paris. https://www.leru.org/news/data-summit-in-paris (accessed 11 April 2020).

7 Association of European Research Libraries (2020). Open consultation on FAIR data action plan – LIBER. LIBER 13 July 2018. https://libereurope.eu/blog/2018/07/13/fairdataconsultation/ (accessed 11 April 2020).

8 Mons, B., Neylon, C., Velterop, J. et al. (2017). Cloudy, increasingly FAIR; revisiting the FAIR data guiding principles for the European open science cloud. *Information Services & Use* 37 (1): 49–56. https://doi.org/10.3233/ISU-170824.

9 Stodden, V., Guo, P., and Ma, Z. (2013). Toward reproducible computational research: an empirical analysis of data and code policy adoption by journals. *PLoS One* 8: e67111. https://doi.org/10.1371/journal.pone.0067111.

10 Wellcome Trust (2017). Data, software and materials management and sharing policy.https://wellcome.ac.uk/grant-funding/guidance/data-software-materials-management-and-sharing-policy (accessed 11 April 2020).

11 Pillai, P. (2020). The role of data in a rapid and coordinated response to infectious disease outbreaks. https://www.rd-alliance.org/sites/default/files/attachment/Role%20of%20data%20in%20ID%20outbreaks_RDA15_Pillai_v02.pdf (accessed 11 April 2020).

12 Mons, B. (2020). Invest 5% of research funds in ensuring data are reusable. *Nature* 578 (7796): 491. https://doi.org/10.1038/d41586-020-00505-7.

13 Allen, R. and Hartland, D. (2018). FAIR in practice - Jisc report on the Findable Accessible Interoperable and Reuseable Data Principles. https://doi.org/10.5281/zenodo.1245568.

14 Christophe, B., Makx, D., and Brecht, W. (2019). Results of an analysis of existing FAIR assessment tools. *Research Data Alliance* https://doi.org/10.15497/RDA00035.

15 Martinez P.A., Erdmann C., Simons N. et al. Top 10 FAIR data & software things 2019. https://doi.org/10.5281/zenodo.3409968.

16 Sansone, S.-A., McQuilton, P., Rocca-Serra, P. et al. (2019). FAIRsharing as a community approach to standards, repositories and policies. *Nature Biotechnology* 37: 358. https://doi.org/10.1038/s41587-019-0080-8.

17 Wise, J., de Barron, A.G., Splendiani, A. et al. (2019). Implementation and relevance of FAIR data principles in biopharmaceutical R&D. *Drug Discovery Today.* 24 (4): 933–938.

6

The Art of Writing and Sharing Methods in the Digital Environment

Lenny Teytelman and Emma Ganley

protocols.io, Berkeley, CA 94704, USA

6.1 Introduction

In 2013, the ambitious effort "Reproducibility Project: Cancer Biology" (RP:CB) was launched to assess reproducibility of 50 high-profile cancer papers. Five years later, the project stopped at just 18 papers [1], in large part due to the difficulty of figuring out the precise details of the methods used in those papers. In fact, Tim Errington, the RP:CB lead from the Center for Open Science, said that not a single one of these 50 papers contained the full details of how exactly the research was carried out [2].

Reporting on the initial results of RP:CB in Nature [3], Monya Baker and Elie Dolgin wrote, "Perhaps the clearest finding from the project is that many papers include too few details about their methods... Replication teams spent many hours working with the original authors to chase down protocols and reagents, in many cases because they had been developed by students and postdocs who were no longer with the lab." And Ed Young, writing in The Atlantic [4], noted that:

> "The hardest part, by far, was figuring out exactly what the original labs actually did. Scientific papers come with methods sections that theoretically ought to provide recipes for doing the same experiments. But often, those recipes are incomplete, missing out important steps, details, or ingredients. In some cases, the recipes aren't described at all; researchers simply cite an earlier study that used a similar technique."

This issue is not limited to biomedical research but cuts across disciplines from physics to psychology. It is very common to read frustrated screams about this problem from researchers themselves on Twitter [5]. The endless "as previously described" trail from one reference to the next in search of the original method has some researchers tweeting gifs of laptops being jettisoned out of the window; sometimes there simply is no definitive account of the original method to be found in any published article anywhere.

Digital Transformation of the Laboratory: A Practical Guide to the Connected Lab, First Edition.
Edited by Klemen Zupancic, Tea Pavlek and Jana Erjavec.

The good news is that the past two decades are full of serious efforts to encourage better reporting of research methods and protocols. These initiatives originate from publishers, academics, and entrepreneurs, with many of them enabled by the development of the World Wide Web. This is particularly gratifying as Tim Berners-Lee specifically invented the Web with the goal of helping scientists share information [6].

When protocols.io was founded, it was a surprise to learn that in 1999, Chris Yoo, a postdoc from the same lab in UC Berkeley where one of the authors, Teytelman, did his graduate studies, cofounded http://bioprotocol.com with the same intent. They raised a million dollars of venture capital [7], and though their effort did not succeed, a dozen years later Dr. Yoo was enthusiastic in advising protocols.io and helping us to avoid their mistakes.

Also in 1999, Protocol Online was started by Dr. Long-Cheng Li to organize disparate life science protocols in a single categorized database [8]. Six years later, in 2005, OpenWetWare opened to the public [9], creating a Wikipedia-like website for crowdsourced sharing of step-by-step protocols, with widespread adoption by the community.

On the publisher side, Veronique Kiermer founded Nature Methods in 2004 to ensure that people focusing on method development could receive credit for their work [10]. Two years later, Nature Protocols and Protocol Exchange launched to further fill in the gap of proper crediting for, and easy sharing of, protocols [11]. Also in 2006, the Journal of Visual Experiments was set up to publish protocol videos, in order to make it easier to learn new techniques [12].

Other publishers have begun to take steps to improve upon the traditional, often vague, narrative, method sections, and to make sure that modifications of the existing protocols are transparent: Cell Press rolled out STAR methods [13], EMBO announced a similar initiative [14], and journals such as eLife have partnered with Bio-protocol [15]. And since 2017, we have seen an increase from 2 to more than 500 journals that include in their author guidelines strong encouragement to authors to deposit in protocols.io a precise step-by-step protocol of their methods and to link to the protocols.io digital object identifier (DOI) from their manuscript.

There are now tens of thousands of protocols available in journals and platforms in a way that was unfathomable even 20 years ago. Funders have recently begun to push for protocol sharing, with the Moore Foundation, Chan Zuckerberg Initiative, and Alex's Lemonade Stand Foundation explicitly expecting that methodological resources are shared in a public repository, and NIH and Wellcome Trust emphasizing protocol sharing in their grant submission guidelines.

It is exciting to see protocol-sharing venues flourishing and the ever-increasing recognition of the value and importance of the precise details behind the research results both for comprehension, sharing, and to increase reproducibility. The following sections review the appropriate digital tools and resources for protocols and the best practices for crafting the protocols themselves.

6.2 Tools and Resources for Tracking, Developing, Sharing, and Disseminating Protocols

Probably the most important advice we can offer is to get into the habit of organizing and keeping track of protocols early, as you are performing the experiments, and to encourage your collaborators to do the same. This practice will save time when you need to repeat the work or report on it; it will ensure that when you graduate or leave the lab for the next position, the critical information remains for the new lab members to build upon your work. Do not forget that as often noted at conferences, your most likely collaborator in the future is you in six months [16].

One of the key barriers to sharing detailed methods is lack of time. When you are trying to publish your results, there is an overwhelming list of requirements for submission to the journal, and you are probably simultaneously continuing with other research, teaching, presenting at meetings, applying for grants and fellowships, and so on. When submitting your paper, having to dig back through one or more paper notebooks for the protocol details and trying to decipher which set of tweaks were the optimal and final ones used is the last thing you want to do. Then needing to transcribe them into an appropriate format for sharing is one more burden and time-consuming task – demanding precious time you do not have. However, if your protocol is already digital and up-to-date, sharing it during the paper submission ought to require virtually no time at all.

6.2.1 Tools for Organizing and Tracking Your Protocols

Needless to say, scribbling details on post-it notes is not the ideal way to keep track of your protocols. Neither though is taking notes in a digital note app or Microsoft Word. This is because these text-processing tools, while superior to floating pieces of paper and hand-written notes in the lab notebook, still fall far short of what you really need for method development and recording. In contrast to research results, methods evolve over time as you hone the approach and optimize the procedures; today you think that room-temperature incubation is best, but tomorrow you realize that actually 20 °C is better. Without versioning, built-in collaborative editing with your colleagues, and easy sharing, relying on your word processors leads to headaches, mistakes, and often permanent loss of the nuanced details of how your method evolved over time and what discoveries about the method you made along the way.

The following is a list of decent-to-great options for organizing protocols.

- GoogleDocs is a far better option than a local word processor as it is in the cloud, has collaborative editing and commenting, versioning, and easy sharing with individuals or groups.

- If you are considering adopting an electronic lab notebook (ELN) or switching to a new one, there is an invaluable resource for evaluating the many digital notebook options from the Harvard Medical School library: https://datamanagement.hms.harvard.edu. The HMS guide specifies for each ELN whether it includes explicit support for protocols. A few ELNs such as SciNote and Rspace have gone one step further and have integrated directly with protocols.io to facilitate dynamic recording, tracking, and editing of methods.
- Dedicated tools for working with protocols include LabStep and protocols.io. These are like a blend of DropBox and GoogleDocs, specifically designed for you to organize and edit your protocols collaboratively, in a step-by-step manner over time.

Whatever your tool of choice, you should make sure to export and backup all of your protocols periodically. Unfortunate events such as lost passwords, stolen computers, broken hard drives, floods, and fires all happen much more often than you expect. You can choose tools that make exporting and backing up easy and set a monthly reminder in your calendar to backup your protocols. protocols.io keeps all submitted protocols securely, and all data is routinely backed up for your peace of mind.

6.3 Making Your Protocols Public

If you have already created digital versions of your protocols in one of the above tools, then when it is time to publish them, the effort needed to make them public will be minimal. However, it is important to choose a proper venue for sharing your method details.

Currently, many authors choose to share protocols when publishing a paper by uploading PDFs as supplementary files. While this is absolutely better than not sharing at all, it is still essentially a protocol graveyard. Supplementary files are unreliable for many reasons; there is a known issue with link rot [17]; also, supplementary files cannot be versioned with updates, corrections, and/or optimizations, and these files are discoverable exclusively to those who are reading the paper. Add to this the fact that some publishers do not archive these files with the rest of the journal's main published content, and longevity is not even guaranteed. Imagine if instead of a central repository of DNA, RNA, and protein sequences in NCBI's GenBank, we shared these as supplementary files or on individual lab websites. You could never run a query for your sequence of interest against the entire database. We therefore strongly recommend that you share your protocols in a purpose-built repository. But which one?

All of the following allow versioning, have long-term preservation plans with backups and archiving, provide unique identifiers for citing the protocol (DOIs), and facilitate discoverability and reuse.

- Open Science Framework (osf.io) from the Center for Open Science, http://figshare.com, and Zenodo are all solid repositories with open access and free sharing of public documents.

- Protocol Exchange from Nature Springer and protocols.io are repositories that have been specifically designed to host and disseminate research methods. Both are open access and free to read and publish. The added benefit of sharing on these platforms is that the protocols are not just PDFs but are formatted and presented nicely on the web, making it easier to reuse and adopt them for other researchers.

protocols.io also facilitates reuse providing the option to dynamically run through the protocol and store experimental notes and details alongside every run. And you can easily update after publication to correct a detail or provide more information about enhancements/tweaks, etc. Finally, other users can "fork" a protocol – create their own copy and modify it for their own purposes.

Whichever platform you choose, once you have shared a digital version of your method, you can place a link to the public protocol's DOI in the materials and methods section of any related research manuscripts or grant applications that you write. This is much better than attempting a narrative write-up of a step-by-step method or referencing a previous article that may itself reference a previous article. Even better, if other researchers make use of your published protocol, they too can cite it easily with the protocol's DOI.

6.4 The Art of Writing Methods

There is a lot more to a good protocol that can be easily followed by others than just the choice of tool or platform for sharing it. Just because you shared code on GitHub, does not mean that your code compiles or is readable and usable by others. In the same way, writing up and sharing a good protocol takes care and thought.

Before we suggest tips and best practices for writing reproducible and useful protocols, let us go through a hands-on exercise. Please take a blank piece of paper and follow the instructions below. After you complete the nine steps, we will reveal what the drawing should represent in the text below. Do not peak ahead of time.

Drawing Exercise

1. Draw a small 8/10th of an oval in the center of your paper with the open part facing to the bottom left.
2. Draw ⅓ of a large circle underneath the oval and draw small perpendicular lines on each end of it.
3. At the center of the ⅓ circle, draw two squares with a small gap between them.
4. Draw a wavy line underneath the squares.
5. On the ends of the larger circle, draw smaller half circles around them with three dots in the middle.
6. Draw two big circles on each side of the small oval from the beginning. They should be more on the top than on the bottom.
7. In the big circles, draw smaller circles, and in the smaller circles draw even smaller filled in circles.
8. Draw a wavy square around everything you drew so far.
9. Draw seven small ovals within the space of the square.

Solution:

We have gone through this exercise many times in workshops, and the results are always variable.
Clearly, the instructions provided are imperfect. The initial idea is to draw a funny SpongeBob character.
Reviewing the many results of researchers following them makes it easy to spot the problems with the instructions, to identify what is ambiguous, and to make a new version that would ensure near-100% reproducibility.
The key takeaway from this exercise is that no one writes a perfect protocol at first. It is an iterative process, and as scientists e-mail you questions about specific steps, over time, you understand what you can clarify and improve. This is why it is so important to share protocols in repositories that make versioning easy. The versioning removes the stress and pressure of trying to report a perfect protocol from the first draft. However, while striving for the perfect in draft #1 is not a good strategy, there are many simple tips to keep in mind when you write your protocol.

Best practices for writing reproducible methods

- Digitize the methods: protocols.io, Google Docs, and/or ELN (not paper or word processor).
- Think of a protocol as a brief, modular, and self-contained scientific publication.
- Include a three to four-sentence abstract that puts the methodology in context.
- Include as much detail as possible (Duration/time per step, Reagent amount, Vendor name, Catalog number, Expected result, Safety information, and Software package).
- Chronology of steps.
- Limit jargon, use active voice.
- Notes, recipes, tips, and tricks.

Reproducible Method Checklist Example:

- ☐ Digital?
- ☐ Abstract?
- ☐ Status?
- ☐ Enough detail?
- ☐ Images/Videos
- ☐ Exact reagent details (vendor, catalog number, etc.)
- ☐ Equipment detail
- ☐ Computational tools
- ☐ Timers, etc.
- ☐ Active voice?
- ☐ Chronology of steps?
- ☐ Data available?
- ☐ Additional tips?

References

1 https://www.sciencemag.org/news/2018/07/plan-replicate-50-high-impact-cancer-papers-shrinks-just-18
2 https://twitter.com/fidlerfm/status/1169723956665806848
3 https://www.nature.com/news/cancer-reproducibility-project-releases-first-results-1.21304
4 https://www.theatlantic.com/science/archive/2017/01/what-proportion-of-cancer-studies-are-reliable/513485/
5 https://twitter.com/lteytelman/status/948289956161581056
6 https://home.cern/topics/birth-web
7 https://www.bioprocessonline.com/doc/bioprotocol-a-web-based-collection-of-protoco-0001
8 http://www.jbmethods.org/jbm/article/view/20/17#ref1
9 https://www.nature.com/articles/441678a
10 https://www.nature.com/articles/nmeth1009-687
11 https://www.nature.com/articles/nprot.2016.115
12 https://www.wired.com/2007/10/video-sites-help-scientists-show-instead-of-tell/
13 https://www.cell.com/cell/fulltext/S0092-8674(16)31072-8?code=cell-site#article Information
14 http://msb.embopress.org/content/14/7/e8556
15 https://elifesciences.org/inside-elife/ab137ad4/working-with-bio-protocol-to-publish-peer-reviewed-protocols
16 https://twitter.com/RetractionWatch/status/1055828725747666949
17 https://www.the-scientist.com/news-opinion/the-push-to-replace-journal-supplements-with-repositories–66296

Part III

Practical

While informative Knowledge Base chapters covered important terms and concepts to understand, the following Practical chapters will give you examples and guidance on defining your lab's digitalization strategy.

7

How to Approach the Digital Transformation

Jana Erjavec[1], Matjaž Hren[2] and Tilen Kranjc[1]

[1] *BioSistemika d.o.o., Koprska 98, SI-1000 Ljubljana, Slovenia*
[2] *SciNote LLC, 3000 Parmenter St, Middleton, WI 53562, United States*

7.1 Introduction

Many of us who spent at least part of our careers in research laboratories also think of them as places where we innovate. Innovations are mostly directed toward the specific research field, for example finding a novel mechanism that will eventually be used as a new therapy to treat a specific disease. In academic research laboratories, such innovations usually end up being published either as research papers or as patent applications. In industrial research laboratories, they become new products or services. Being 100% focused on your research, you would rarely think of laboratories as places where you could actually innovate the laboratories themselves, including the processes and tools which are used to generate research results. Consequently, people working in laboratories often end up using very old and sometimes outdated approaches to how they are planning, executing, documenting, and reporting their work. One might even go that far and conclude that the way we do science and research has not changed for decades, while we observed significant technological progress.

Many laboratories are already using digital solutions to some extent: the majority of the instruments now have digital data outputs, which are stored, analyzed, and published. We read a lot of digital content such as online research publications, and we prepare reports and research papers in a digital way. However, there are still many practices left from the analog era such as extensive use of paper to plan, document work, and even manage results. Majority of laboratory data (including digital data) is not organized and is fragmented across multiple locations, which leads to temporary or permanent loss of data. Science is also facing data integrity challenges, where organizations are often unable to assure accuracy and consistency of data. Being unable to find, access, and reuse data or assure data integrity causes serious bottlenecks and can slow down the entire research process, making it extremely inefficient. This is where digitalization can help.

If we compare research and industrial processes, we can conclude that the scientific process is centuries older than the industrial process. The run for profit was a

Digital Transformation of the Laboratory: A Practical Guide to the Connected Lab, First Edition.
Edited by Klemen Zupancic, Tea Pavlek and Jana Erjavec.

very important driving force behind all four industrial revolutions, which brought automation, robotization, and digitalization into industrial processes, making it more efficient. On the other hand, the research process, or scientific method, if we call it like that is somewhat lagging behind. Industrial or commercial laboratories that are financed entirely by their immediate output and are profit oriented, such as service laboratories (e.g. laboratories analyzing clinical, environmental, samples, materials, oil, and gas), recognize efficiency as very important as it translates into profitability. Therefore, these laboratories are more motivated to automate and digitalize than academic and other nonprofit laboratories. Having a research laboratory digitalized will also improve data management as well as increase the quality of data as it can become Findable, Accessible, Interoperable, and Reusable (FAIR) and in the end contribute to overall efficiency of how research is being conducted [1]. In the end, efficiency and productivity in academic research is also becoming a more and more frequent topic, usually from the point of view of either a concrete researcher or science funding bodies [2].

A shift in organizational culture is the key aspect related to digitalization of a laboratory, in addition to digital tools (software tools and laboratory instruments that are being used to generate, modify, or store data in a digital way) and processes (how you do research including how you manage the data). Whether your team members will accept the new technologies or not is a vital ingredient of digitalization and will be your legacy for the upcoming years or decades. That is an important reason why digitalization needs to be planned well, and the goals related to it need to be clear and understood and accepted by team members. A shift in organizational culture is likely to be the most challenging part of your digitalization strategy.

Some predictions estimate that digitization will be so disruptive or transformative that four out of top 10 leading businesses across all industries will be displaced by digital transformation within next five years [3]. This is also true for laboratories, and therefore, any effort toward digitalization of laboratories is not just beneficial but will soon become vital. That is why digitalization and digital transformation need to be planned well.

7.2 Defining the Requirements for Your Lab

7.2.1 Digitization Versus Digitalization Versus Digital Transformation

First, let us have a look at the following definitions and the main differences between three crucial terms: digitization, digitalization, and digital transformation.

Digitization is defined as a process that changes information from analog to digital form [4].

Example *Converting handwritten notes into digital notes.*

Digitalization, according to Gartner, is defined as a process of employing digital technologies and information to transform business operations.

Example *Implementation of a LIMS system into a diagnostic laboratory to manage samples, analysis, and reporting. Implementation does not only increase efficiency but also minimizes errors, automates certain tasks, and improves information exchange internally as well as externally.*

Digital transformation is a process that aims to improve an organization by triggering significant changes to its properties through combinations of information, computing, communication, and connectivity technologies [5]. Digital transformation only happens when leadership is recognizing the strategic importance of making profound organizational changes to the company that is customer driven, rather than technology driven.

Example *A company may decide to run several digitalization projects, for example they will implement a LIMS system in their analytics laboratories and ELN into their research laboratories. But just implementation of new systems will not lead to a digital transformation of a company. The management will, however, decide that the company will continue to innovate in the digital space and that this will be the company's no. 1 priority. Soon, the company will use newly digitalized processes to gather data for advanced analytics. They might decide to develop additional algorithms that will enable them to develop better products much faster or offer their customers a better service. They might go further in investing in young start-ups that complement their vision to stay competitive and create value.*

To summarize, we digitize information, we digitalize processes that are part of business operations, and we digitally transform the business and its strategy (Figure 7.1).

7.2.2 Defining the Approach and Scope for Your Lab – Digitization, Digitalization, or Digital Transformation?

When we are speaking about digital technologies, laboratories and businesses often struggle to define what they really need, want, and how much they can invest.

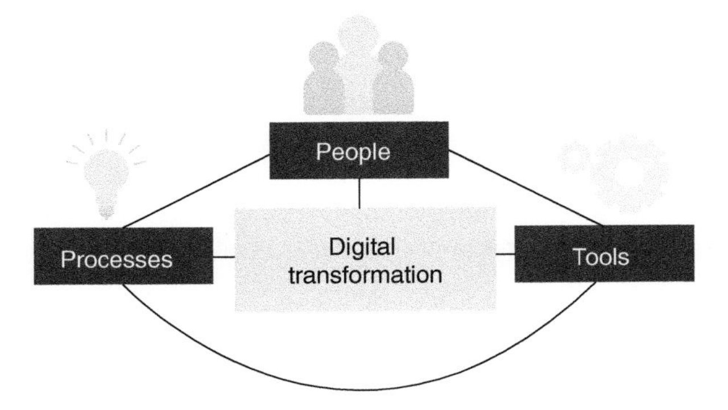

Figure 7.1 Digital transformation encompasses at least three aspects: people, processes, and tools. Source: Courtesy of Janja Cerar.

We have set up a list of very important questions that can help you identify your lab's needs and keep your decision-making process in focus.

1. Which challenges do I have now?
2. Which challenges need my immediate attention?
3. Which challenges do I see in the future?
4. How could the changes affect my current business?
5. What is my long-term business strategy?
6. How will I manage legacy data?
7. How will I get people to cooperate?

Your answers will depend on many things, such as company or laboratory size, your position within the organization, your experience in change management, and your long-term strategy.

Let us address each question separately.

7.2.2.1 Which Challenges Do I Have Now?

Identifying which challenges you are facing at the moment might sound easier than it actually is. They can be related to the technology you are using, people you work with, or regulatory standards and guidelines that you need to follow. Write them down and provide enough detail which will help you make decisions or present your proposals to decision makers.

Example *We have reported three false-negative results from qPCR analysis in the past 3 months and overall 10 this year. The reason was an error in manual data transcription from the analysis report into a final diagnostic report by different operators. The interpretation of the analysis report was correct, so the operators are well trained to interpret the results but sometimes sloppy when rewriting them. The mistakes are costing us money and reputation. We need to find a fast solution to prevent this from happening in the future.*

While this example describes a very pressing problem, others might be less obvious, but nevertheless important. Many laboratories still use a combination of paper laboratory notebooks, raw data files that are stored in folders on PCs that are connected to instruments, and other documents on a file-sharing system. Although the folders are well organized, the information is still hard to find, and even harder to aggregate, analyze, or compare it. This becomes virtually impossible if a person leaves the laboratory, because the data is saved in so many places and more often than not, not properly annotated. Although the data is there somewhere, the organization has lost it (and hence the money). As a consequence, people are often reinventing the wheel, cannot compare the data and seek for the trends, or cannot successfully repeat the experiments, which lack detailed information. As a result, you lose even more money, with wasted working hours, reagents, and overall poor efficiency. This is actually not only a single-organization problem, but a global problem.

7.2.2.2 Which Challenges Need My Immediate Attention?

After listing the challenges, you will need to put a priority on which ones to solve first. If you are reporting false-negative results to your customers this is something that you solve as soon as possible as it is causing serious business damage. In short term you will probably decide to double check everything, which can marginally improve your problem but also increase overhead. In parallel, you can start looking for tools (e.g. software solutions) that are able to automate this process and reduce human errors.

On the other hand, having research data scattered and hard to find is not something that is causing immediate damage to your business. This does not mean, however, that you should not plan how to solve it. The fact that you are not extinguishing a fire does not make it less important. Not leveraging the existing data midterm or long term is also costing you serious money. The ultimate goal should be to achieve data integrity through careful planning and execution of digitization and digitalization projects, which would result in a complete, accurate, and consistent data to build your future work on.

7.2.2.3 Which Challenges Do I See in the Future?

Predicting what the future brings is difficult but critical. It will help you make the right and sustainable decisions in the present and avoid repeating the same process in a few years.

Example *We are currently running only one laboratory, but in the next 5 years, we will grow in numbers and we will have at least three units, with an estimated number of employees growing from 10 to around 50. I would like to keep data and information flowing smoothly between units and would like to keep everything traceable and compliant to the required standards.*

If you believe that you will grow in size, you would like to implement solutions that save your immediate problems but are also scalable to support your growth in the future.

Making the transition from one system to another can be extremely painful, so making the right choice from the start is extremely valuable. That being said, in the process of digital transformation you will inevitably make some wrong decisions, so be ready to change them, if some do not work.

The important idea here is to understand that the technology trends will change in the following years, as well as your lab's growing business. So, defining your main challenges well will help you evaluate software solutions later on that can support you in addressing those challenges. Specifically, lab scale-up, growth in the number of lab personnel, infrastructure, and lab instruments interoperability. It is not so important to choose the software by its name or current popularity but by the agility of the vendor to adapt to the technological advances and support your growth in the future.

7.2.2.4 What is My Long-term Business Strategy?

How will my laboratory or organization work in 5, 10, or 20 years? Who will be my customers, and which services and products will I offer? Will we be introducing a new technique, instrument, or technology? How fast will we grow in terms of employees? Having a vision setup for your organization will help you incorporate digital strategy into an overall business strategy that will help you pursue the vision. That means that digitalization projects will be equally important as any other project that you will set in a strategy. This will have a positive effect on company culture and an overall perception of digital projects within the organization.

If you do not have a long-term strategy you may only see the benefit in digitalizing processes rather than completely transforming your organization. This is a legitimate approach that might work for you short- or midterm. It is highly likely that sooner or later market demands will force you to digitally transform the company, and at that time you will need to take time to think about your long-term business strategy.

7.2.2.5 How Will Changes Affect My Current Business?

Every change to the existing processes and toolset causes disruption. In a transition period, the workload will very likely increase, which increases the likelihood of mistakes that you cannot afford to make, prolongs the adoption period, and may even negatively affect the company culture.

Example *You are using paper reports which you send to your customers* via *e-mail. You have the new system running for 3 days, and the reporting part is done automatically. However, people are not used to using the system, and one of the operators has problems sending the report from the system. Since time is pressing, he uses the "old way" and sends the PDF* via *e-mail. There is no log in the new system that the report was sent, so everyone considers this a pending job. Only after a week and quite some investigation, you find out that the report has been sent on time. You would probably like to remind your team at this point to use the new system only and report any troubleshooting.*

It is very important that you plan ahead and consider all the effects the changes will bring. There are many questions you can ask yourselves, depending on the changes you are making. And it is better to ask more questions than too few. We have listed some examples. It is very important that this information is transparently communicated to employees.

How will a transition period be executed?
Will people spend more time using the new system at the beginning?
If yes, what will be the consequences?
If the new system will automate a lot of work, will I need to let some people go?
If not, how will their everyday work change, will they get new assignments?
If the new system will change communications with customers, will they accept them?
Who will train the employees?

7.2.2.6 How Will I Manage Legacy Data?

If you decide to digitalize processes and introduce new tools, you need to have a plan on how to deal with legacy data. There are a few options: migrate everything, archive everything and access when needed (clean-cut approach), and hybrid approach, which means that you migrate certain information, while other is archived. Very often a hybrid approach or partial migration is being recommended, as full migration might not be feasible (extremely demanding, time-consuming, and expensive) and is not always possible. With only archiving all old information you might lose some relevant data. The hybrid approach enables you to migrate important data (i.e. sample information) and archive data that you are predicted to use only very rarely (i.e. analysis reports).

Example 1 ***Complete migration***

Our company has data in a LIMS system that I started using 2 years ago. We were not too happy with a system and have found a better provider. I am able to export the data from old LIMS and import it into a new one. I will need some support from the new LIMS company to do a complete migration.

Example 2 ***Archive everything***

So far our laboratory has been paper based. We are now implementing a new ELN software to keep track of our R&D work. We have decided to archive all notebooks and data files and grant certain people long-term access. We will start completely fresh with the new system to avoid confusion (clean-cut approach). Optionally, we could grant a grace period and instruct laboratory members that they are obliged to document every new project into the ELN, but for the ongoing projects, they have a choice to either complete them in paper notebooks or transfer them to ELN and continue working there.

Example 3 ***Hybrid approach***

Our company has been using a combination of paper-based processes, supported by software systems, to keep track of our R&D work. We have protocols and sample databases saved in MS Excel documents, final reports in PDFs, experiment plans on paper, and raw data on local drives. We will be able to import all PDFs, protocols, and samples into a new ELN system, but we have decided to archive all paper notes and raw data to a dedicated location on a shared file server. We have decided who will have access to the archive.

Answering the following questions will help you decide how to choose the right option for legacy data:

How does the new system support migration?
How much information do we have?
Which information is critically important to migrate?
Which information is duplicated?
Where do we currently keep the information?
How will we access the information that we do not migrate?

How long will we need to access archived information?
Who will have access to archives?
How do we ensure data integrity?

7.2.2.7 How Will I Get People to Cooperate?

Introducing new digital tools in the laboratory often causes some resistance within the team or organization. Changing existing habits is hard, and there are many reasons behind the objections you will get from people. User adoption of new digital technologies is critical for the success of digitalization projects and digital transformation strategy in general. The chapter *Addressing the user adoption challenge* is dedicated to this matter.

7.3 Evaluating the Current State in the Lab

Whenever we are making a change, we always first need to understand where we are now. When changing the core concepts of how an organization lives and breathes, it is even more important to understand the current state in detail. This is clearly needed to plan a transition to a future state. But besides that it also helps us to understand what can go wrong and to eliminate the unknowns.

Digitalization heavily affects how you manage your data. One of the positive side effects is an increase in data integrity. Data integrity means that the data cannot be changed between the source and destination. In practice, this means eliminating manual tasks from managing data. Automated data management also brings more metadata, which is data about data. These are usually information such as who created the data and when, which instrument was used, and how it was transferred. Such metadata are helpful when you are searching and filtering databases.

7.3.1 Defining the Overall Goals of the Digitalized Laboratory

7.3.1.1 Example

Goals of a digitalization project can be defined in the following way:

Goal 1: Improve the data management by implementing digital tools (e.g. electronic laboratory notebook (ELN)).
Goal 2: Increase the efficiency of the laboratories by 25%.
Goal 3: Improve data integrity by eliminating manual steps from data flows.
Goal 4: The acquisition of new technologies should be 100%, and it should be easy for the users to start using the new tools.
Goal 5: Project should be finished in 12 months.

Goal 1: Improve the Data Management by Implementing Digital Tools

Digital tools, such as ELN or laboratory information management system (LIMS), can significantly improve the data management in your laboratory, because they provide a framework for implementation of good data management practices.

The most significant effect is data management efficiency, because you can eliminate all paper forms from your workflow, perform automated calculations, report generation, and report overview. Besides, many instruments used in routine labs today integrate directly with LIMS, and your lab staff do not need to worry about data transfer and saving anymore. The other effects are better security and data integrity. You can measure the improvement through improved efficiency and less procedural errors.

Goal 2: Increase the Efficiency of the Laboratories by 25%

Increasing the performance starts by analyzing laboratory processes and the amount of time they take. Then you can identify the bottlenecks and look for improvements. Some simple improvements might be digitalizing the equipment booking, maintenance, and remote access to instruments. There are many tools already available for this and can significantly simplify the daily operations in your labs. Then you can measure the performance by comparing the time spent on laboratory tasks before and after optimization.

Goal 3: Improve Data Integrity by Eliminating Manual Steps from Data Flows

This goal extends Goal 1 by integrating the laboratory equipment with the digital tool you are introducing. This helps you to eliminate manual data management, which improves data integrity and efficiency. You can measure the success by tracking efficiency improvement (see Goal 2) and error rate (see Goal 1).

Goal 4: The Acquisition of New Technologies Should Be 100% and Should Be Easy for the Users to Start Using the New Tools

One of the common problems with digitalization projects is user adoption. That is why we want this to be an important and equal part of our digitalization strategy and included it as a requirement. Most often the digitalization projects fail because they fail to implement the digital culture within the organization. Or, in other words, implementing technology is easy. But if people do not use it, it is a failed investment. User acquisition should be carefully planned and regularly tracked. One good approach is to survey the users to learn about their experience with the new technologies and get suggestions for improvements.

Goal 5: Project Should Be Finished in 12 Months

The digitalization project will affect your whole organization, and therefore, it is important to stick to the planned timelines. You should be clear and transparent about how much the disruption will be present in the organization and what will happen after the project is finished. Make sure to treat this project as any other project within the organization, which has a project manager and a dedicated project team. Do not forget to follow the progress and address any problems immediately.

7.3.2 Defining the Data Flows

Data is generated at multiple points in the laboratory and then transferred to one or more endpoints and possibly enriched or processed in between. This is called

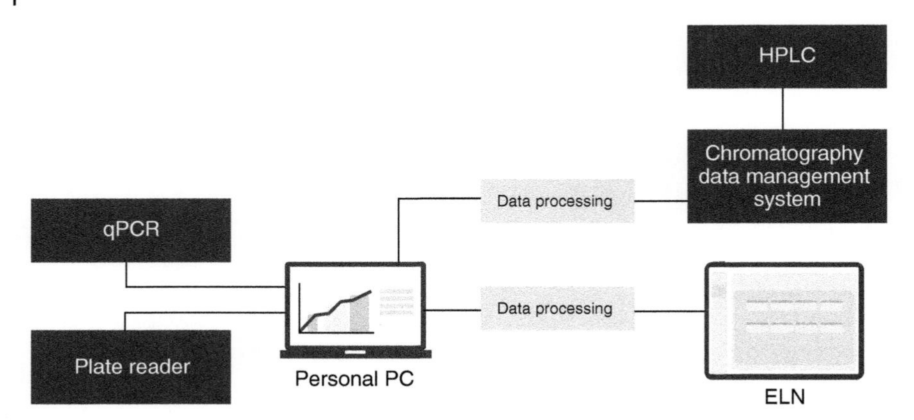

Figure 7.2 An example of a simple data flow of a laboratory. Source: Courtesy of Janja Cerar.

data flow – the process of moving, enriching, and processing data between different stations in the laboratory. You can describe the data flows in diagrams or by physically drawing the channels on the graphical layout of the laboratories (Figure 7.2). Drawing the data flow diagram will help you later when you will be identifying the bottlenecks in your data management process.

Figure 7.2 shows the instruments HPLC, qPCR, and plate reader that generate data, which are then transferred to the PC. The data from a PC is then uploaded into ELN. Data from HPLC are first stored in the chromatography data management system and then transferred to the PC and ELN. This is a typical setup, where users still transfer all data first to their PCs and then to ELNs. The connections between instruments and computers (dark lines, also called edges) can have additional data linked to define characteristics of the connection.

The analytical process starts by describing the data flows in the laboratories. While doing this, think about data in the most general terms possible. Each laboratory experiment or process generates data from the beginning to the end. Therefore, it is not just the final readout but also the details about how you perform the experiment or the environmental factors during it. Another aspect you could include here are materials (e.g. samples). Materials can be considered as data in physical form; therefore, they would also fit into the data flow.

During this process you are designing a graph representation of data flows. The data generators and acceptors will be nodes, and data channels will become edges. For each node, you will add some descriptors of the data channels:

- The type of the node.
- The node can be connected to the local network (yes/no).
- The communication protocols enabled by the node. This includes the manual data transfer which you might be using now (USB stick/CD, DVD/network/wireless).
- The software and hardware used on the node. This is helpful when finding bottlenecks and opportunities for improvements.

- Naming rules for the samples (name of patient/code of patient/barcodes).
- The labeling technique of samples (writing/barcodes).

You can define different types of nodes that will be used in your graph. Do not be too specific, because this usually makes the graph too complex. Focus on a few functional ones, such as:

- Analytical instrument
- Digital tool, such as ELN or LIMS
- Sensors
- Digital service/application programming interface (API)
- Samples

For each edge you should also add some additional parameters:

- The type of data that travels in that channel.
- The type of communication channel represented by the edge (USB stick/CD, DVD/network/wireless/physical).
- The amount of data that is transferred/the required bandwidth of the channel.
- The modification/enrichment process happening on the edge.
- The digitization of sample name (manual/photo/barcode).

You can use different colors to better visualize the different node and edge properties. You can start by organizing this data in a spreadsheet, one for nodes and one for edges. Node rows should contain at least a unique name of the node and all the properties. Edge rows should contain the names of both nodes and all other properties. Then you can use software tools, such as Dia, draw.io, or Graphviz, to draw and visualize the network.

7.3.3 Describing the Processes

Laboratory processes are most commonly the bottleneck when trying to improve the efficiency of the laboratory. We often see that the minor administration processes, such as scheduling, equipment booking, report generation, and logging, require quite some resources, are repetitive, or simply affect the efficiency of the laboratory. That is why you will often find improvement opportunities in such simple tasks.

Laboratories of any size have an enormous number of processes in place; therefore, it makes perfect sense to use the existing resources. These are usually standard operating procedures (SOPs) or similar documents that you already have in the organization and contain all the required information you need.

In case you do not have your processes described, or it is not adequate, then you would need to spend some time describing them. This is essentially similar to writing SOPs. The most basic form of a SOP are steps of the process, with some metadata, such as number, title, and author. You might want to separately describe the needed materials and tools. Then write down the steps. The steps can be sequential or hierarchical. Simple processes usually follow a sequence of steps, while complex ones might cover multiple different scenarios. You can also use a diagramming software, such as DIA or draw.io.

7.3.4 Identifying the Bottlenecks

At this stage we should have all the information collected. Now we analyze the collected information on the current state of data flows and processes, identify opportunities for digitalization, shape them into one or more projects, and define digitalization goals for each one. This is probably the most difficult process and requires quite some experience with digitalization and optimization of the processes and data flows.

7.3.4.1 Bottlenecks in Data Flow Optimization

The main bottlenecks in data flows are manual tasks, which affect both the efficiency of the laboratory and the data integrity. By automating the manual tasks through device integration and data automation you can increase the efficiency of the laboratory. This normally also makes users happier, since they spend less time on tedious tasks. Regardless, you should take into consideration their opinion when introducing changes.

7.3.4.2 Efficiency and Integrity of Data Flows

One of the sources of inefficiencies in the current laboratory setup, processes, and data flows are redundancies. You can ask yourself what opportunities exist to eliminate redundancies. Many data management tasks are still manual which requires resources and are error prone. You can identify the opportunities by looking at the data flow graph and find all manual data management tasks, such as manual data transfer from the instrument, recording data on the paper, and then transcribing into report and manual generation of reports. You can then search for solutions on how to avoid manual data transfer and assess each one, based on the complexity and the return on investment you are expecting. We will look into a few examples, which can be digitalized and/or automated.

Example *Save data directly into digital tool or shared drives*

Most laboratory instruments today have a built-in network interface and can be set up to store the data on a shared network drive. Some instruments also have an API built in the software that allows it to directly communicate with the digital tool, such as ELN or LIMS. A good starting point is therefore to directly import the data to the digital tool or at least save it to a shared network drive. If the instrument does not have a network interface and is not connected to a PC you can use a special middleware device that can do that.

Using a proxy data medium, such as paper (either handwritten or printouts), can take a significant amount of time to digitize and transcribe. Laptops, tablets, and mobile phones are becoming widely present in the laboratories, so you can use them to access digital tools and enter data directly. This will save a significant amount of time and help you maintain the data integrity.

Example *Automated report generation*

Data reporting is the task that usually requires the most manual input. The good news is that it can often be automated, which improves efficiency and integrity. Data reporting essentially means conversion of raw data into a human-readable document. Reports frequently start with data analysis, usually done in a spreadsheet software. This is easy but highly error prone – one wrong operation in a spreadsheet might change the whole dataset. A better approach is to use automated data analysis scripts, which can be written in a scripting or statistical software, such as R, Python, or Bash. Such scripts take the raw data as an input and generate all the required data needed for the report without any human interaction. To generate the report you can use a templating system or just extend the data analysis scripts to insert the data into the report template. So just by importing your raw data into an ELN you could trigger an automation sequence that would do the analysis, generate the report, and send it by e-mail to the person who needs it. Automated report generation will save a significant amount of time to your scientists, since manual report generation mostly means copying numbers into data tables, which can be efficiently done by the computer.

Example *Scientific paper writing*

When it comes to academic research laboratories, a research paper is the ultimate report. Writing a research paper is more complex and difficult to automate. However, advances in artificial intelligence (AI) can already help you with these tasks, such as SciNote Manuscript Writer that outlines a draft of your manuscript from the data in your ELN. It can also find relevant publicly available references online and write the introduction. The draft serves as a good starting point but still needs a significant amount of editing. However, this tool will still save you a big part of the time you initially invest into drafting.

7.3.4.3 Example: Make Data Machine Readable

Often the data is not machine readable. By this, we mostly refer to data that are saved in proprietary file formats that cannot be read outside of the software provided by the vendor. However, such data can often be exported as open file formats and used by other software solutions. Although it is a good idea to use open file formats when possible, beware that proprietary file formats sometimes store additional data that get lost during the export.

You can go beyond machine readable and implement good data management guidelines, or shorter – FAIR. Acronym FAIR stands for Findable, Accessible, Interoperable, and Reusable. It is a guideline for maintenance and generation of scientific data with sustainability in mind. Many data generated by laboratory instruments will not be FAIR by default. To make such data FAIR you would need to add some metadata and descriptions of the experiment. Sometimes this can be done through the digital tool by properly structuring the experiments and projects. Check with your digital tool vendor for advice on how to achieve this.

Example *Use the ELN as a primary data repository.*

When implementing an ELN the scientists might not be aware of how it can simplify their work. Scientists often e-mail the results from the instrument-connected PC to themselves, and then they would do the analysis on their office PC. They might keep the same practice even when using an ELN. It would be much simpler to log into ELN on the PC connected to the instrument and upload the files containing results directly into the ELN.

7.3.5 Opportunities in Process Optimization

Bottlenecks in data flows are a good starting point to begin with process optimization. But the main contributions toward improved laboratory efficiency are likely to be found in other bottlenecks, not necessarily linked to data flows. That is why we are looking at the bottlenecks in processes that take a lot of time and are repeated often.

We will use the process description we prepared earlier, which is usually based on SOPs in your lab. Finding bottlenecks in the processes takes much more time and dedication, mostly due to a large number of processes in the laboratory. Processes that take a long time or are frequently repeated on a daily basis contribute most to improved efficiency. You can use this information to prioritize the processes that need immediate attention rather than the ones where optimization would have less effect. In the following sections we discuss some guidelines and good practices that you can use as starting points. You might find something else to be more relevant for your organization.

7.3.5.1 Time-consuming Processes

Some optimizations can be done in experimental processes that have a lengthy execution, but can be parallelized or multiplexed, i.e. by performing multiple such tasks in parallel. This is particularly important for assays with long incubation times, such as immunoassays, or difficult pipetting schemes, such as qPCR.

For example, you could optimize qPCR using a more high-throughput format of 384 well plates instead of 96 well plates and/or consider using automated liquid-handling instruments (pipetting robots or smaller devices that guide you through pipetting). Such devices do not only reduce human effort but also increase repeatability and can often be connected into laboratory data flows as they can read and output data.

Another example are assays with many washes and long incubation times, such as ELISA or Western blot, that take a lot of time to perform. The main reason is that washes are usually short (5–15 minutes) and need manual handling. A washing process can therefore take one or two hours of manual work, most of it being changing liquids and short incubations. Effectively, the scientist does 15% work and 85% wait. Using appropriate automation techniques you could perform more such assays, but with a huge efficiency boost.

When you are having long incubation times you can try to do many assays at once. Instead of doing individual ELISA assays, you could perform a few together. This might not improve the efficiency that much but will get you the results faster.

Automation is not relevant only in the laboratory but also for data analysis. Spreadsheet tools are still the most commonly used tools for data analysis in science. Some assays have a complicated data analysis process that might take quite a lot of time as well. Instead of using a spreadsheet software you can automate the data analysis of such processes with data analysis scripts (R or other statistical software or scripting languages). You could also use template spreadsheets; however, they are more error prone because the user could unexpectedly change data points without noticing.

7.3.5.2 General Laboratory Processes

Some processes in the laboratory are executed very often, they constitute the general operations, and are used by all staff. Some examples are equipment booking, remote access to the network and equipment, procurement, document management, record keeping, and training. These processes are not specific to laboratories, and therefore many tools already exist that can help you digitize and optimize them. Although time savings seem small, you would be surprised how much time it saves to the whole team in a year.

For example, laboratories use printed calendar sheets, placed next to the lab instrument, for booking the equipment. This means that the person needs to be physically present in the laboratory to book or change the booking. Moreover, if this needs to be coordinated with multiple people, they need to be available at that time to discuss the alternative arrangements. Instead, such laboratories could implement a shared online calendar. Such a tool would allow you to book the equipment from the office and easily coordinate the booking with other people.

Another great *example* of process optimization is training of new users. It is actually one of the most time-consuming tasks in life science laboratories. These are normally done individually for each new member and take days or weeks to complete. This can be improved by recording the training sessions and using the recordings to train the new members. When thinking about optimizing laboratory processes, it is important to think broader than just about what is happening with the data. Optimizing general processes can save a lot of time to a lot of people.

7.3.6 Gap Analysis

A gap analysis is a process where you will identify the differences between the current state and the state that you anticipate to achieve by the end of the project. The information and analyses you gathered up until now are important to build a gap analysis that you will use in the digitalization strategy planning phase.

It is important to document this analysis properly. You can use a large sheet of paper and divide it into four columns: Current state, Future state, Gap description, and Solution. You can add extra fields to do some rough resource planning, although this can also be done later and differently. Most data you can just fill in, since you gathered it during the current state description and while searching for opportunities.

7.3.6.1 Example

Let us fill out the columns for the example of a digitalization project in the laboratory (Table below).

One of the objectives of the digitalization project is to implement an ELN into the laboratory. In the current state analysis you identified that most data are saved on the shared network drive. In the future you would like to save all data directly into ELN.

You also identified that most users still transfer data on USB sticks. In the future, you would like them to access ELN directly from the instrument's PC and upload the data directly.

The adoption of ELN in the department might also not be smooth, so you need to find a solution for this as well.

The gap-description column in the table can be used to put more details about the gap.

You will probably have some ideas about how to improve this, such as performing multiple tests in parallel or automating the liquid-handling tasks. You can write this under the Solution column.

The table below gives you an example of the gap analysis, describing the current state, future state, gap description, and solutions.

Current state	Future state	Gap description	Solutions
Data is stored on a shared network drive.	Store data directly in ELN.	Shared network drives do not provide adequate access control and project management.	Implementation of ELN.
Equipment booking is taking too much time and is not digitalized.	Digitalize equipment booking.	The booking arrangements cannot be managed remotely and are therefore suboptimal.	Implement a shared calendar for equipment.
Data is transferred with USB sticks.	Automated data transfer between devices and data storage.	Manual data transfer is a risk for data integrity and is inefficient.	Implement automated data transfer middlewares.
The adoption of digital tools in the laboratory is low.	The adoption of digital tools should be 100%.	Laboratory staff involved in digitalization pilot tests did not remain using the digital tools after the study.	Define the technology adoption strategy.

There are also several tools available to help you with gap analysis. The two most common ones are McKinsey's 7S framework and SWOT analysis (Figure 7.3). Although these tools are tailored for analysis of businesses, you might find some useful aspects in it.

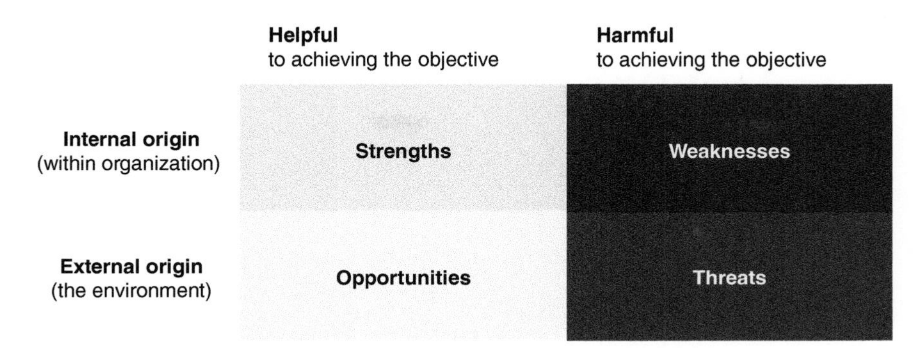

Figure 7.3 SWOT analysis: Strengths, Weaknesses, Opportunities, and Threats. Source: Xhienne, https://commons.wikimedia.org/wiki/File:SWOT_en.svg.

Strengths and Weaknesses are attributes of the organization, while the Opportunities and Threats describe the environment the organization lives in.

References

1 Elsevier (2020). 10 aspects of highly effective research data. Elsevier Connect. https://www.elsevier.com/connect/10-aspects-of-highly-effective-research-data (accessed 1 June 2020).

2 Forero, D.A. and Moore, J.H. (2016). Considerations for higher efficiency and productivity in research activities. *BioData Mining* 9 (1): 1–4.

3 Technology Networks (2018). Preparing laboratories for the digital revolution. Informatics from Technology Networks. Technology Networks. https://www.technologynetworks.com/informatics/blog/preparing-laboratories-for-the-digital-revolution-304663 (accessed 1 June 2020).

4 Digitization (2020). Gartner. https://www.gartner.com/en/information-technology/glossary/digitization (accessed 1 June 2020).

5 (2019). Understanding digital transformation: a review and a research agenda. *The Journal of Strategic Information Systems* 28 (2): 118–144.

8

Understanding Standards, Regulations, and Guidelines

Matjaž Hren

SciNote LLC, 3000 Parmenter St, Middleton, WI 53562, United States

8.1 Introduction

Standards and guidelines as defined by International Organization for Standardization (ISO) are documents established by consensus and approved by a recognized body that provides, for repeated and common use, rules, guidelines, or characteristics on the design, use, or performance of materials, products, processes, services, systems, or persons. Standards can be developed by national, regional, and international standards developing organizations and also by businesses or other organizations for their own internal use. They can also be developed by consortia of businesses to address a specific marketplace need or by government departments to support regulations. Standards are not mandatory, they are voluntary.

Regulations or codes can be described as rules of order having the force of law, prescribed by a superior or competent authority, relating to the actions of those under the authority's control. Regulations are issued by various government departments and agencies to carry out the intent of legislation enacted by the legislature of the applicable jurisdiction. Compliance with regulations is mandatory.

Examples of standards applicable to life science laboratories are ISO 9001 (general quality management standard), ISO 17025 (for testing and calibration laboratories), ISO 15189 (for medical and diagnostic laboratories), and Good Laboratory Practice (GLP). For example, a laboratory that is performing quantitative detection of genetically modified organisms in food can be accredited under ISO 17025 in order to demonstrate the quality of their services and their technical competence to the customers. Sometimes such laboratories decide on their own to become accredited against a standard as their competitive advantage on the market, but lately the national regulatory bodies also recognized such standards as good practices and require the service providers to be accredited.

An example of a regulation is Title 21 CFR Part 11 (mostly known simply as 21 CFR Part 11 or even Part 11). This is a regulation put into force by the US Food and

Digital Transformation of the Laboratory: A Practical Guide to the Connected Lab, First Edition.
Edited by Klemen Zupancic, Tea Pavlek and Jana Erjavec.

Drug Administration (FDA) and regulates digitized processes, specifically electronic records and electronic signatures by defining the criteria under which electronic records and electronic signatures are considered trustworthy, reliable, and equivalent to paper records. As such this regulation is law in the United States. On the other hand, the EU counterpart of the 21 CFR Part 11, the EU good manufacturing practice (GMP) Annex 11: Computerized Systems (frequently known simply as Annex 11), is not a legal requirement but rather a strongly recommended set of guidelines for the interpretation of the principles and guidelines of GMP when computerized systems are used as part of GMP-regulated activities.

Another example of regulation is General Data Protection Regulation (GDPR), a regulation in EU law on data protection and privacy in the European Union. A similar example is a Health Insurance Portability and Accountability Act of 1996 (HIPAA), which is a US legislation that provides data privacy and security provisions for safeguarding medical information.

It may seem that standards are only targeting industrial, commercial, or service laboratories; however, there are many guidelines and initiatives applicable to academic research laboratories. Some are more general such as findable, accessible, interoperable, and reusable (FAIR) guiding principles for scientific data management and stewardship, which provide guidelines to improve the findability, accessibility, interoperability, and reuse of digital assets [1]. Some are specific to techniques, such as MIQE guidelines: minimum information for publication of quantitative real-time polymerase chain reaction (PCR) experiments [2], or The digital MIQE guidelines: minimum information for publication of quantitative digital PCR experiments [3], and good chromatography practices [4], providing information on good practices to be considered in the analysis of samples when chromatographic systems are used. Since many of the laboratory instruments now have digital data outputs and can be controlled digitally, some standards are covering these aspects such as AnIML, an open standard for storing and sharing any analytical chemistry and biological data and SiLA, setting standards for open connectivity in lab automation.

8.2 The Need for Standards and Guidelines

From the definitions and examples of standards, guidelines, and regulations, we see that in addition to covering processes in laboratories related to products or services (e.g. ISO 9001, ISO 17025), they also cover quite specific methods (e.g. good chromatography practices), products, and instruments or instrument-related aspects such as data-exchange formats (e.g. AnIML and SiLA).

If we look at all these aspects in the context of a laboratory, we see that standards and guidelines ensure quality in laboratories, or as ISO puts it, they aim at achieving the optimum degree of order in a given context. Their choice of the term "optimum degree of order" is very interesting and important, as they do not use the term "maximum degree of order." This means that standards are not imposing a rigid and perfect set of rules on how to organize activities in the laboratory but rather help laboratories find a way to get to an organized state that best fits their processes and organizations.

Standards that are covering processes tend to be agnostic of technology used in the laboratory and whether the processes happen in physical (paper) or digital or in a hybrid way. Their main aim is to standardize processes and make them reproducible and robust. Standards also assure traceability, introduce risk management into processes, and introduce processes for detecting, correcting, and preventing errors, which ensure self-improvements. Moreover, they ensure that the methods used in laboratories are validated and introduce good practices for laboratory instruments used in the processes such as regular maintenance, calibrations and validations, and so on.

In a simplified way, standards increase the quality of work in the laboratory and the quality of laboratory outputs, whether these are certificates of analyses or published research papers. If you imagine that you would get two certificates of analysis from medical diagnostic laboratories and one would bear a stamp ISO 17025 and ISO 9001 and the other one not, you would more likely trust the first one.

There is a difference when a laboratory works "in compliance" with a standard, or if a laboratory is actually accredited against a standard. In the first case, the laboratory simply adopted the standard and works according to it, and in the second case, the laboratory went through a formal accreditation process with assessment bodies that are in turn accredited by national accreditation bodies. Accredited laboratories usually go through periodic internal audits, where trained internal auditors check compliance with the standard, and external audits (e.g. once a year), where they are independently checked against the standard by independent external expert auditors who often have domain expert knowledge. They review how your work complies with standards, not specifically what you do, but how you do it.

8.3 How Does Digitalization Relate to Standards and Guidelines

There are a few aspects of how digitalization relates to regulations, standards, and guidelines, and we are discussing them below.

8.3.1 Standards Should Affect the Selection of the Tools for Digitalization

When you are preparing your digitalization strategy, e.g. in the step of assessing current data flows, identifying bottlenecks, and defining the overall current requirements for the digitalized laboratory, make sure you add to that list the current regulations, standards, and guidelines you adhere to, and try your best to predict if you will need to comply with any additional ones in the future. This is often related to the short- and mid-term vision of the laboratory or organization, so it might be useful to involve people or management responsible for strategy in your organization into the decision-making process to select the right digital tools.

Before you start searching for the providers of the digital tools, you should list requirements for the tools that derive from a specific standard or regulation. However, if you are not able to do so, include a list of names of standards and regulations that apply to your use-case. This will be harder to do for standards you did not implement yet, but it is still a useful and very important exercise. Having a list of requirements will help you determine whether digital tools are supporting the standards when you will be researching the tools on your own or when you will be requiring more information from vendors (e.g. does your tool have electronic signatures and audit trails; we are performing some of our studies according to GLP, will your tool support that; is your tool GDPR compliant; and we have to comply with HIPAA regulation, does your tool support that). When a tool complies with a certain standard or regulation, vendors make sure they have this information easily available. In some cases companies need to go through a tender process when they are purchasing new tools. A common step in the tender process is issuing a Request for Information (RFI) to different tool vendors. Having a list of requirements as part of RFI that also includes standard and regulation aspects will make your decision process more efficient.

It is also important to note here that the software itself normally is not "compliant." It can support you in meeting the compliance requirements, but being compliant normally includes all actions that are taken into account within the lab and lab's operations in general.

Therefore, not all the tools will address all requirements, so you should be prepared to make some level of adjustments to your processes when you will be digitizing them. However, proper research in advance can save a lot of effort.

Example 1 *Our lab is digitizing the analytic process for the quantitative detection of genetically modified organisms in food. We are transferring the paper-based analytical process, where we are using a set of forms and SOPs into an ELN. As a part of the process, we have a review step where the principal analyst needs to review and sign-off on the results generated by a technician. We have selected an ELN vendor and transferred all the necessary forms and SOPs into it. During the pilot phase, we realized that the ELN does not have adequate capability where principal analysts would be able to use electronic signatures to confirm or reject the results prepared by the technician. We now have to find a workaround in the ELN, pay for customization to the vendor, or possibly find a new tool. If we identified that the tool needs to support e-signatures (for our lab to be compliant with 21 CFR part 11) in the phase of collecting requirements for the tool, we could have avoided this complication.*

8.3.2 Digital Tools Promote Good Practices

Digitalization will inevitably lead to the development of internal guidelines and best practices which will raise the level of quality of processes in laboratories.

For example, digital tools can bring in among other things:

- *Traceability* of actions that users or systems perform on records, data, etc. is often a core requirement of many standards and regulations and ensures trust between the regulatory body and the laboratory and generally reduces the chances of bad practices such as data manipulation

- *Improved findability of information* generally increases efficiency
- *Improved control of access to information* is a must in accredited laboratories, where certain actions pertain to specific roles in a laboratory, such as confirmation of analyses is limited to laboratory analyst and in cases where some data (e.g. personal data and customer data) should have access limited to specific people or roles in the laboratory
- *Easier access to information* generally increases the efficiency of processes
- *Ability to enrich data with metadata and tags* helps users group search pieces of data
- *Structure, organization, and hierarchy of data* are sometimes enforced by digital tools; in the absence of a standardized hierarchy of data in laboratories, any agreed organizational structure of data is a good practice that improves findability and goes in line with the "optimum degree of order" promoted by standards
- *Good data management practices* such as storing data in one place, data integrity, data encryption, data backup strategies, and security, which are all aspects that are much more challenging to implement in a paper or a hybrid paper–digital system. In order to have a backup/restore or encryption in place, a certain level of IT infrastructure and expertise is needed, which not every laboratory has. Therefore, it is beneficial if any of these features are brought in by digital tools.

These benefits vary greatly from tool to tool and also depend on the tool's main purpose and flexibility; however, any of these aspects will increase the quality of work and will be a step in the right direction. Sometimes some of these features may be perceived by end users as limitations at the beginning; however, imposing order and rules are always limiting compared to total flexibility, which sometimes borderlines chaos.

In addition to increased quality of work and outputs, digital tools may also improve efficiency, especially in cases when tools automate or replace manual or labor-intensive tasks that are prone to human errors. Benefits are especially noticeable in certified/accredited laboratories that are required to manage risks in their processes. In these cases, digital tools may reduce the number of quality checks that were required in the manual or paper-based system in order to control the risk of human errors. For example, if an analytical laboratory had to manually transfer the data from instruments into a LIMS system or to a spreadsheet, there was an additional check in place performed by an independent analyst who would verify that the data were transferred without errors. If the new tool would be able to transfer the data automatically, this would make the additional quality check obsolete and make the process more efficient. Another example is saving time due to faster access to raw data due to better traceability brought in by digital tools or reducing the cost of archiving printouts of raw data or certificates of analysis, which is sometimes outsourced to professional archiving specialists.

Data exchange in laboratories (data flows) is a very important and sensitive aspect and is often at the core of digitization efforts. There is a clear lack of unified data-exchange standard; however, there are some competing attempts that look promising and are being increasingly adopted by instrument manufacturers (e.g. AnIML, SiLA2, and Allotrope Foundation). They are working to improve interoperability, which can be defined as the ability of different information systems, devices, and applications to access, exchange, integrate, and cooperatively use data in a coordinated manner. As such, interoperability strongly promotes digitalization of laboratories not simply by making data flows more seamless and improving data integrity but also making the entire processes more efficient and enabling the use of artificial intelligence algorithms and big data analytics [5].

Perhaps one of the most literal and the most obvious advantages of digitalization is the reduction of use of paper. In accredited laboratories, the use of paper becomes even more obvious due to retention policies for recordkeeping in regulations or standards. This means that a laboratory is required to keep all records in a way that they remain legible and can be stored and retained in such a way that they are readily retrievable in facilities that provide a suitable environment that prevents damage, deterioration, or loss. Although many times standards leave the retention time policy up to the laboratory, the times are commonly measured in years (e.g. five years is a common retention time). In laboratories that have a high throughput of sample analyses this can become a significant and a costly challenge. Some laboratories even outsource archiving of records to professional services. Although archiving and access of digital records also have their own challenges, it is much more convenient and efficient than physical archiving.

To summarize, the advantages digital tools bring may shorten the time and save effort required for certification or compliance with regulations, standards, or guidelines. Sometimes tool vendors provide information with which regulations, standards, and guidelines the tool is compliant with. However, it is important to understand that if a tool is, for example labeled to be GLP compliant, this does not mean that you will be automatically GLP certified using the tool – you will still need to implement the tool, integrate it with your processes, set up the quality management system, and go through formal certification or accreditation process. But this path should be shorter and easier.

8.4 Challenges Related to Digitalization in Certified Laboratories

Regulators and accreditation bodies are to various levels adapted, equipped, and familiar with auditing digitized environments. In case of regulations that specifically cover digitized environments, such as 21 CFR part 11, the regulators are much more likely to be ready, especially when such regulations have been around for quite some time (21 CFR part 11 was initially published in 1997). When it comes to standards and guidelines that cover general processes such as ISO 9001, ISO 17025, and GLP, which can be perfectly well established in nondigitized as well as in digitized environments, then regulation bodies have various levels of experiences with auditing digitized environments. This is especially true in industry segments that are not well digitized. Although the level of digitization should not be in any way affecting the audit process, it sometimes can make the auditing process a bit more tedious because the auditors might not have as much experience with digitized environments as with "paper-based environments" or are simply not that technology savvy. However, this should be less and less of a challenge as digitization is expanding and is also bringing several benefits to auditors.

Sometimes the bodies that prepare and maintain standards and guidelines need some time to adapt to digital solutions, and sometimes specific guidelines for digital environments are prepared to help auditors and consumers of the standards. One such example is the document OECD Series on Principles of GLP and Compliance Monitoring No. 17: Application of GLP Principles to Computerized Systems.

8.5 Can Digital Strategy be Implemented without Certification?

Absolutely. As we saw earlier, digitalization will inevitably raise the level of good practices in laboratories independently of certification or accreditation.

Every laboratory has good practices, possibly without even knowing. They might have been gradually introduced in time based on experience, sometimes by trial and error, and may not even be written down, for example having all protocols or standard operating procedures used in the lab by various people stored in one location and storing all raw and analyzed data on a central location on a laboratory server (and not locally on laptops).

Example *A good example of evolution of good practices is protocol management. An organization or laboratory may decide to store laboratory protocols in one location, starting this by making it obligatory for every researcher to save the protocols on the central location and check whether there are any similar protocols there. This will help them to avoid reinventing the wheel or using a very similar protocol with slight modifications. One of the next steps could be an introduction of codes used for versioning protocols, so that every user would know which is the latest version of the protocol or tags or a set of standard (semantic) expressions (controlled vocabulary) to describe each protocol and make it easier to find a related protocol. The ultimate level would be to upgrade this system into a document management system with a formal document approval process and official versioning of protocols.*

Following the line of thought that the majority of businesses believe they will be out of business or become marginal players if they do not digitally transform [6], any digitalization effort is a step in the right direction and should bring in good practices, which are the cornerstone of many regulations, standards, and guidelines.

References

1 Wilkinson, M.D., Dumontier, M., Aalbersberg, I.J. et al. (2016). The FAIR guiding principles for scientific data management and stewardship. *Scientific Data* 3 (1): 1–9.

2 Bustin, S.A., Benes, V., Garson, J.A. et al. (2009). The MIQE guidelines: minimum information for publication of quantitative real-time PCR experiments. *Clinical Chemistry* 55 (4): 611–622.

3 Huggett, J.F., Foy, C.A., Benes, V. et al. (2013). The digital MIQE guidelines: minimum information for publication of quantitative digital PCR experiments. *Clinical Chemistry* 59 (6): 892–902.

4 World Health Organization (2020). Good chromatography practices. https://www.who.int/medicines/areas/quality_safety/quality_assurance/QAS19_791_Rev1_Good_Chromatography_Practices.pdf (accessed 1 June 2020).

5 Lehne, M., Sass, J., Essenwanger, A. et al. (2019). Why digital medicine depends on interoperability. *NPJ Digital Medicine* 2 (1): 1–5.

6 Infor (2020). Designing tomorrow: the digital transformation journey|Infor. https://www.infor.com/blog/designing-tomorrow-the-digital-transformation-journey (accessed 1 June 2020).

9

Interoperability Standards

Sören Hohmann

Department Molecular Biotechnology and functional Genomics, Technical University of Applied Sciences Wildau, Hochschulring 1, 15745 Wildau, Germany

The past 50 years have been characterized by major advances in instrumentation and computer modeling, which are changing the way scientists work. This trend toward digital data management is certainly a good one and offers as many challenges as it offers advantages, depending on how you handle it. One of the key challenges that bring benefits on a broad scale is to connect the hardware and the software in the lab. In the past this was only applicable with custom-made solutions, e.g. if you buy a whole ecosystem from a vendor. As those custom-made solutions are not affordable for the vast majority of labs, multiple initiatives started to enable connections to the lab in a more flexible and affordable manner. The keyword for connecting the lab is interoperability.

Interoperability is a characteristic of a product or system, whose interfaces are completely understood, to work with other products or systems, present or future, in either implementation or access, without any restrictions. To give a fairly simple example, a mouse with an USB cable might be used on every PC (hardware) and work immediately, without bothering which device it is or which operating system (software) runs on the hardware.

As data and algorithms increasingly shape the world, the data becomes more valuable than ever. Getting your data interoperable increases its value.

If one strives to achieve interoperability for himself, we recommend watching out for already active initiatives that pursue the goal. Most of them welcome new members, and some may take an entrance fee. On the one hand, there are consortia dedicated to interoperability, such as the Data Interoperability Standards Consortium (DISC). There are also paths via an alliance of same interest partners for specific fields, such as the alliances behind Allotrope and SiLA. Additionally, the standardization organization offers paths toward creating standards for the field and also offers more viable short-path procedures such as the DIN-SPEC to reach a lightweight standard.

Digital Transformation of the Laboratory: A Practical Guide to the Connected Lab, First Edition.
Edited by Klemen Zupancic, Tea Pavlek and Jana Erjavec.

If interoperability on a high level is achieved, the benefits are:

- flexible data that is longtime sustainable.
- time saved in case you add a new compliant soft- or hardware to your interoperable lab.
- reduced cost by creating automated data exchanges and therefore safe work hours that are spent on dull and error-prone manual transfer.
- increased data transparency as a modern interoperable format will surely consider the FAIR principles or FDA 21 part 11 guidelines.
- improved control of access to information is a must in accredited laboratories, where certain actions pertain to specific roles in a laboratory, such as confirmation of analyses is limited to laboratory analyst and in cases where some data (e.g. personal data and customer data) should have access limited to specific people or roles in the laboratory.
- structure, organization, and hierarchy of data as interoperable formats are well defined and usually even come with an ontology (a explaining vocabulary) for each used term.

As interoperability quite certainly is one of the key components for a better digitalization, it is worth presenting more information about three good examples. The aim is to learn from them how to successfully create and maintain a standard.

9.1 SiLA

The SiLA standard handles instrument communication in a harmonized way. SiLA's vision is to create interoperability, flexibility, and resource optimization for laboratory instrument integration and software services based on standardized communication protocols and content specifications. SiLA 1.0 was published in 2009, and the overall improved SiLA 2.0 was released in early 2019. SiLA is owned, maintained, and improved by the SiLA Consortium. This Consortium is a nonprofit organization that can be joined by both sides, life science experts and industry. SiLA 2.0 is developed to be a state-of-the-art technology base centered around easy-to-implement, outstanding, and elegant concepts that can quickly grow with the help of the community.

To achieve this, the SiLA Consortium set multiple goals:

- SiLA must map to a base technology & communication standard that will survive for a considerable amount of time. It needs to be accessible, robust, and open.
- The base communication protocol must be capable of surviving traversal over common Internet infrastructure.
- SiLA standards are free and open; SiLA enables and supports open-source efforts with licensing that facilitates and does not impede adoption.
- Purposeful handling of metadata.
- A communication protocol is only part of a working infrastructure, and extensions should be possible without losing the original spirit and the reason for SiLA as a standard.

- Furthermore, the aim is to keep the right level of simplicity for a maximum of user-friendliness.

On a technical level the Consortium favors multiple technologies as they are commonly used and robust:

Technology	Description
HTTP/2	As a base communication layer. HTTP enables the retrieval of network-connected resources and can deliver fast, secure, and rich medium for digital communication.
"REST"-like communication paradigm	A REpresentational State Transfer (REST) is an architectural style for providing standards between computer systems on the web, making it easier for systems to communicate with each other (https://www.codecademy.com/articles/what-is-rest).
ProtocolBuffer data structures	Protocol Buffers are widely used for storing and interchanging all kinds of structured information.
Interface description language (IDL)	IDLs describe an interface in a language-independent way, enabling communication between software components that do not share one language, for example, between those written in C++ and those written in Java.

9.2 AnIML

The Analytical Information Markup Language (AnIML) is a vendor-neutral cross-technique XML standard for analytical and biological data and is maintained by the ASTM International E13.15 Subcommittee. The first release was in 2013. The goal is to improve exchange, analysis, and retention of analytical and biological data in an open-source approach. AnIML focuses on raw data handling, with priority on ease of handling. Therefore, the standard is machine searchable and human and machine readable. This concept is suitable for many different measuring techniques, as well as for collecting the same technique from different devices. Moreover, AnIML openly states what it expects from the different stakeholders and why they should contribute:

- *Instrument vendors*: AnIML aims to be usable for every type of analytical instrumentation, so instrument companies will need to integrate export functionality in their software.
- *Analytical laboratories*: Laboratories will want to move toward storing analytics data in the AnIML format because of its features for data integrity and archiving.
- *LIMS vendors*: Laboratory information management system (LIMS) currently in use in more extensive laboratories will need to integrate options to store and process AnIML files as companies move to consolidate and archive instrument data in one format.
- *Government agencies*: Reporting of instrument data to different government agencies is likely to increase significantly as requirements become more stringent, and AnIML is likely going to be the format of choice.

- *Researchers*: Scientists currently deal with a variety of instruments producing multiple file formats and have to keep software to read these data files – forever! Conversion or export to AnIML will significantly improve this situation as generic viewers will always be available for an open standard. In addition, AnIML can be used to store data from research equipment where software can be designed specifically to write the specification. Researchers can write their technique definitions for in-house built research equipment if they wish (although these would not be an official part of the AnIML specification).

Today, AnIML offers standards for multiple techniques (UV/Vis and chromatography), with multiple more in the pipeline under development (IR, MS, nuclear magnetic resonance [NMR], etc.). Even though AnIML is open source and can be implemented on its own, the company BSSN Software specializes in implementing AnIML for customers. This company was bought by one of the big players in the pharma market in 2019, which may be interpreted as an indicator for the rising importance of interoperability.

9.3 Allotrope

The Allotrope Foundation emerged in 2012 and focuses on raw data handling by developing a single universal data format to store linked data that connects people with the raw data, results, and evidence. This is somewhat similar to AnIML, but instead of eXtensible Markup Language (XML) they use Resource Description Framework (RDF) as a base format. The Foundation is an international consortium of pharmaceutical, biopharmaceutical, and other scientific research-intensive industries. In comparison to AnIML they are a closed community that is membership based and demands an entrance fee.

Allotrope's approach is divided into three categories:

- The Allotrope™ Data Format is a vendor, platform, and technique agnostic format adaptable for any laboratory technique to store the data, contextual metadata, and any ancillary files.
- The Allotrope Taxonomies and Ontologies form the basis of a controlled vocabulary and relationships for the contextual metadata needed to describe and execute a test or measurement and later interpret the data.
- Allotrope™ Data Models provide a mechanism to define data structures that describe how to use the ontologies for a given purpose in a standardized way.

Today, multiple Allotrope Data Models are available (Calibration, Cell Counter, GC, Raman, etc.), and some more are in the pipeline for 2020.

Even though progress is visible from work in recent years, there are still unfilled standardization gaps in the management of data. Research showed that either non-profit consortia or networks effectively are able to host standards (apart from the standardization organizations). Most of the interoperability standards are around 10 years old and are still actively worked on, yet some are relatively unknown among the scientific community – indicating that promoting a standard is also part of getting it to be successful. Furthermore, all the consortia are keen to include all sides and create a broadly consultative and inclusive group. The FAIR principles are widely considered to be the key to modern science and therefore also must be considered for a modern standard. On a technical level it seems purposeful to use common and open formats that are long-term sustainable. The standards observed use one file format for the core data and multiple other formats to enable transfer and communication of the data. Each of the offered standards comes with an extensive Ontology or Model description and obviously aims for a maximum of user-friendliness.

9.4 Conclusion

Even though digital data is considered the oil of the twenty-first century, this seems quite contradictory to many people. It will take time until it is realized, and with a high chance, the awareness will just rise after severe issues are discovered, which frankly will be quite late and cost substantial amounts of money. Therefore, it is highly important to create max value data, and interoperability is one of the keys to reach this potential. Standards, guidelines, and digital tools that promote good practices offer solutions that enable data to be harmonized, and we strongly recommend to implement them sooner than later.

10

Addressing the User Adoption Challenge

Jana Erjavec

BioSistemika d.o.o., Koprska 98, SI-1000 Ljubljana, Slovenia

10.1 Introduction

User adoption is successful when a user transfers from an old system and adopts a new system that is usually better, faster, and more efficient. However, during the process of implementing changes it is highly likely that you will face resistance where users will want to maintain the status quo and insist that there is no need for a change. There are some common user objections that can be predicted and managed.

"I do not want to change because":

1. The current system works well for me.
2. The current system works well because I helped to establish it. There is no need for a new one.
3. I do not have time to change, my schedule is already full.
4. The new system looks complicated and will take a lot of time for me to learn it.
5. I do not trust people who are making the change; they do not understand how we work.
6. How will I benefit from the new system? I really do not see the added value for me.

Before we dig into some best practices on how to tackle the user resistance, let us understand first, why user acceptance of new technologies represents a problem in the first place.

One possible explanation is being offered by the famous Maslow's Hierarchy of needs (Figure 1). Maslow stated that people are motivated to achieve certain needs and that some needs take precedence over others. Basic needs, such as food, water, and safety, need to be fulfilled before a person is motivated to fulfill their psychological needs, such as having friends and having a feeling of accomplishment. Last are self-fulfillment needs or achieving one's potential [1]. For example, you achieve your full potential with curiosity and exploration, willingness to try new things instead of maintaining the status quo. So the reason why people do not like big changes is supported by a theory that puts self-fulfillment needs at the bottom of

Digital Transformation of the Laboratory: A Practical Guide to the Connected Lab, First Edition.
Edited by Klemen Zupancic, Tea Pavlek and Jana Erjavec.

the priority list. Companies that understand this motivate their employees through great work environment conditions that address the basic needs (i.e. salary, health insurance, and great infrastructure) and psychological needs (i.e. great teamwork and an opportunity for career development). Such companies will be more likely to succeed when they start to introduce big organizational changes such as digital transformation because people will be more motivated to change and are more likely to perceive changes as an opportunity for self-fulfillment rather than a nuisance.

It is important to note that Maslow also suggested that the order of needs might be flexible based on external circumstances or individual differences. Most behavior is multimotivated, that is, simultaneously determined by more than one basic need.

Acceptance and use of new technologies have been subject to several studies. One of the best known is research done by Venkatesh et al., which resulted in a unified theory of acceptance and use of technology (UTAUT) [2]. Many other studies have been done to further update the UTAUT theory, including one done by Dwivedi et al. who have found that attitude played a central role in the acceptance and use of new technologies. For example, attitude was also influenced by whether I believed that my working environment supports the new change and how others feel about the change. Attitude also had a direct effect on behavioral intention, meaning that if I have negative feelings toward change, my intentions to accept and use new technology will be weaker [3].

Reaction to change perceived as negative can also be explained by the Kübler–Ross Grief cycle [4]. Let us explain each phase through the following example.

Sarah has been working with the company for five years. She is a great employee and an appreciated colleague. She understands the processes and gets the job done. Sarah is being briefed that a new software system will be implemented, which will significantly improve her efficiency but also change the paper-based system she was used to working with. For Sarah, this is not great news or at best, she has mixed feelings about it. She is surprised and is unable to react to the news immediately ***(immobilization)****. After taking some time to think, she really believes that this is not the best idea, and she tries to ignore the news (****denial****). Anger starts to build, and Sarah resents that managers did not ask her for her opinion before the change has been made, also the change will take a lot of time of her already busy schedule (****anger****). She approaches people in charge and tries to negotiate a different solution or postpone the deadline for implementation (****bargaining****). She finally realizes that the change is inevitable and does not feel great about it (****depression****). After a while, Sarah starts seeking solutions on how to best move forward (****testing****), which ultimately leads her to accept the change (****acceptance****).*

Since Sarah is an appreciated coworker, her initial negative attitude toward the new software system will likely influence other coworkers, who will also perceive the change as negative and will go through the stages of the grief cycle. It is the duty of every management to minimize the risk of user rejection *before* the change is being announced. This will more likely lead to positive perception of a change and faster technology acceptance or at least shorten the grief cycle intervals.

You will be more likely to succeed if you have previously successfully implemented major changes, which also means that you were able to recognize and address resistance. Success also depends on your current culture, what you set as goals, and whether you have resources available. But most importantly, you have increased the chances of success if employees trust people who are making changes.

Let us have a look at some of the good practices that increase a chance of successful user adoption, whether you are implementing only one solution (i.e. electronic laboratory notebook [ELN] or laboratory information management system [LIMS]) or digitally transforming your laboratories.

10.2 Identify Key Stakeholders and Explain the Reasons for Change

In order for digital transformation to succeed, management needs to be not only onboard with the proposed changes, it needs to have a clear vision of the desired culture and manage the execution of it as one of its top priorities. There are usually many reasons that incentivize management to digitally transform the company, for example, process and cost optimization, data integrity challenges, and compliance issues. These are very rational reasons for change, and you should pass them on to the rest of the company. Support your findings with data to further support your plans. Regardless of your reasons why, it is important to clearly set and communicate goals and expectations of the change.

Example *In the past 10 years we have had 300 noncompliance issues because of manual data entry, which could be automated but is currently still paper based. This has cost our people 1500 working hours, and the total costs of managing the issues were USD 3 million. Besides that, we have multiple blind sports, processes that are not tracked, which makes it hard to manage risks or assess if any improvements are needed. And on top of everything, it also turned out that information exchange between silos is not great, and poor communication leads to misunderstandings, mistakes, and slows progress. With future company growth, the lost hours, mismanaged processes, and poor communication will cost us more money if we decide not to change anything.*

People understand facts, which is especially true in research. It is highly likely that they will have mixed feelings about your proposed changes, but at least they will understand the reasoning behind it. Sooner or later, they will also accept it emotionally.

Make sure to also identify key stakeholders. In the organization with laboratories, these could be individual departments, functions, or positions, for example: laboratory technicians, researchers, R&D managers, and QA managers (Figure 10.1). Now identify how they currently feel about the change (state A) and how would you like them to feel about it in the future (state B). Are they against the change, neutral, or in favor of a change? Which stakeholders are influencing which? Think about how you could bring them from state A to state B. Understanding this will help you address fears and objections of every stakeholder and help you plan actions that you need to take.

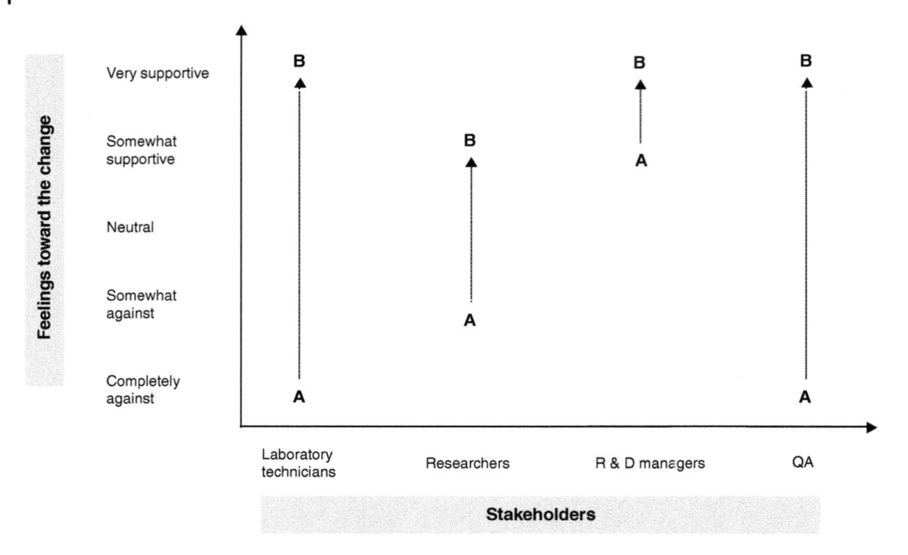

Figure 10.1 An example of a graphical representation of stakeholder analysis. Draw arrows and write down a strategy on how to move them from A (current state) to B (desired state).

Clear, honest, transparent, and continuous communication from the management will help stress the importance of the digitalization process, raise awareness, and motivate the people to collaborate with the project team which is in charge of digitalization. And ultimately, change the company culture.

10.3 Establish a Steering Committee

A steering committee will be in charge of carrying out the digitalization projects or strategy. The project team should involve representatives from all stakeholders within the company, including end users. Everyone should have a voice in the process. Typically in larger organizations, the key members are chief digitalization officer (CDO), chief information officer (CIO), or any other board member and person in charge for process management. In smaller organizations, these functions can be represented by one person, for example CIO. The roles of each steering committee member are different, and duality of roles can sometimes represent a problem. For example, CIO needs to take care that everything is running smoothly, while CDO should promote innovation. If both are represented by one person, the digital transformation may sometimes fail.

Many successful use cases show that involving end users into the process early significantly increases a chance of success. But how do you identify the right people?

Talk to them. Identify potential ambassadors, based on whether they are digital enthusiasts and/or would like to prove themselves outside of their everyday tasks.

Larger organizations would often involve only managers or senior employees, who are usually too busy and could perceive the steering committee obligations as additional work on the top of an already busy schedule. Because they might fear that the workload and the change of well-established processes will cause many issues, they might take a risk-averse stance. Their perception will be carried down to other employees, and the general perception of the digitalization project will be negative. Having a steering committee of people who are too busy and risk-averse will lead the project nowhere. However, seniority and the experience these people bring to the table are valuable. And it is absolutely crucial that leadership is either part of a steering committee, as mentioned earlier, or at least clearly supports the digital transformation. Without management support, there will be no transformation.

On the other hand, individuals who are more inclined to the new technologies and the benefits that they bring would spread positive connotations of the word "digital" and "change." They will act as ambassadors and will be willing to invest time and effort because they believe in the technology and the fact that the changes will bring improvement. That being said, having a committee with only enthusiasts onboard will not make great progress either. They are likely to make changes too fast, convinced that everyone understands their goals. In spite of acting in good faith, people will not be able to grasp changes fast enough; they might feel misunderstood or even ignored. Resistance would be inevitable. It is also a good practice to create new positions within the organizations for people that get new responsibilities, preferably with financial incentives (salary raise), training, and certificates. This way they will see it as an opportunity to grow their career.

So the truth is in the middle. A steering committee needs people with different views and backgrounds, representing different stakeholder groups. However, everyone should recognize the problems they are having and agree on the solutions they are planning to implement. Together they should be able to engage the right people at the right time, which should bring your company or laboratory closer to establishing a digital culture.

10.4 Define the Project Objectives, Expected Behaviour, and Timeline

Once you have assembled the steering committee, the team can identify high-priority projects (within the digitalization strategy) as well as define project plans or charters, milestones, and timeline. It is absolutely necessary that digitalization strategy is not just a piece of paper with high-level goals. Projects need to be planned, prioritized, and executed. An example of a project is a product trial, which is a smaller scale implementation of a larger project. It is used to identify any issues and risks before starting a larger project and is limited to a smaller group of people. Once the product trial has ended, a steering committee assesses the outcomes and decides whether they should roll out the solution or not. The execution of the main project is then also carefully monitored and checked against milestones and KPIs (Key Performance Indicators) that were set.

10.5 Check for Understanding and Encourage Debate

It is crucial that everyone understands *why* changes are being planned, which projects are being prioritized, and how they will be executed. At this point, make sure to be ready to answer any questions and address any concerns that you will get as a response. This is also a good moment to encourage debate. It might give you some valuable insights into how people are perceiving upcoming changes. Identify potential ambassadors that you could bring onboard as superusers. Also, recognize skeptics who are less inclined to make any changes. Understand the reasons for resistance and build an implementation strategy that will be able to address this.

You might encounter different types of resistance to change from different personality types, and you will have to be prepared to handle them differently, giving them different arguments. For example, a person who is always busy will likely react that they do not have time for changes. Such a person needs to understand that there will be some activation energy necessary at the beginning and when the benefits will arrive and that they can get some help with prioritizing the changes with their current tasks that make them busy. Engineer type of researchers, who perceive only their ideas for change as the best ones, will require detailed information about the approach and benefits of the new system. In extreme situations, the management might also need to be involved directly to help explain and reinforce the necessity for changes.

Example *Bob has been with the company for 20 years. He understands the processes inside-out, including strong points and weaknesses. He is used to coming to work on time, writing down instrument measurements, filling out QA documentation on paper, and submitting it for approval. He knows that he is waiting for approval for too long and that documentation management takes too much of his time every day. He needs to double-check whether he has entered all the data correctly, and he is under stress as he is responsible for correct information.*

The proposed new solution will automate Bob's process. Bob will come to the office, instrument data will already be stored in a Software. Bob will quickly review the data and the electronic QA information and with a few clicks electronically sign and send the data for approval.

Suddenly a task that would require an hour of work has shrunk to a few minutes of work. If Bob has already too much work, it is highly likely that he will welcome the solution, which might give him an opportunity to focus on more important tasks. However, if Bob can manage everything in eight hours, the new solution will take a significant part of his work off the table. He might be afraid that the new solution will replace him and that he will be laid off. Without knowing, he will feel resentment toward new technology and people who support the change. At the meeting, he will probably speak against the changes.

If you are hearing what people say, you will be able to understand the barriers to successful implementation and appropriately address them. You might decide to reach out to Bob for opinion more often and engage him in the project. It is also highly likely that other people share the same opinion as Bob. So you might want to bring this topic forward at your next group meeting and explain how changes will affect their everyday work. Again be honest about problems, but do not forget to emphasize the positive effects. Ask the skeptics for help and feedback. They will be less afraid of new technologies and will feel that they are significantly contributing to the project.

What management also needs to understand is that resistance is a function of disruption and cannot be avoided, it needs to be managed. Management needs to communicate that objections do not eliminate the need for change and should not allow exceptions or agree to any ultimatums. Consensus may never be reached, but the need is still there and the management needs to lead the way, hold people accountable, and enforce changes that will benefit the company.

10.6 Acknowledge Ideas and Communicate Progress

Stay in contact with the employees throughout the project. You might be surprised by how many great ideas you will receive. A steering committee's duty is to evaluate each idea and use it as long as it helps the overall strategy. The danger here is that too many different ideas might defocus the project and delay the overall delivery. It needs to be clearly communicated that ideas are being taken into account, but that the steering committee decides which to implement. You might decide to use some software to manage this, such as Microsoft Excel spreadsheets or Trello.

Once the project is running, also do not forget to communicate progress. If you are proud to be reaching certain milestones, people should learn about them. An example of a milestone would be the completion of a product trial. Remember that with consistent communication you are also slowly changing people's perception of the digitalization projects and hence the company culture. Next time they will hear you are planning to implement another digital tool, they will already be familiar with the process, less skeptical of new technologies, and hence less likely to resist your suggestions.

10.7 Provide a Feedback Mechanism

Getting feedback is absolutely necessary at any stage of the project. However, it is even more important to get it once you start implementing solutions. You would often run a product trial before you roll out the solution on a larger scale. At this point, you would like to know everything about how the solution is performing, whether the vendor support team is helpful, and whether users are using the solution. Make it clear that initial mistakes are expected and will be tolerated. You will notice that some employees will be giving very relevant feedback and suggest improvements, while some comments might be based on anger or fear. It is important to distinguish between both and take into account the relevant. Do not be afraid to acknowledge the effort of those who are very cooperative. But rather than recognizing individual effort, support and/or reward process improvement which has a direct positive effect on your operating costs.

There are different ways to set up feedback mechanisms, from a simple spreadsheet to which a steering committee has access to or a dedicated person who gathers all the feedback and reports to the project owner and steering committee. Depending on the project complexity, you will decide which option works best. In order for the feedback mechanism to work, make sure you have a process well defined and that any issues are being resolved on a regular basis.

10.8 Set Up Key Experience Indicators and Monitor Progress

Large corporations, such as Google, have developed frameworks (i.e. HEART framework) that help understand user-centered metrics which measure user experience of their products [5]. But how does this have anything to do with laboratory digitalization?

Actually a lot. You can look at your employees as new users of, let us say, a new Software system that you are implementing. After a product trial kickoff, you would like to be hands-on on the progress of implementation, you would want to check in regularly and receive feedback from key stakeholders. The first thing that will come to your mind is setting up KPIs that will measure business value which a steering committee needs to report to the top management. This is valuable but will not be very informative short term. Taking into account an onboarding period, it is very unlikely that business KPIs will talk in favor of a change in a very short period of time. But you are aware that if an implementation is successful, significant business value will follow.

Therefore, you should be focusing on user-centered metrics or *key experience indicators (KEIs)* and measure only those that are critical for implementation. I will focus on user *happiness*, *engagement*, and *adoption* which in my experience are most important to measure when implementing new technologies or tools in laboratories.

10.8.1 Happiness

In relation to products and services, happiness is joyful or satisfying user experience. Understanding people's attitudes toward a new product is extremely helpful in identifying the strengths and weaknesses of that product or an implementation service. It can also help understand the effect of a change after it was introduced and over time. One way to measure happiness is with the percentage of satisfied users. For example, if you have 100 users who have responded to your question about how happy they are with the new software system, 20 were unhappy, 36 undecided, and 44 indicated that they are happy, the score is 44%. You should aim to increase the number as your goal should be that a large majority of users are happy with the new software. Speaking with users will help you to identify the reasons for dissatisfaction and act accordingly. Sometimes just poor onboarding is a problem; in other cases, general resistance toward change might play a role.

10.8.2 Engagement

Engagement measures whether users find value in the product or service and whether it is meeting their needs. You can measure the frequency and intensity of interaction between the user and the product. Engagement is a very reliable measurement because it is unbiased.

Different engagement metrics can reveal the frequency of usage and how much people are using new products. When employees are forced to use a bad product, they limit engagement to the minimum necessary, avoid engaging deeply, and do a lot to find bypasses. This behavior can be reflected in engagement metrics and supported by happiness measurement. One way to measure user engagement is by measuring the percentage of users who used new software products during a given period of time (out of all users who had the opportunity to use it). The time period that you measure can vary, depending on the complexity of the software and implementation process; for example, 5% of users used the software in the past day and 20% used it in the past seven days. Also, ask Software providers if they are able to provide you with user-specific engagement metrics for their software system and/or onboarding process. This can save you a lot of time.

10.8.3 Adoption

Understanding people's adoption behavior toward a new software is extremely helpful in identifying whether or not the new solution has value. High adoption numbers mean that the new solution has promise.

A good idea would be to measure the adoption rate, which is the percentage of new users of a new product or service. The formula for calculating the adoption rate is: Adoption rate = number of new users/total number of users. For example, if you have a total of 50 users, of which five are new, then your adoption rate is 10%. The adoption rate should always be calculated for a specific time period, which will help you understand the trend, whether more users have adopted a new system, or the implementation has stalled.

Only collecting numbers is not enough though. They only serve as a basis for an in-depth interpretation and analysis. Say that user adoption is not increasing and user happiness is stuck somewhere in the middle. You would like to learn why this is happening by interviewing stakeholders. You will be drawing some conclusions and making adjustments that should improve the metrics. For example, perhaps the vendor onboarding process is not great and you will request a more thorough approach. Perhaps managers are not encouraging team members to engage in the new project, so you might organize a meeting and set expectations for them. With each iteration, you should reflect on the positive and negative outcomes. Everyone needs to take responsibility for the successful transition from the old to the new system. Remember, this is not negotiable and everyone needs to actively participate!

10.9 Gradually Expand to a Larger Scale

The pilot phase will provide you with important insights into what works and what should be improved. Although you might feel confident enough to implement the strategy on a large scale, it is a good idea to pace yourself. One option is to have an agile approach, a gradual expansion of the strategy in waves or sprints. The typical phases of the agile approach are plan, design, develop, test, deploy, and review. This means repeating the pilot phase and reflection with different departments or business units. Each wave will get you a better understanding of the process. Each wave can also be larger as you feel more and more confident with the implementation approach.

Figure 10.2 shows waves that can represent business units' departments or even software features as they get adopted. Wave zero usually represents a product trial, where the system is being adapted to the laboratory requirements for the first time, onboarding plans have been set up, and security measures addressed.

You might wonder how to prioritize which departments to use in the pilot and early expansion phases. A good idea is to prioritize based on the level of the cultural change – the more digital-oriented and digital-embracing units should be the first ones to make the transition to a new system. This is important because your first adopters will become the ambassadors of the new culture and a sort of a reference site for digitalization. They will and should have an important role in the expansion phase. You can empower them to help you train new users, provide support, and be role models. However, these additional ambassador roles of pilot departments/groups need to be clearly communicated to them, and they should be a part of the pilot phase strategy. Another approach would be that you start with units where there is a pressing business need that a new system can address. You should always seek to improve processes and seek the best return on investment.

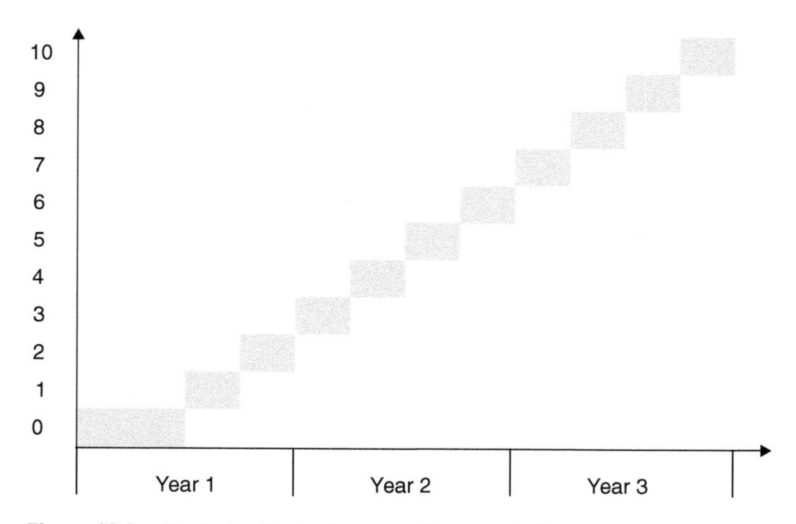

Figure 10.2 Gradual rollout of a new software solution.

An important fact to note is that you will be educating the multigenerational workforce on the new digital technologies, which presents an enormous challenge for internal education. One tool that is worth mentioning is gamification. By definition, it means that you are applying typical elements that you would find in a game playing to other areas where you would like to increase engagement. Gamification promotes friendly competition, engages learners to achieve desired behavioral change, motivates them to advance through the course and take action, and provides a sense of achievement. When applied to corporate learning programs, gamification can help stimulate enjoyment and engagement transforming tedious, boring learning experiences into a fun and rewarding experience. Why does gamification work? It has been proven that by giving virtual rewards for achieving learning goals, a dopamine is released, which makes us feel good. Since gamification focuses on giving instantaneous feedback, which makes us feel good, learners begin to associate learning with positive emotions, which encourages them to repeat it [6]. There are many successful corporate gamification strategies already described, and I would recommend reading through at least one of them, and if possible incorporating at least some gamification elements into an overall digitalization strategy.

Regardless of the approach, however, you should continuously communicate the importance of digital transformation and which business goals you are trying to achieve. Everyone needs to agree that the goals make sense and that digital transformation is the solution. Remember, you would like to establish a digital culture and establish good practices in change management, so that next time a new system or process will be implemented more smoothly.

10.10 Conclusions

We have learned that digital transformation utilizes technology and software to transform the way laboratories operate.

You can expect some level of resistance as it is not an easy task to rethink and redefine how a laboratory or organization has been operating for many years. Humans are creatures of habit, and people in organization get comfortable with the way processes are set up with predictable revenue streams and predictable planning very fast. We also wonder whether now is a good time to make changes, especially if things are stable. Though understandable, such a way of thinking will leave your laboratory in the status quo. The businesses that distinguish themselves in success know that taking calculated jumps is necessary to improve their operations and better serve their employees and customers. Therefore, a written digitalization strategy that is appropriately communicated well to all employees and leadership becomes very valuable.

References

1 Google Books (2020). Motivation and personality. https://books.google.com/books/about/Motivation_And_Personality.html?hl=sl&id=DVmxDwAAQBAJ (accessed 1 June 2020).
2 Venkatesh, V., Morris, M.G., Davis, G.B., and Davis, F.D. (2003). User acceptance of information technology: toward a unified view. *MIS Quarterly* 27: 425. https://doi.org/10.2307/30036540.
3 Dwivedi, Y.K., Rana, N.P., Jeyaraj, A. et al. (2017). Re-examining the unified theory of acceptance and use of technology (UTAUT): towards a revised theoretical model. *Information Systems Frontiers* 21 (3): 719–734.
4 Mayers-Elder, C. (2008). On grief and grieving: finding the meaning of grief through the five stages of loss. *Journal of the National Medical Association* 100: 865–866. https://doi.org/10.1016/s0027-9684(15)31384-5.
5 Rodden, K., Hutchinson, H., and Fu, X. (2010). Measuring the user experience on a large scale. Proceedings of the 28th International Conference on Human factors in Computing Systems – CHI'10. http://dx.doi.org/10.1145/1753326.1753687
6 Reiners, T. and Wood, L.C. (2014). *Gamification in Education and Business*, 710. Springer.

11

Testing the Electronic Lab Notebook and Setting Up a Product Trial

Blazka Orel

SciNote LLC, Middleton, MI 53562, USA

11.1 Introduction

When investing into a new digital lab solution for your laboratory it is important to test the product before the purchase. Not testing the product could bring you a myriad of problems that you could easily avoid from the beginning.

We may take the analogy of purchasing a car, where you would not likely buy one without taking it for a test drive first. Without testing the car, you could find out later that it has bad brake pedals, a stiff steering wheel, or a small cargo space. Some of these later realizations can have big consequences that are potentially very damaging. Therefore, you should perceive the product testing an important step before selecting a solution. Once you shortlist the potential vendors who could meet your needs, you should set up a product trial to systematically test and evaluate their solutions.

11.2 The Product Trial

A product trial is a preset testing environment designed for systematic testing of different solutions, let it be in parallel or one by one. The testing environment should emulate the real conditions of the working environment since a product trial is considered as a pilot, an experiment done before the actual implementation of a solution.

We can also describe a product trial as a small-scale project conducted for the purposes of evaluating a product performance, quality, security, and compliance with established standards. However, besides testing the product, testing the service is also imperative. After all, having great customer service can bring so much added value to the table even if the product does not tick all the boxes of your requirements.

On a second note, a product trial can also be viewed as a preliminary phase which is constructed before the performance of a full-scale implemented solution. In other words, a product trial takes place prior to the solution implementation, where the service is being expanded on a larger scale as a result of a successful pilot.

Digital Transformation of the Laboratory: A Practical Guide to the Connected Lab, First Edition.
Edited by Klemen Zupancic, Tea Pavlek and Jana Erjavec.

11.3 The Importance of a Product Trial

You may be wondering what the benefits of a product trial are. The main advantage is that you are able to detect the potential risk factors and mismatches with your requirements. This way, you are able to eliminate the obstacles, renegotiate the status of expected service, discuss custom feature development options, or even prevent a misfortunate investment if the product or the service does not fit your needs.

For example, if you are looking to invest in an electronic lab notebook (ELN), the risk factors worth considering may be safety reasons (e.g. data access and permissions, data protection, hosting, and backups) or the reasons related to the commercial contractual aspect (e.g. the contract type, period of performance, and intellectual property use and ownership). As for the general or custom-developed ELN features, the requirements vary between different types of laboratories. For example, a chemical laboratory might be considering an ELN that enables the drawing of chemical structures and reactions, and molecular genetics laboratories might prefer a digital tool for constructing plasmids.

A product trial should mirror the real conditions of a working environment in such a way that testing the solution helps you understand and predict what implementing a new solution would look like in reality and what would be the overall impact of the implemented solution. Let us say you work in a lab and you need to coordinate with your colleagues about booking a certain laboratory instrument. In this case you should assess if the communication in the testing environment such as text messages within the ELN application is equally or more efficient, as the communication in a real working environment via e-mail, for example.

Additionally, setting up a product trial can help you reach several goals before the full implementation occurs and thus ensure a better and easier transition from an old platform to a new one. Some of the goals that are reached during the product testing do not need to be achieved again during the implementation, for example, you do not need to go through the initial cloud setup of the ELN again once it is done. Thereby, you can actually reduce the adoption friction of a new solution in the implementation process, make the implementation a better experience for everyone involved, and save a lot more time during the learning phase, which means saving on the expenses as well.

If we are looking from the vendor's point of view, enabling a pilot or a so-called trial period for customers to test the product plays a huge role in the customer's purchase decision process, and in a positive way as well. A study shows that customers who had a good product experience during a trial period are more likely to purchase the product in comparison with those who had a bad experience or those who did not test the product at all [1].

Although the good product experience is largely achieved through design and quality control, the product also needs to educate, engage, and adapt to users' needs. You could have all the ELN features a lab's heart desires like an inventory management tool, but without a proper training and good support service your lab personnel could stop testing the product right from the start, before even getting to the "good stuff" and seeing the real added value of the software. So, take your time to organize a product trial, you never know how this might increase the product adoption and user satisfaction after the trial is done.

11.4 Setting Up a Product Trial

11.4.1 Phase I: Planning

You should start planning a product trial shortly after you acquire a list of products that you want to test. Once you determine what are the final solutions for evaluation, only then can you decide on your testing strategy and the entire process of a product trial. In order to do this, you need to take inventory of all your resources, such as time, people, and money.

First, you need to decide if you are going to test one or multiple products, and if you are going to analyze one product at a time, two at a time, or several products simultaneously. After that, you need to define the scope of the product trial, meaning, you need to think about the product trial timeline, its members, and any other resources you would need to run a successful pilot.

When it comes to setting a project timeline, estimate how many hours it would take to test all the solutions and how long would the pilot last from the very beginning until the finalization (e.g. days, weeks, or months). You should ask yourself when you want to complete the pilot stage with respect to the start of implementation of the new solution. Also, you should take into account all the days off, such as employees' vacation and holidays.

How many people and who you include in the pilot is also very important. In turn, to assign the right participants, you need to identify who is going to use the product, who are the main end users (i.e. super users), and do they have any kind of experience with such tools. Knowing all of this can help you create targeted groups of people who can test and evaluate solutions in a more structured and objective manner.

Besides planning human resources, you also need to define other ones that could help you meet your testing objectives. Here, we are referring mostly to financial and intellectual resources. How much expense could you afford for testing solutions is completely up to your organization; however, what and how you learn about the solution is part of the provider's responsibility. Some of the vendors offer user training and learning documentation (e.g. manuals, tutorials, and knowledge base), others do not, so make sure to make a query in advance, and this might help you in testing the product in the right way.

The pilot milestones or goals are the most essential aspects of a product trial that you need to establish. The goals you need to achieve are very much intertwined with the timeline of the product trial and the activities carried out by people who are testing the products. In short, timeline, goals, and activities should represent the backbone of your entire product trial, and you should not disregard them when planning a product trial.

Now, how you measure the success of a product trial entirely depends on the metrics you define prior. The metrics help you measure various product features and the service that comes along. It is crucial that you set clear goals and establish methods for gathering metrics, so in the end, you get a nice clear picture of the pros and cons of the solutions you would have tested.

11.4.2 Phase II: Conceptualization

One of the most challenging parts of the product trial is to form a concept, the idea of how the testing would look like. The concept of a product trial might be defined before or after the planning phase. If you create a product trial concept before the planning phase, you can better plan the entire project when having a viable testing strategy already in place. Otherwise, you can define a testing approach after the planning, once you learn what resources are available to you.

We might say that the conceptualization phase is the part where you basically design a product trial, a detailed examination process for testing the potential solutions. Based on identifying your key problems that you want to solve, you are analyzing the solutions with the same evaluation method to select the best one among them. Added to an examination process, there is another side of the conceptualization phase which is designing the testing environment for your trial participants, where they analyze different solutions in the same way. The key is to construct such testing conditions that enable smooth, controlled, and unbiased testing of digital lab solutions you are interested in.

To design the best possible testing environment, you should be familiar with the testing conditions first. Since we are discussing the testing of digital lab solutions, their features and functionalities are the baseline for constructing the testing conditions. Besides product features and functionalities, you should also acknowledge the capacities of your work environment where you will be setting up a product trial. It might happen, for instance, that you are unable to test the digital lab solution because of your low server performance when having intensive workloads. Therefore, it is beneficial to do some research on your work environment and capacities before jumping right into the solution testing.

Now, let us take a look at the example of designing a testing environment for examining an ELN, a digital lab solution used for research data management in a laboratory [2]. When considering the testing conditions for examining an ELN, you should take into account multiple factors, starting with device access and connectivity. You can access the electronic laboratory notebook offline, in case it is a desktop application, or online if the ELN is web or cloud based. In either case, access to the application should be easy and convenient but still secure to prevent privileged account abuse and third-party data breaches. You should also think over testing the connectivity with other digital lab solutions or instruments you are using in a laboratory. Sometimes, antivirus software and firewalls can prevent access to the application, so you need to keep this in mind as well.

When talking about data access and security, you need to establish individual user permissions having user identifiers (user IDs), where people log into applications with their unique username and password. Additionally, you need to review if there are any data security protocols in place to ensure secure data transmission between your browser and application server. Usually, it goes without saying that all communications between the browser and the application need to be encrypted, but there is no harm in asking the ELN provider about the data encryption. The same applies for data backup; inform yourself about the possibilities of data backup, such as the data backup servers, where are they located, how are they secured in the event of natural disasters, and how frequent are the backups.

Given the testing of an online solution, you should consider the Internet connection as well, that is the Internet speed and quality. The server location plays an imperative role here, so think carefully where you would like the ELN to be hosted so you can test the solution smoothly, undisturbed by the Internet connection issues.

We previously mentioned that solution features and functionalities serve as the baseline for constructing the testing conditions. This is especially true when we are talking about file formats, file uploading capacity, and size of storage space. The information about the file formats, size, and uploading can help you prepare for testing the application capability, that is if the application can uphold your work requirements. For example, if you are working with large electron microscopic images or big datasets for machine learning research, you should test whether the solution supports these types of files, if uploading of such large files is even possible since there could be a limit for individual file size per one-time upload, and if there is sufficient storage space in general. Another alternative solution to this is that the software can integrate with your existing file storage system and only references the files.

In order to test the parameters above, you first need to identify what kind of data you input, process, and produce (text, tables, images, etc.). After that, you should prepare some files that contain dummy data, data that does not hold any information of higher value and servers for testing purposes only. Alongside, you need to identify different occasions or events when you are working with data and where you are pulling it from; in other words, identify your data sources so you can replicate your environment for data management in a product trial.

Even if you set up the entire testing environment, you still need to design an examination process suited for different users who would potentially use the new digital platform. Usually how it goes is that there are several people participating in a product trial, who hold various job positions in a multidepartmental organization. For that reason, it is advised that you write down all the work processes performed by different people or teams, working in different environments with different modus operandi, so to speak. This can help you organize numerous testing teams that can examine digital lab solutions in a more structured and tailored way according to their work expertise.

If we once more take an example of testing an ELN and look at the examination process in an academic laboratory full of researchers and a production laboratory owned by a pharmaceutical company, they would be examining a completely different set of ELN features. In case of an academic lab, students would be interested in testing data-sharing functionalities since they are often working on common grant projects, whereas employees in a pharmaceutical lab would be far more interested in testing e-signing of reports since they need to follow certain regulations on electronic records and electronic signatures [3].

To conclude, examination processes and testing conditions can vary between different roles of people working in the same organization. With this in mind, when designing a concept of a product trial, try to think what type of organization you have, how big is your organization (in terms of number of people, teams, or departments), what type of work you do, what kind of data would you manage, and how would you prefer to access the application (via offline or online). Once you clarify all of this, then you are ready to start testing the solutions.

11.4.3 Phase III: Testing

The timeline of your product trial is set, the milestones that need to be achieved are known, and all your groups of testers are familiar with the examination process. But here comes the hard part – the actual testing and evaluation of potential digital lab solutions. Although there is a common belief of *trying before buying*, the testing itself may cost you more than expected if you and your team are not well prepared. Therefore, the cost of the evaluation needs to be proportional to the overall cost of the initiative.

First, you need to specify what aspects of the product are truly important to your organization, then you need to prepare criteria for assessing product suitability, a checklist of specifications your testers would need to analyze, and to evaluate potential solutions. Besides product requirements, one thing not to disregard is testing the service that comes with the product. Many vendors offer free service of setting up a testing environment or the so-called sandbox, where they create several testing user accounts for your organization. It is in the vendor's interest to immediately provide you the first product experience and establish a relationship between you and their customer service. The service can also include consulting on how to use the product for your specific purposes, advising on how to set your priorities, conducting the user training, and so on.

To assure a valid and regulated testing, you need to establish the same evaluation method for all the testers. They should be able to analyze different solutions in the same way. For example, they should have the same amount of testing time for each solution and the same evaluation form or questionnaire to acquire easier comparison of testing results.

In case of testing the ELN platforms, the evaluation form can represent a simple table covering examination of four different segments which can present four different table sheets: performance, security, quality, and compliance.

Example of an evaluation table:

Each of the abovementioned four segments can have a list of requirements written in the first table column. While testing each of the requirements individually, the testers should fill out the second column by entering yes (Y), no (N), or not applicable (N/A). The "Y" should be entered if the ELN platform is meeting the requirements, the "N" should be entered if the requirements are not met, and finally the "N/A" should be entered if the tester cannot assess if the requirements are fulfilled. Creating an extra column to enable evaluators to write comments for each specific requirement is also a good idea. Sometimes, the vendor is already planning to add the specific functionality to the software in one of the future releases, or they can even offer you custom development service and develop it for you, depending on your budget.

The choice of performance requirements to test in an ELN is completely up to you. You can start with examining how easy and intuitive it is to manage projects and have an overview of their progress, uploading data (e.g. ELISA images and standard operating procedures (SOP) documents), finding the information you are looking for (e.g. using search function), etc. Under the security requirements you should be evaluating if the ELN can be hosted on a cloud or local server, how do they explain the data protection practices they implemented, if the ELN enables individual user permissions having user IDs, and so on. The quality aspect of the ELN could represent the loading speed of the ELN application, the overall user experience (UX) design, etc. Lastly, under the compliance you should evaluate if the ELN enables good laboratory practice/good manufacturing practice (GLP/GMP) compliance for different lab activities from protocol versioning to data traceability (e.g. audit trail and activity history).

If we look at the product performance and quality they are defined entirely from your subjective perspective; on the other hand, product security and compliance should not be evaluated subjectively; either the product has security measures and compliance in place or it fails to do so.

Additionally, the participants should have the testing conditions that imitate the conditions of the real work environment to get the most accurate testing results possible. At the same time, the people who are participating in a product trial should be the actual end users who are going to use that digital lab solution (e.g. if the product is designed for laboratory usage, then it should be tested by people working in a lab).

Example In the table below are some of the questions labs would often ask when testing the ELN. The answers will be explained on the example of SciNote ELN. Please note that in most cases, software is being updated regularly so many functionalities are being upgraded as you read this example:

Question	Yes/No/NA	SciNote ELN – comments
Is there an integrated search tool in the software? If yes, on what level(s) can search be performed (project/experiment/attachments/etc.)?	Yes	Search can be performed on the majority of data entities in the software (projects, experiments, and tasks) as well as on comments and attached files (full-text search). The general search function will find all hits on all levels of SciNote. It also searches through attached documents and inventories.
Is the software suitable to handle big files/attachments?	Yes	The default file size limit on files is 50 MB, but this can be expanded upon request. There is no limit to the number of files. Any type of file can be stored in SciNote. Alternatively, files can also be stored on a local server or in a designated cloud storage and linked to SciNote via hyperlink.

Question	Yes/No/ NA	SciNote ELN – comments
Version tracking should allow the user to maintain a history of changes made in the notebook as well as who made them. Version tracking should be in place for library items (such as protocols) as well as project notes.	Yes	SciNote allows you to track all changes within the ELN on a general level (exportable audit trails with user id, time stamps, and content change details) as well as on every specific level (trail of every change made to a protocol, comment, and result). Changes to a particular protocol must be saved with versioning and stored to the library.
System shall allow for custom reports to be generated.	Yes	SciNote automatically generates custom reports. The functionality allows the user to choose which level (project, experiment, and task) and which associated content/data should be included in the report. Custom reports can be exported in a PDF as well as Word (.docx) format.
System includes repositories/libraries and/or inventory management (protocols, samples, reagents, equipment, etc.)	Yes	Yes, inventory management is the important part of SciNote functionalities. It enables you to create custom inventories and manage all your samples, reagents, instruments, protocols, etc. Each table is customizable, so it can support your repository/library needs as well. In the columns, you can include all needed data, even barcodes and files with thumbnails. Also, every item from every inventory can be linked with the tasks (experiments) in which it was used.
Is it possible to create templates of experiments and/or projects?	Yes	Yes, experiments and workflows can be saved and reused as templates. Also, SciNote offers default templates that include detailed protocols from today's most widely used life science experiments.
Archiving and deleting	Yes	In SciNote, we prioritize the traceability of your data and all actions in the lab, so primarily, you can archive data. If you really want to permanently delete something, you can open the archive and then delete permanently.
Does the software comply with the 21 CFR Part 11 requirements? Does the system support electronic signatures?	Yes	Yes, SciNote supports electronic signatures, electronic witnessing, locking of tasks, advanced user permissions, system log records, and audit trails. All electronic signatures are tracked and have their own audit trail with time stamps, user id, and comment section.

Question	Yes/No/ NA	SciNote ELN – comments
Is it possible to make exports? If yes, state export formats.	Yes	Yes, if you need to export everything, there is the Export all functionality in SciNote. It will export all your data, organized in folders (by project). When you open a project folder, you will see that experiments within each project have their subfolders. You can also see the folder that contains files from your archived experiments and your inventories. Also, here you will be able to find the full project report file. Project report file is a ready-to-print and completely readable file in a .html format that can also be opened in Word. In addition, and apart from exporting everything, you can export inventories in the form of .csv files (Excel readable), and sciNote protocols in the form of .eln format. Also, you can use the reporting functionality to export your protocols, or other project data, as a .pdf and .docx (Word) file.
Support communication (comments and annotations) between users for in-progress experiments	Yes	In SciNote, you can invite members of your team to collaborate on projects, assign them permissions and roles, and notify each other about the latest progress. It enables you to post comments, tag, and notify team members to speed up lab management, correspondence, and delegate tasks.
What kind of support is provided with this ELN product?	Yes	SciNote team prides itself in having the top-rated customer support service. Support specialists have a scientific background and are fully qualified to offer consultancy for every lab. They offer a full range, from email and online meetings, to fully personalized onboarding for the entire lab management and lab employees. The scope of support you receive depends on the SciNote plan of your choice.
How could we separate or identify the contributions of multiple users inside a project or an experiment?	Yes	Every person logs in to sciNote with their username and password. This means that all activities are traced by unique user ids and can be seen throughout the software. Basically, you always know who did what and when. Also, in-depth filtering of all activities is available. All users have their roles and a determined set of permissions which makes sure that only authorized personnel can perform certain actions.

Question	Yes/No/ NA	SciNote ELN – comments
Does this ELN support different levels of access to the system?	Yes	Yes, all users have their roles and a determined set of permissions which makes sure that only authorized personnel can perform certain actions.
Are there requirements for computers/tablets that are needed to run the ELN on?	Yes	SciNote runs in a browser. Its responsive UX design allows authorized personnel to access their accounts via different smart devices (smartphones, tablets, laptops, or stationary computers).
What file types are supported (i.e. .txt, .doc, .xls, .jpeg, .pdf, Microsoft Project)?	Yes	SciNote supports a multitude of file types, such as .txt, .doc, .xls, .jpeg, .pdf, .csv, .png, and ODF Documents (.odt, .ods, .odp).
Data protection practices, backups, storage, encryption, etc.?	Yes	To learn more about SciNote's data protection practices, please download this white paper and read more: https://scinote.net/blog/scinote-data-protection-whitepaper/
Does the product support multiple web browsers?	Yes	SciNote supports Chrome, Firefox, and Safari browsers.
References and user stories?	Yes	Available on SciNote Reviews webpage: https://www.scinote.net/user-reviews/

While evaluating the product, you can also schedule product demos, meetings, or questions-and-answers sessions with the vendor's product specialists. This approach often helps clarify some overlooked functionalities, functionalities that should be used in a specific way to gain the most benefit, or even discuss the overall implementation according to your requirements, on your own use case.

Also, while doing that, you include the support team or the sales team from the side of the vendor and test their responsiveness, willingness to help, communication, and overall collaboration that you will rely on in the future.

11.4.4 Phase IV: Reporting

When the testing is finished, the feedback from evaluators should be collected and the findings concluded in the evaluation report.

Now, in order to collect all the information obtained during the testing phase, you should have in mind to set the metrics before the testing, so that you could measure the product capability and the satisfaction with the service. If the digital lab solution has data export functionality, you could also retrieve particular data out of the platform, so you could use it for your evaluation report. In that case, you do not be shy to reach out to the provider to consult with you about options for exporting data and automatic report generation.

Collecting the feedback from the testing participants can be done by the team lead or the person who is leading the product trial. The best way to collect the data is in a table form, so you are able to make calculations and draw statistical graphs if needed. Besides the numeric information, the evaluators also give qualitative feedback on the product they tested, which could be later on revisited and summarized in a group discussion of all the participants. Usually in these kinds of discussion sessions, evaluators help the team lead to draw final conclusions from the product trial, which are then presented in a final report in front of the decision-makers (e.g. lab managers, purchasing managers, and company owners).

The report should contain a summary of the most important aspects of the evaluation, from the main objectives and questions used to direct the evaluation, quantitative and qualitative data obtained during the testing, and finally to conclusions and recommendations [4]. It should be evident, in the evaluation report, how the products performed during the testing and whether they are meeting your requirements and to what extent.

Besides the evaluation findings it is equally important to present the report in a visual pleasing manner (e.g. in a short PowerPoint presentation), so the people who were not involved in a product trial can easily understand the final results and conclusions. The evaluation report can show only the top three ELN vendors and show their comparison, where it is justifiable to which degree a specific ELN is or is not meeting the requirements. You can display if the ELN requirements are met numerically (e.g. in percentages) or by word (e.g. fully, partially, or not at all).

As mentioned before, the evaluation report should be presented to the decision-makers or key stakeholders. This means that they hold the power and the responsibility of making decisions at the high level in an organization. A lot of times, the key stakeholders are not directly involved in a product trial, meaning, they are not familiar with the technical aspect of the tested products. Consequently, it is better to present the key stakeholders with a few recommendations and report to them about the risks and benefits that might happen by selecting any of the solutions.

Once the key stakeholders have all the main findings and recommendations in one place, they can make a final decision that would lead to the implementation process of the selected digital lab solution. The decision can be based not only on the best value for money but also on the business relationship with the vendor according to the common vision and values. After all, this could be a long-term commitment considering you would not want to be switching from one digital lab solution to another (too often) since it could be very time consuming and costly. Not to mention, it could also be quite stressful for the entire organization, especially for the end users.

11.5 Good Practices of Testing a Product

All is good and well when discussing the setting up of a product trial in theory, but in reality, many things can go wrong, some even having more serious consequences than anticipated. The aftermath of a failed product trial can have various negative impacts on your UX, one, for instance, leaving your organization hesitant to try any new digital lab solutions ever again. Therefore, a well-executed product trial can present an ancillary step toward a more successful implementation of a new solution later in time.

In this subchapter we focus on some of the good practices your organization can follow before and while testing different solutions. We believe that learning about the good practices of testing a product can be very beneficial to you even if you do not decide to implement any of the solutions you would test. In addition, we address the most common hurdles you may experience during the testing of digital lab solutions and how to overcome them.

11.5.1 Taking the Time for Planning

It may seem unimportant to you to plan a product trial carefully, or to plan it at all, but as with planning any other project, good preparation can save you in the end a lot of time and money. Even though you must invest some time for getting valuable information, you can easily lose even more time in the long run if you do not take time to plan. Additionally, if you do not have a plan in place, testing of solutions can become very unsystematic and can lack focus. This may cause some chaos and can become very frustrating, especially for testers who need to be guided through different phases of a product trial.

The plan can be very simple, without going into much details, but it should include clear goals and a solid timeline of activities that would bring you to achieving those goals. Besides goals and a timeline, you should also envision who would carry out specific tasks and what would be their main responsibilities. Again, keep it simple, the assignments do not have to be like case studies. Testers just need to be aware of their testing group's focus so they test the product features and functionalities they are supposed to test.

11.5.2 Having a Bigger Picture in Mind

Before trying out the product, one of the things that potential customers forget to do, or they do not put much effort into, is acquiring a list of existing and future product functionalities. The other subject they forget to address in a conversation with a vendor is the possibility of a product custom development. If your intention is to purchase a digital lab solution, for which you plan to be a part of your organization for a longer time, you should ask yourself if and how you can enable your organization a tailor-made platform that would be evolving with the time according to your organization's growth and user requirements.

The heart of an issue, however, is not the customer's unawareness about the possibility of modifying a product, but the fact that potential customers do not always know exactly what their requirements are or they have too many requirements. In that case, the provider's role is to help the client to figure them out. This way, the client and the provider can define the current versus desired state together and create a clear list of must-have requirements. In addition, it is also important to address the possible technical challenges and how they can be solved.

11.5.3 Keeping Your Testers Motivated

Testing products can be a very tiring process requiring a lot of mental focus. Therefore, try facilitating a good testing environment for your pilot participants so they can feel comfortable and motivated enough to put their technical mindset into the practice. You can also reach out to the provider to help you with setting up a testing environment.

Moreover, you might experience some hesitation or even disinterest from people willing to test the products. They need to understand why they are doing this, what is in it for them, or how they can benefit from it. If the people who are going to use the product are not aware of their personal gain, the resistance and sabotage to test the products might appear. Many employees feel that testing products is a waste of time because they are already too busy and do not want to get an extra workload. The other reasons might also be hesitation of trying new solutions and using them once they are implemented because changing people's habits is hard. So, the key is to find good reasons to aspire pilot participants to test the products.

Although you can get a lot of help from a provider to set you up with the sandbox, the product testing is still driven in-house by your project leaders or team leads. They are the ones responsible for creating a strategic pilot plan; however, they are also often not part of the testing process. This does not mean that leaders need to be omnipresent, but they should be involved at key steps along the entire product trial and be in charge of leading groups of testers and motivating them.

11.5.4 Systematic Evaluation of Products

To ensure accurate, relevant, and consistent evaluation results, you need to have a systematic approach when it comes to conducting a product testing. There are a couple of things to have in mind, from testing conditions, evaluation methods, focus groups of testers, to possible reexamination in the near future.

We already talked a lot about arranging a testing environment, considering the testing conditions as well. The main message to remember here is to enable such a testing environment that easily imitates the real working conditions. It is also advised to use dummy data that is resembling the real data as much as possible, so you can properly test how the data behaves within the platform, and how you can manage and extract it.

The most important part of the evaluation process is to have a unified evaluation method. This involves having the same evaluation form or questionnaire designed for testers. They should be testing the same set of features in all products, in the same or at least very similar way. When people are following the protocol or guidelines, less mistakes can be made and more accurate results can be collected.

It might be necessary to create various evaluation forms for different focus groups of testers. This is absolutely valid since you need to test the usage of digital lab solutions by different workers anyway, users who have different work responsibilities and work in different environments. However, the same people should be testing the products so in the end you can compare the results from the evaluation reports.

Finally, after the testing is done, save the evaluation reports for later, in case you would want to revisit certain products and reexamine them.

11.5.5 Cooperating with Vendors

Assembling a testing environment for a product trial does not have to be a one-person task. The best way to go around it is to talk with the product provider and set the pilot together. After all, having an expert onboard can make your work easier.

Nevertheless, asking a vendor for assistance with the product trial might seem a bit discouraging at first, but keep in mind that it is in vendor's interest to offer you a trial period with free customer support, so they can persuade you into buying their product. At their worst, they are learning new information about their customers and how to improve their products and services.

So, connect with those providers who offer this kind of benefit and use their knowledge and assistance in setting up a product trial. They might help you with setting up the sandbox, consult you on how to use the product functionalities, organize a special training for your pilot users, etc. Besides, they can direct you to their knowledge base that is rich with various useful blog posts, user documentation, frequently asked questions, video tutorials, online courses, etc.

Now, purchasing a digital lab solution for your organization can be a long-term commitment; therefore, you should do a bit of research about the vendor as well. What is their vision, what are their values, what is their company culture, how do they do business, and so forth. The decision whom you are stepping into a business with is a big one, and it should not be taken lightly. If the two of you share the same values and the way you do business, this is a good basis for a successful business cooperation.

11.6 Conclusions

In this chapter we covered multiple aspects of what a product trial represents to people working in a laboratory in case of testing the digital lab solutions. The pilot is conducted on a smaller scale for the purposes of evaluating a product performance, quality, security, and compliance with established standards. Additionally, it represents a great opportunity to connect with different vendors and to try their services.

The main purpose of running a pilot is to try the product before the purchase in order to detect any potential risk factors and mismatches with the client's requirements. Furthermore, it can have a great impact on the adoption success of the selected solution in later stages of the implementation; thus, it should not be disregarded.

If we look at the product trial setup closely, it has many similarities to starting any project in general or even correlations to designing a laboratory experiment. For that reason, the product trial should be planned in advance just like any other project, it should also have clear goals, and a systematic evaluation concluded with key findings.

It is very important that the evaluation of digital lab solutions is done by the actual end users and in a systematic and unified way. However, there are no strict rules on how to run an examination process, only that the people who are testing the products need to be familiar with the guidelines.

To conclude, try before you buy, but keep in mind that the cost of the product testing needs to be proportional to the overall cost of the initiative.

Good luck with testing and finding the best digital lab solution!

References

1 Lu, X., Phang, C., Ba, S., and Yao, X. (2018). Know who to give: enhancing the effectiveness of online product sampling. *Decision Support Systems* 105: 77–86.

2 Murphy, A.J. (2019). Maintaining an effective lab notebook and data integrity. In: *Success in Academic Surgery: Basic Science* (eds. G. Kennedy, A. Gosain, M. Kibbe and S. LeMaire), 31–41. Cham (Online ISBN 978-3-030-14644-3): Springer.

3 Wingate, G. (ed.) (2010). Electronic records and electronic signatures. In: *Pharmaceutical Computer Systems Validation*. Chapter 13, 248–275. Boca Raton, FL: CRC Press.

4 BetterEvaluation International Collaboration Project (2012). Final reports. https://www.betterevaluation.org/en/evaluation-options/final_reports (accessed 24 January 2020).

Part IV

Case Studies

In the following section we present different case studies and expert comments on the subject of going from paper to digital.

You will be able to read how different laboratories and professionals approached the subject and put it into practice, what are their conclusions, advice, and lessons learned.

12

Understanding and Defining the Academic Chemical Laboratory's Requirements: Approach and Scope of Digitalization Needed

Samantha Kanza

Department of Chemistry, University of Southampton, Faculty of Engineering and Physical Sciences, University Road, Southampton, SO171BJ, UK

12.1 Types of Chemistry Laboratory

There are many subdisciplines of chemistry, and therefore, it logically follows suit that not all chemistry laboratories look the same. When many people think of chemists, they have an image of a scientist in a white coat and goggles, mixing some chemicals together in a conical flask. These laboratories do indeed exist and are the more traditional "bench chemistry" laboratories that contain the expected chemicals, fume cupboards, workbenches, etc. and have a hard-and-fast safety goggles and lab coat policy. However, there are also a wide range of other types of laboratories: crystallography laboratories typically contain large diffractometers, microscopes for crystal preparation, and several computers hooked up to various machines in the laboratory. Physical chemists often work in laboratories akin to (or indeed the same as) physicists, with laser equipment, that operate as partial or full cleanrooms with strict regulations about what can and cannot be taken into the laboratory. Sometimes a chemist's laboratory could be full of computers linked up to different machines, or even containing machinery with strong magnets that computers need to be kept at a safe distance from. The point here is that we cannot consider a one-size-fits-all approach to digitizing a chemistry laboratory as each one has different characteristics and different requirements.

12.2 Different Stages of Digitalization

Maintaining the record is a fundamental part of scientific practice; it is a serially numbered collection of pages that is used to record the mental and physical activities of the scientific experiment [1]. It plays many important roles: It creates a legally binding record that serves as a historical artifact, protects intellectual property, and provides a useful communication device to other scientists [2]. It also provides the ability to disseminate information and to create order [3]. However, if we take a minute to decouple the scientific record from scientific research as a whole, it is

Digital Transformation of the Laboratory: A Practical Guide to the Connected Lab, First Edition.
Edited by Klemen Zupancic, Tea Pavlek and Jana Erjavec.

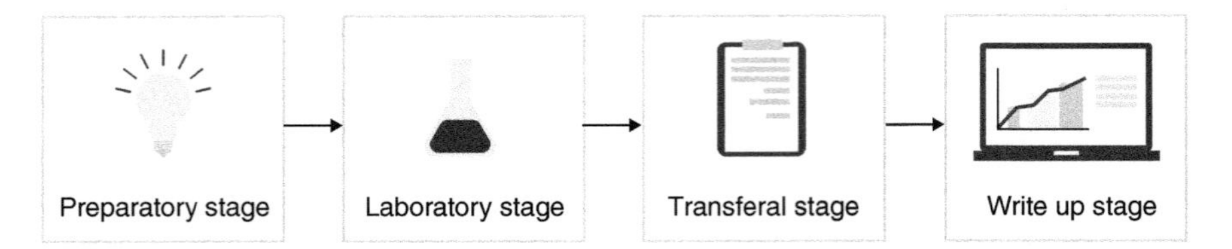

Figure 12.1 The different stages of scientific research. Source: Courtesy of Janja Cerar.

important to remember that scientific research embodies an entire life cycle of work. It has been questioned whether a daily scientific record could be seen to be solely the contents of a laboratory notebook, or whether it could stretch further to encapsulate data files, additional documents, and other items associated with that lab entry [4, 5]. Scientists may differ on the perception of what constitutes a specific record, but the scientific research life cycle as a whole covers many different stages which all need to be considered with respect to identifying the digital requirements of a laboratory. The different stages of scientific research include writing notes, conducting experiments, making calculations, analyzing data, reading literature, and the final write up stages of producing a scientific paper or report. These can be broadly broken down into four main stages (Figure 12.1):

- *Preparatory stage*: This encapsulates thinking about experiments, any preplanning of experiments, and filling in health and safety information (Control of substances hazardous to health [COSHH] forms).
- *Laboratory stage*: This covers all work that takes place in the laboratory, from setting up to conducting your experiments.
- *Transferal stage*: This stage deals with transferring the information from the laboratory (whether that is notes in a lab book, or handwritten data, or data collected electronically from an instrument) to a computer.
- *Write up stage*: The final stage deals with work done on a computer already, once scientists have sat down to write up their experiments or papers or thesis, etc.

The following sections detail the different stages, noting their digitalization requirements, barriers to adoption, and some initial solutions to consider. This will be followed by some suggested solutions for the entire laboratory life cycle, finishing with some overall considerations of digitizing a laboratory.

12.3 Preparatory Stage

This initial stage is very important, as how scientists set up their work from the beginning will impact how they conduct the rest of their work going forward. If they plan their experiments on paper, take it into the laboratory on paper, and get into the habit of doing everything on paper, then there is a high chance that they will continue using paper and get stuck in these habits. There are several pieces of work that will be conducted before scientists enter the laboratory to run their experiments, and these all form part of the digitalization requirements.

12.3.1 Digitalization Requirements

There are several different aspects of digitalization requirements for this initial stage:

- Experiment plans and protocols
- Experiment notes
- COSHH forms
- Literature reviews

These are typically done with a mix of a lab book, paper, and a computer. Literature reviews often have a paper-based component, as it is common to print out research papers, highlight them, and make handwritten notes; however, the use of reference managers to manage literature is also very common, and formal literature reviews tend to be written electronically and are conducted as part of a scientist's research. However, the rest of the activities often take place on paper.

It is common for COSHH forms to be paper based, especially in universities. Further, as scientists need to be able to see their experiment notes, plans, and protocols when they are actually conducting the experiments, they often create them in their paper lab books so that they have easy access to them in the laboratory.

This raises several digitalization requirements:

- Software to capture these preplans, notes, and protocols
- Software to handle COSHH forms and preferably link them into experiment plans
- Hardware to display these in an accessible usable format in the laboratory

12.3.2 Issues and Barriers to Adoption

This may sound simple, but there are some key barriers to adoption for these requirements, namely hardware accessibility in the lab and available, usable software. Scientists' experiment plan serves as a guide for them in the laboratory. If there is no easy way for scientists to access this information electronically in the laboratory, they will take it in on paper, meaning they will probably create it on paper, particularly, as a lot of scientists are in the habit of recording all aspects of their experiments in their lab book. Further, with respect to COSHH forms, these are often quicker to fill out on paper and may involve going into the laboratory or the chemical stores to look up specific details about the equipment or to check the hazards of the chemicals the scientists are planning on using.

12.3.3 Suggested Solutions

There are several potential solutions to aid with digitizing these processes, and they need to involve both adequate hardware and software. The first step would be to identify the software needs of the scientists, as even if there is suitable hardware for accessing electronic material in the laboratory, it will not matter if the software is not sufficient.

Investigate what type of software is needed to write up experiment plans and protocols. Are there bits of the plan that are easier to write on paper, e.g. diagrams? Or are these plans purely being written in lab books because they are easier to take into the laboratory and there currently is not suitable hardware for accessing electronic plans? If the latter is the case, then this is a hardware issue which will need to be addressed in the laboratory stage, whereas if it is the former, then this would require more work to understand what features the scientists are looking for to create their plans, and then either identifying an off-the-shelf solution to this, e.g. a specific electronic laboratory notebook (ELN), laboratory information management system (LIMS), or other software package, or if this is not sufficient, considering commissioning something custom built. Similarly for COSHH forms, if they are currently being filled out on paper, is this because they are not available electronically? Or because there is a perception that it would be harder to fill them in electronically? If it is the former, then it is worth looking into the electronic versions of the forms or specific COSHH software.

12.4 Laboratory Stage

This stage covers all work conducted in the laboratory, including any additional notes that are made while setting up an experiment, the recording of any observations or values during the experiment, and writing up notes about experiments or any other work conducted inside this environment. This is likely to be the last digital stage in most institutions due to the high adoption barriers of using electronic solutions at this stage. However, if these issues can be overcome, facilitating digitalization at the laboratory stage is the cornerstone in the solution to fully digitizing a laboratory.

12.4.1 Digitalization Requirements

The main digitalization requirement for this stage is having some hardware and software to interact with seamlessly during an experiment, both to view experimental plans and setup, and then to note down observations and values during the experiment. This can take different forms, depending on the type of laboratory and the type of work it has been designed for. Some laboratory work will involve working in a fume cupboard to conduct an experiment and needing to have a digital solutions in specific places, others will involve taking data readings from different pieces of equipment and needing to be able to consolidate these in one place, whereas some laboratories are designed for quality control and assurance and will need to be able to fill in forms while making quality checks. These are all contrasting setups, but the overall need remains the same:

- Having ready and available access to hardware in the laboratory

- Software that already displays any preplanning information
- Software that is easy to use and does not disrupt a scientists workflow
- Software that can integrate with the different pieces of laboratory equipment to read data values.

12.4.2 Issues and Barriers to Adoption

This is arguably the hardest stage to digitize, both because there are significant barriers to taking hardware into many laboratories and because scientists find transitioning from conducting an experiment to using a piece of hardware intrusive. The main barriers have been categorized as such:

Accessibility of hardware: There are differing rules and concerns linked to taking hardware into laboratories. The laboratory is a hostile environment, and many chemists may not wish to take their electronic devices into the laboratory in case chemicals are split on them or they get knocked off a workbench. There are also some laboratories that house equipment with powerful magnets that could wipe a hard drive clean if it was brought into close range. Equally, many cleanroom laboratories often will not allow external hardware to be brought into their laboratories in case of contamination. Work often takes place in areas that are not suited to a computer either because of the equipment around or space to put down laptops, and not all laboratories have Internet access.

Ease of use: How users perceive the ease of use of a solution impacts on their willingness to use it. Scientists consider the idea of using an ELN or similar to digitize their work inside the laboratory as difficult and disruptive to their workflow. If they are in the middle of conducting an experiment, they do not necessarily want to have to take off their gloves to type out some information they could quickly jot down, and find it easier to just move their lab book around with them.

Data compatibility: Even if a laboratory has not been fully digitalized, it often still houses digital components, such as pieces of equipment that are controlled by computers and produce electronic data or readings. Unfortunately, these different pieces of software do not always play well together, they sometimes produce readings or data in incompatible formats, which cannot be read by other pieces of software. There is an important distinction to make between a digital laboratory and a fully connected digital laboratory, and it is important to make sure digital solutions are compatible with one another in different areas of the laboratory.

Attitude to digital laboratory solutions: As noted previously in this chapter, digitalizing a laboratory is not just a technological solution, it is a human-based one also. If the scientists who are going to use your laboratory are against the idea of digitalization, then this can prove just as big a barrier as hardware and software issues. In industry, this is potentially less of an issue, as new policies can be mandated as part of a companywide policy (although it is still a good idea to get your staff onboard!), but in academia where there is typically more freedom allowed with respect to laboratory procedure, this can pose a big problem.

12.4.3 Suggested Solutions

In order to digitize the laboratory stage, there needs to be accessible hardware in the laboratory. The experimental areas should be assessed, considering when scientists would typically access their lab book during these experiments, coupled with where is a suitable place to put the said lab book (and if this would be safe for hardware). The main options are to set up desktop computers in accessible safe places for laboratory users to use, to provide protected portable devices, or a mixture of both. There is not one correct answer; it will depend on the setup of your laboratory and the nature of the work conducted in it. It is worth noting that there are many different types of hardware solutions for this.

There is the obvious desktop option, which has the advantage that these could be stored in the laboratory full time, they could be adjusted to a height above the work benches or pieces of equipment, and have moveable keyboards and mice with protective covering so that they could be used during experiments. There are portable devices such as laptops and tablets, which can also have protective cases put around them. It would be ideal to invest in laboratory-specific portable devices so they could be left in the laboratory in their protective cases, and be left charging so they would be readily available.

While scientists are often reluctant to take their laptops into the laboratory, the same hesitation often does not apply to their mobile devices. Mobile phones are often used to take pictures to aid with experiment observations or to provide images for papers or presentations. This development in itself demonstrates that there has been a shift in the attitude to at least mobile devices in the laboratory. This is worth considering and exploring with respect to hardware solutions in the laboratory. Why will scientists take mobile phones into a laboratory but are potentially more hesitant toward tablets or laptops? Is it the size, ease of carrying, and the fact that it can fit into a pocket? Is it deemed a necessary device to have nearby at all times? Are there solutions that could make use of smaller mobile devices such as smartphones or small tablets – such as taking notes alongside pictures or finding equipment that can be controlled and synced with a mobile device.

Similar to the preparatory stage, the software aspect also needs to be considered. Once appropriate hardware has been set up, there needs to be suitable software to access the preparatory information and fill in the required information at the laboratory stage. Having accessible hardware will break many of the barriers to adoption down; however, there is still a high perception that paper is quicker and easier to use with respect to jotting down notes. There are a few ways this could be addressed; most tablets facilitate data entry via a stylus, so notes could still be made in a similar fashion to paper but recorded electronically and potentially converted into typed notes using Optical Character Recognition (OCR). There are also electronic paper notebook replacement devices such as the remarkable [6], which functions as a digital notebook and can convert handwritten notes into typed form. Alternatively, it is worth considering voice-based systems whereby results and observations could be dictated directly into an ELN or word-processing software.

12.5 Transferal Stage

This stage covers taking the information recorded in the laboratory and converting it into a form useful for the write-up stage. This is the stage that also has the greatest potential for data loss as scientists frequently do not digitize the things that did not work and have a tendency only to digitize what they perceive as necessary for the write-up stage.

12.5.1 Digitalization Requirements

The main digitalization requirement is to be able to move work and potentially data from the lab book and machines into one place for the final write-up stage. With respect to data retrieval this means that the different pieces of equipment need to produce data that is in a usable compatible form that can be easily imported into any software used for data analysis or writing up.

12.5.2 Issues and Barriers to Adoption

There are four quite contrasting main barriers to adoption for this stage, which would all need to be considered and addressed separately.

Duplication of effort: This barrier is heavily linked to lack of digitalization in the laboratory stage. If the laboratory stage remains paper based and information comes out of the laboratory in the paper lab book, then at some point it will need to be made electronic. This leads to duplication of effort as scientists would then have to essentially reenter information such as observations and results that they had already noted down in their paper lab book.

Intellectual property and patenting: There are two separate barriers here. The first dates back to one of the original uses of the paper lab book which was to have a signed and dated record of any work conducted in case of a patentable invention later on, which if necessary to the scientists research would need to be reflected in any software-based solution. The second is with respect to privacy and security. If all laboratory work is electronic and potentially cloud based to be synced between the different work stages, then there is the potential for data to be hacked, or potentially mined by the software providers.

Data compatibility: This barrier relates both to equipment and any software for capturing data (whether from equipment or manually) in the laboratory. It is commonplace for pieces of equipment to be hooked up to a computer and for the data to be read off them electronically; however, this data can be in multiple different forms and is often in a locked down proprietary form that may not play well with other software packages for data analysis or ELNs. Similarly, ELNs can also make it difficult to export data out of them in a format that is useable with other software.

Stigma in science: The final barrier is very person based rather than technological. There is a stigma attached to writing up any negative results, and often the pieces of work that do not get digitized are the bits that do not work. If a scientist runs an experiment and it is unsuccessful, then chances are the main record of that experiment will remain in their paper lab book rather than being written up electronically as they will not be looking to publish it or put it in any report.

12.5.3 Suggested Solutions

In a fully digital laboratory, this would barely exist as a stage. Transferal from the laboratory to the write-up stage would be instantaneous as everything would be recorded electronically, and data files could be copied in a usable format straight from the computers linked to the scientific equipment into the necessary user's filestore. Anyone looking to digitize their laboratories should be implementing solutions, which means once a scientist leaves the laboratory all their experimental notes and data should already be placed in a suitable place to be immediately accessed for the write-up stage. Additionally, when purchasing any new equipment as part of this digitalization, these should be evaluated with respect to what data formats they support and how compatible they are with the software the scientists need to use, to ensure that the data transfers between the laboratory stage and the write-up stage are as seamless as possible.

There also needs to be an overall shift in attitude with respect to writing up experiments that do not work. It is immensely useful to know what does not work; otherwise, experiments can be needlessly duplicated. Finally, as noted above this is the stage with the highest potential for data loss as typically substantially less concern is given to backing up paper-based work, meaning that if a lab book is lost or a fire happens in the laboratory, then the contents of that lab book could be lost forever. This needs to be addressed and made clear as a benefit to digitizing laboratories; while there may be an initial time cost in converting to using digital methods, this does have the strong advantage of reducing the potential for lost work and the need to repeat it.

12.6 Write-up Stage

This is the final stage and encompasses the follow-up of work outside the laboratory for dissemination and presentation and involves writing up experiment reports, thesis, and papers and creating resources to present one's research.

12.6.1 Digitalization Requirements

This stage can often be disregarded as it is already fairly digital, scientists typically do already use computers at the dissemination stage as all papers and reports typically need to be delivered electronically, and work is often presented using resources such as PowerPoint to make presentations or through the medium of posters or documents that area also created electronically. However, despite the fact that it is already a relatively digital stage, there are still key digitalization requirements to consider.

- Domain-based software
- Software for data analysis
- Software for creating resources for dissemination and presentation
- Knowledge-management tools to manage and link together a wide range of different electronic resources.

12.6.2 Issues and Barriers to Adoption

Given the already technological nature of this stage, there are fewer barriers to adoption for the idea of digitalizing this stage, as typically scientists perform the write-up stage electronically. However, just because this stage is already quite digital does not mean it is optimized. This stage, while typically the most digital of the stages currently remains quite disjointed. Data is often stored arbitrarily in different places when it is read from pieces of equipment, and related notes, presentations, and papers are often stored all over the place, which can make finding specific pieces of information and producing write-ups after the fact more time consuming. Further, scientists often require specialist software for parts of their write-ups such as software that makes it easy to write equations or software to draw molecular models such as ChemDraw. A common reason cited for not using ELNs at this stage is that they do not contain all of these specific domain-based features (although of course neither do Microsoft Word or LaTeX), so other more generic notebooking software is used to produce write-ups with domain-based software, where necessary. Having this spread of disparate electronic resources can cause a lack of knowledge management, which was identified as a considerable issue by Kanza et al.'s [7] research into the digitization requirements of academic laboratories.

12.6.3 Suggested Solutions

There is a need for knowledge management solutions for scientists to link and store their data such that they can easily identify all necessary files (including plans, health and safety, data, write-ups, presentations, and posters) about different projects and experiments. This can be addressed in both a structural- and a software-based way. Research has demonstrated that scientists organize their notes in a very personal way [4]; however, it would be useful to have structures and procedures in place for how to store and link experiments and data together for retrieval. This could be as simple as having specific folders for specific projects and experiments, making sure there is sufficient storage space to put all of the different types of files in one place, and perhaps having model folders for new experiments or projects with the appropriate expected subfolders such as plans, data, results, and write-ups. Alternatively, there could be knowledge management software solutions. This could involve using an existing cloud-based storage platform such as Google Drive or Microsoft Teams where different folders can be shared with different groups, files can be indexed and searched, and there can be specific conversation threads associated with different pieces of work. Or, this could involve implementing software such as ELNs, LIMS, or a custom knowledge management system that facilitate tagging and linking documents together such that they can be easily searched for, or intelligent systems that suggest where documents might be linked to one another, or that intelligent queries can be performed on such as finding all experiments that have used specific compounds.

12.7 Conclusions and Final Considerations

Overall, digitizing a laboratory is as much as a human-based endeavor as a hardware and software one. There is no one magical solution that will work for every laboratory; each team and laboratory will have individual needs and these need to be taken into account. There are also several other factors that need to be considered:

Digitalization aims: If you are digitizing your laboratory, why? Is it to preserve the scientific record? To facilitate reproducible work? Is digitizing the laboratory indicative that these things will happen automatically? Digitizing a laboratory opens up new possibilities, but these cannot be taken advantage of unless other procedures are put into place. For example, if by digitizing your laboratory you are purely aiming to have electronic records of all your work to prevent losses, then once you have implemented a secure backup procedure you will have fulfilled your aim. However, if you are aiming for knowledge to be shared across groups, departments, or even your entire institution, then collaboration and sharing aspects need to be considered such that the appropriate people can access and accurately search the system. If digitalization is to facilitate reproducible research, then digitizing a laboratory is only a very small step in this process, as there will need to be new regulations and minimum standards for disseminating and publishing work such that data, data collection methods, and analysis methods are all published alongside work such that research can actually be reproduced.

Terminology: Something that has become very clear in several years of researching different aspects of digitizing scientific research. Scientists do not like the term ELN. It is associated (potentially unfairly or incorrectly) synonymously with giving up their paper lab notebooks and disrupting their current workflow. When embarking on a digitalization project, it is worth considering different terminology and making it clear that you are looking to improve the workflow as a whole, step by step, and will not be needlessly disrupting their workflow by making their lives harder. Instead, focus on why you are looking to digitize the laboratory and emphasize the benefits.

Play to technologies strengths: Paper still has affordances over electronic data entry; it is quick and easy to use, can be easily transported around, is cheap, and you can make squiggles and notes in any direction or placement. However, there are some tasks that computers can just objectively do better. A big one is searching if you are looking for a piece of data or notes about an experiment, you can find a document or multiple documents much faster on a computer than in a paper lab book. Electronic resources are also substantially easier to backup, collaborate on, and transfer to different places. It is important to focus on the positives of digitalization and also consider the time-saving effects in the long run. For example, producing a resource like an article about the return on an investment when implementing an ELN published by SciNote [8], for your laboratory to demonstrate how digitizing it can save time or solve current issues or likelihood for error could make the difference in getting adopters onboard. Further, case studies and prototype studies could be run with a group of early adopters to aid with these resources.

Ultimately, digitizing a laboratory requires more than just good technology, it requires a change of attitude and organization, and a full consideration of the human elements of the laboratory, as well as the hardware and software requirements.

References

1. Kanare, H.M. (1985). *Writing the Laboratory Notebook*, 145. Washington, DC: American Chemical Society.
2. Shankar, K. (2004). Recordkeeping in the production of scientific knowledge: an ethnographic study. *Archival Science* 4 (3–4): 367–382.
3. Latour, B., Woolgar, S. and Salk, J. (2019). Laboratory life: the construction of scientific facts (Paperback and Ebook). Princeton University Press [Internet]. https://press.princeton.edu/titles/2417.html (accessed 19 September 2019).
4. Shankar, K. (2007). Order from chaos: the poetics and pragmatics of scientific recordkeeping. *Journal of the American Society for Information Science and Technology* 58 (10): 1457–1466.
5. Kanza, S. (2018). What influence would a cloud based semantic laboratory notebook have on the digitisation and management of scientific research? University of Southampton. https://eprints.soton.ac.uk/421045/ (accessed 19 September 2019).
6. reMarkable (2019). The paper tablet [Internet]. https://remarkable.com/?gclid=Cj0KCQjwzozsBRCNARIsAEM9kBNNQAqdjn8cOL_Z4cPMXpYUmscy-5lHxHud6l71Km1wWGJnV8Vic-waAgnfEALw_wcB (accessed 19 September 2019).
7. Kanza, S., Gibbins, N., and Frey, J.G. (2019). Too many tags spoil the metadata: investigating the knowledge management of scientific research with semantic web technologies. *Journal of Cheminformatics* 11 (1): 23.
8. SciNote (2018). Return on investment of an electronic lab notebook [Internet]. https://scinote.net/blog/return-on-investment-electronic-lab-notebook/ (accessed 19 September 2019).

13

Guidelines for Chemistry Labs Looking to Go Digital

Samantha Kanza

Faculty of Engineering and Physical Sciences, University of Southampton, University Road, Southampton, SO17 1BJ, UK

The first step to digitalizing your lab is accepting that the transformation between a paper-based lab and a digital lab will take time and is not a quick simple process. Furthermore, this is not solely a technological solution. It is as much about the people in the laboratory and the organizational practices as it is the technology. This section covers some main guidelines and advice points for those looking to digitize their laboratories.

13.1 Understanding the Current Setup

Understanding the laboratory conditions will enable a higher level of comprehension of the documents created as part of this process [1]. Creating the appropriate tools and environment for scientific record keeping is in itself part of the record creation process [2]. Equally, when considering the wider context of how scientists could digitalize their processes, it is important to be aware of the limitations of their environment. It is important to understand the current processes and pipelines that exist in a laboratory (and assess different options for improving them) before digitalization can be considered. Often it can be tempting to try and find a piece of software that can emulate the current paper-based or semiautomated processes. However, this is not always the best way forward and could prove an enormous upheaval to the laboratory and current working practices, not to mention a costly exercise both in terms of money and staff training.

It is also absolutely imperative that different laboratories are not all treated in the same way. A crystallography laboratory cannot be digitalized in the same way as an organic chemistry laboratory. Even if the perfect solution was identified for one laboratory, it almost certainly will not align with a different laboratory, because it will contain different processes, different equipment, and will have contrasting needs.

Take the time to map out the current laboratory workflow, understand which processes are paper based, which (if any) are electronic, and most importantly understand the data flow and how information gets passed around the lab. It is

Digital Transformation of the Laboratory: A Practical Guide to the Connected Lab, First Edition.
Edited by Klemen Zupancic, Tea Pavlek and Jana Erjavec.

also important to remember that digital is not always equally efficient. Often, labs have multiple digital processes, but they do not play well with one another, or they even merit the additional use of paper, such as needing to use four different pieces of software with incompatible export files that require noting down specific values on paper to enter into the next piece of software in the process. Look at the entire system and work out where pieces can be automated.

The current workflow, processes, and software usage need to be formally identified in order to properly assess the situation and move forward. As part of mapping out the workflow, it is important to identify the main pain points of the current processes. If there is a significant blockage in one part of the workflow, or a particularly slow process, it might be worth considering the different options to streamline this first, rather than immediately trying to replace the entire workflow with a new system. Furthermore, the level of errors should be considered both when mapping out the workflow and considering which processes to address first. If a particular process leads to lots of errors, for example because data is hand transferred from one system to another, or one form to another, these processes should be moved up the priority list to address.

This process is not just about introducing an electronic lab notebook (ELN) to your laboratory or replacing paper-based processes. As noted in the previous chapter, there are many stages of digitalization. It is imperative to remember that just because something is currently in a digital form, it does not mean that it has been produced in an efficient manner. It is just as important to ensure that the software and environment a scientist has on their computer once they sit down to do digital work is optimized, as bringing in digitalization to the lab.

It is also worth noting that there are different reasons for digitalizing a laboratory. It might be to facilitate better communication between the laboratory users, and to create shared repositories of digital work within an organization, or to enable more reproducible research. If this is the case, then solely digitizing the laboratory (converting information from paper to an electronic format) is not enough. Think about WHY you are digitizing your laboratory, and what you are hoping to get out of this process, and try to ensure that the work done toward this endeavor aligns with these aims. It cannot be emphasized enough that this is not just a technology-based solution, nor is it purely a technological-based problem; digitizing a laboratory and all of the aims that coalesce with this process requires a human element also.

13.2 Understanding Your Scientists and Their Needs

Scientists organize and conduct their work in a very personal way, as demonstrated by Kanza et al. [3–5]. Industrial labs often have standardized practices, although these can differ between companies. Academia, however, is a very different kettle of fish as it is typically a much freer environment. Universities will have numerous different types of chemistry labs, housing different types of equipment, dedicated to different subdomains of chemistry, and potentially more importantly in this context, supervised by different professors or academic members of staff.

Experience of conducting focus groups and observations of chemistry labs at the University of Southampton (as detailed in Kanza et al. [4]) demonstrated exactly this high disparity between lab practices. These studies illustrated that PhD students and postdoctoral researchers typically followed similar practices if they worked in the same group and had a strict supervisor, but practices and regulations typically differed greatly between each group, and sometimes even within the same group if the supervisors were particularly relaxed, or if they supervised multidisciplinary groups who studied different areas of chemistry. If you are looking to digitize a laboratory, it is important to understand the different types of scientists who work in that laboratory and be aware of the different work that is conducted in the laboratory. For example, which groups of scientists use different pieces of equipment? How is data currently collected from this piece of equipment and in what format? How do the scientists then obtain it? What formats do they actually need it to be in and what pieces of software do they use the data with? It would be counterproductive to replace the existing equipment or processes if the new result ends up not meeting the needs of one set of the scientists who use it.

Further, scientists can often see ELNs as a disruption to their current workflow and that is why they are reluctant to use them. Talk to your scientists and try to understand what their objections are to potential plans to digitizing the laboratory. For example, a scientist might find it very disruptive to stop mid experiment to take off their gloves and type an observation out on a keyboard, whereas they might have a special pen that they use in the laboratory and can quickly use to take notes while not disrupting their workflow as much. In that case perhaps additional computers are required for the laboratory with protective casing, or a voice-based system needs to be considered as a way of recording information.

It is also a common misconception that scientists who do not wish to use an ELN are opposed to digital labs or new technology, and yet this is often not the case. They might be very onboard with the idea of introducing new software into the laboratory but have had a previous bad experience with an ELN and found it very disruptive. Or they might actually already digitize a lot of their work and would benefit vastly from new equipment that enables them to seamlessly collect data in a form that can be immediately analyzed by them or that is compatible with the software they need to use for it. Ultimately, laboratories would not run without scientists, and therefore their buy-in cannot be underestimated.

13.3 Understanding User-based Technology Adoption

There is more to asking a group of people to adopt a technology than just creating a new shiny product and hoping that it catches on. Much research has been conducted on the area of technology adoption: the individual uptake or acceptance of a new idea or product. Some theories seek to quantify adoption rate based on aspects of the technology, postulating that adoption is purely based on its performance, for example the technology acceptance model (TAM) [6]; or some other wider social theories argue that society (or in this instance the laboratory users) shape changes in technology [7]. However, in reality, getting scientists onboard with technological changes in the laboratory is a sociotechnical issue that requires a sociotechnical solution.

A highly pertinent theory to this is the *unified theory of acceptance and use of technology* (UTAUT). This theory was conceptualized by Venkatesh et al. in 2003 [8]. This theory builds on former technology acceptance-based theories such as the TAM [6] and other social and technical theories to try and explain what factors would influence a user to use a piece of technology. It postulates that the four main variables that will influence the usage are:

- Performance Expectancy – how much an individual believes that using this technology will improve their job performance (similar to perceived usefulness in TAM).
- Effort Expectancy – how easy the technology is to use (similar to perceived ease of use in TAM).
- Social Influence – how much somebody more important than an individual thinks that they should be using this technology.
- Facilitating Conditions – how much an individual feels that the technology is well supported with a suitable technological and organizational infrastructure.

These factors are all very important when it comes to accepting new technology, as is the acknowledgment of the fact that this is as much about the people as the software. While many areas of laboratories have gone digital in places (e.g. it is quite common to find equipment linked to computers to facilitate digital capture of data), many scientists are still struggling with the idea of giving up their paper lab notebooks and some of their paper-based procedures. One of the many affordances of paper is how easy it is to use; therefore, any digital replacement for a paper-based procedure or indeed the introduction of any new digital system needs to be easy to use. Similarly, users of these systems need to be convinced that using them, rather than maintaining the status quo, would genuinely enhance their performance. Additionally, social influence is an important factor to consider, as figures in authority (either supervisors/mentors for academia or bosses/managers both in academia or industry) could have an impact on whether an individual makes an effort to adopt a particular technology or in how they perceive it in the first place. Finally, the support network behind a system is also worth consideration; scientists find creating their records a very personal endeavor and have their own unique ways of organizing them on paper exactly how they want. It therefore follows that they would need a well-supported system that facilitates recording and organizing their notes and is more stable than their current paper-based practice, which would not be possible if the software was not well organized and did not have the appropriate technical support in place.

Further, several innovation theories discuss the importance of early adopters such as Rogers [9] who created the Diffusion of Innovations Model. Early adopters are coined in this theory as possessing the greatest degree of "opinion leadership" out of all of the adopter categories. It can be very useful to get a group of early adopters onboard who can lead the way for others to adopt new technologies. Consider identifying a group of those who are most interested in digitizing the laboratory and run some discussion sessions to generate new ideas and help understand the laboratory processes.

13.4 Breaking Down the Barriers Between Science and Technology

Getting the other members of your lab to buy into new digital solutions can be hard and as described above often requires several different elements to persuade them that a new digital solution is a better one. However, in addition to being concerned with ease of use and efficiency, etc. a big issue in technology adoption in the laboratory is lack of understanding of the technologies and why they are useful.

It has become common practice to throw around words such as "artificial intelligence (AI)" and "smart solutions," but what does that mean? What can feasibly be digitized and improved? How will this be achieved? It is advisable to provide the other members of your laboratory with some training (whether in-house if you have the appropriate expertise or outsourced if not) to help them understand the different types of technologies, their limits and capabilities, and how and where they can be used.

13.5 Making Your Laboratory Team Understand Why This Is Necessary

There are obvious reasons for digitalizing laboratories! Share these with the laboratory users and try to bring them on side. For example, if there are currently dangerous or costly errors happening as a result of a paper-based system due to manual transcription errors, highlight these problems and ensure that your scientists understand why the current systems are not working and why changes need to be made.

A key barrier to potentially adopting an ELN or bringing new technologies into the laboratory is concern that it will disrupt the current workflow and make a scientist's life harder. A way to combat this is to show scientists where digital is better. Paper-based systems also do not lend themselves to collaboration, as sharing information requires scientists to make manual copies of their work and send it over to the requesting scientist. This makes it very difficult to track changes and preserve an audit trail of the work [10]. Digital systems have many advantages over their paper counterparts. They provide the ability to quickly enter, retrieve, locate, and share data, in addition to facilitating long-term storage by being able to create backups and archives with ease [11].

13.6 Working with Domain Experts

In order to truly achieve a successfully digitized lab, both technical and scientific expertise are required. The scientists need to be able to fully explain and detail the different aspects of their current systems and demonstrate where they would like digitalization improvements to be made, and what procedures these need to conform to. Equally, a technical team is needed to determine what is feasible, and where small infrastructure changes could be made, or where small pieces of code could be written to automate a process or link two processes together as an interim solution, and to understand both the hardware and software requirements of the entire laboratory workflow.

13.7 Choosing the Right Software

Digitalizing a laboratory is not necessarily about finding one ELN or laboratory information management system (LIMS) system that works for the entire setup. ELNs and LIMS can be very useful for a specific subdomain, and if there are suitable tools out there that fit your laboratory setup, then that is an obvious solution. However, for companies or universities that have multiple different types of laboratories, it might be worth considering other alternatives. In some instances perhaps using One Note or Word and Excel combined with specific domain-based software and a sensible hierarchical folder structure would be more suitable, if the main requirements are writing up experiments and using some specific pieces of software to analyze data or draw graphs or diagrams.

Equally, you may already be using multiple software packages and digital pieces of equipment in your laboratory, and the real issue is that a lot of it does not play well together which introduces the need for paper-based processes as it is not possible to link all the different workflow processes together seamlessly. In this instance, working out how to address this issue and identify the problem areas would be a better start than introducing more software. Often people like familiarity and are comfortable with what they are already using, build on that.

Further, when it comes to digitizing a laboratory it is important to make sure that new solutions do not break existing solutions. When a software engineer writes a piece of code, they would typically write tests for that code and upon adding new functionality would run those tests again to check that the new functionality has not broken the old functionality. Adding new digitalization processes should be treated in the same way, check that they are not going to break anything existing.

13.8 Changing Attitude and Organization

Digitizing a laboratory requires more than just good technology; it requires a change of attitude and organization. This is a human-, software-, and laboratory-based issue. The individual needs of the laboratory need to be considered and the different workflow processes understood before any digitalization can be commenced. Scientists need to be brought onboard with these processes, and those in higher positions need to help with this by consulting them, understanding them, and making sure they are involved in the digitalization process.

Finally, digitalized does not always have to mean fully typed out; a picture of a lab notebook page is still better than it only existing in paper form and still provides more of a backup and improves access to material outside the lab. These simple changes need to be both considered and promoted to make slow progress to further digitizing the scientific record, and thus laboratories.

References

1 Shankar, K. (2004). Recordkeeping in the production of scientific knowledge: an ethnographic study. *Archival Science* 4 (3–4): 367–382.

2 Shankar, K. (2007). Order from chaos: the poetics and pragmatics of scientific recordkeeping. *Journal of the American Society for Information Science and Technology* 58 (10): 1457–1466.

3 Kanza, S., Willoughby, C., Gibbins, N. et al. (2017). Electronic lab notebooks: can they replace paper? *Journal of Cheminformatics* 9 (1): 31.

4 Kanza S. (2018). What influence would a cloud based semantic laboratory notebook have on the digitisation and management of scientific research? [Internet] [phd]. University of Southampton. https://eprints.soton.ac.uk/421045/ (cited 19 September 2019).

5 Kanza, S., Gibbins, N., and Frey, J.G. (2019). Too many tags spoil the metadata: investigating the knowledge management of scientific research with semantic web technologies. *Journal of Cheminformatics* 11 (1): 23.

6 Davis, F.D. (1989). Perceived usefulness, perceived ease of use, and user acceptance of information technology. *MIS Quarterly* 13 (3): 319–340.

7 Bijker, W.E., Hughes, T.P., and Pinch, T.J. (1987). *The Social Construction of Technological Systems: New Directions in the Sociology and History of Technology*. Cambridge, MA: MIT Press. 428 p.

8 Venkatesh, V., Morris, M.G., Davis, G.B., and Davis, F.D. (2003). User acceptance of information technology: toward a unified view. *MIS Quarterly* 27 (3): 425–478.

9 Rogers, E.M. (2010). *Diffusion of Innovations*, 4e. New York: Simon and Schuster. 550 p.

10 Machina, H.K. and Wild, D.J. (2013). Electronic laboratory notebooks progress and challenges in implementation. *Journal of Laboratory Automation* 18: 264–268. https://doi.org/10.1177/2211068213484471.

11 Voegele, C., Bouchereau, B., Robinot, N. et al. (2013). A universal open-source electronic laboratory notebook. *Bioinformatics* 29 (13): 1710–1712.

14

Electronic Lab Notebook Implementation in a Diagnostics Company

Casey Scott-Weathers

Intuitive Biosciences, 918 Deming Way, Madison, Wi 53719, USA

14.1 Making the Decision

The decision to move a scientific lab from a physical notebook to an Electronic Lab Notebook (ELN) is not easily made. There can be conservatism in science labs that manifests as a tendency to continue with processes that have worked in the past, an "if it ain't broke, don't fix it" mentality. There are also costs, both time and money, associated with moving to an ELN: researching the appropriate notebook, training users, and the cost of the notebook itself. The upfront monetary costs associated with paper notebooks are much lower than ELNs, and almost every scientist or technician has been trained on how to use analog notebooks. The combination of conservatism and the perceived simplicity and low cost of paper notebooks mean that many labs are content to continue to use them, regardless of the many possible benefits of migrating to an ELN.

14.2 Problems with Paper Notebooks

Despite the reasons to continue using paper notebooks, Intuitive Biosciences decided to explore the possibility of moving to an ELN. Personnel changes made finding historical data more difficult, as new employees did not know where to begin searching for data. The historical data was organized only temporally, and if one was not sure when an experiment was performed, it sometimes meant manually searching multiple paper notebooks to find the appropriate data. We were also hoping to expand in the future, and the new personnel would have even less idea of where to find critical historical data in our physical notebook system. Even the scientists who had been at the company from the start were spending more and more time searching through notebooks for previous results.

Because of the nature of our experiments, not all the data for every experiment could be printed and stored in a paper notebook. So, once the historical data was located, it was then necessary to find the appropriate folder on our internal server to

Digital Transformation of the Laboratory: A Practical Guide to the Connected Lab, First Edition.
Edited by Klemen Zupancic, Tea Pavlek and Jana Erjavec.

see the raw data, which just added to time spent looking for historical data. Beyond this time, recording current experiments was time consuming as well. The variable nature of our work meant that templates were of limited utility, and even basic templates needed to be updated regularly, resulting in lots of time typing up new procedures to be printed out and taped into a physical notebook. As "wasted" time spent searching old notebooks and recording new experiments grew, we determined that the path we were on was not sustainable.

Advice: My advice on approaching the problems inherent in paper notebooks is to begin solving them as soon as possible. The difficulties in managing analog notebooks do not get easier as a company grows.

14.3 Determining Laboratory's Needs

Of course, making the decision to investigate ELNs did not mean we were going to move forward with purchasing an ELN though. There was a wide variety of ELNs to choose from, and finding the appropriate one for our lab and situation would not be an easy process. Additionally, not all the personnel felt that the move to an ELN was worthwhile; they felt that the upfront costs of switching, both in money and labor, outweighed the increased labor costs of searching for data and entering notebook entries by hand, which meant that the ELN we chose would need to convince skeptics that the efficiency and utility gained by the ELN justified the upfront money and time to transition.

As we moved into comparing ELNs, and justifying the potential move to various parties, we determined that there were three major points that were of critical importance. These points are likely similar to points that most labs will find important if they move toward an ELN: price, ease of use, and flexibility. The pricing of systems is a relatively obvious point of interest, and we quickly found that there was a wide range of prices and pricing structures including one-time purchases or annual fees. Ease of use is another factor that is likely important to any lab preparing to switch to an ELN. Our lab, like most labs, has users with a wide variety of technological aptitude, and whatever system we chose needed to cater to that wide range of skills. Additionally, as the company grew, every new user would need to be trained on our specific ELN. Since we are a small company, we did not want that to be a time-intensive process for the designated trainer. The third and final major point that we considered is one that seemed simple on the surface but removed several ELNs from our consideration almost immediately – flexibility of the system. Almost none of the work that we do is DNA-based molecular biology, and very little of our R&D process is rote, which means that ELNs that specialize in DNA work, or have built in DNA tools, added very little value to us and tended to have limited flexibility to accommodate our R&D experiments. There were, of course, other more minor concerns and considerations. For example, we needed a notebook that was compatible with a wide range of data files, since we produce images, Excel files, PDFs, and more esoteric file types that all need to be stored in an accessible location.

Advice: Once the decision is made to change to an ELN, be thorough in documenting what exactly is needed and what is just wanted. It can seem time consuming, but spending the time up front to really understand your needs and wants means you are much more likely to find an ELN that works for your situation.

14.4 Testing

After determining our major and minor needs, we narrowed the field to a few free ELN options and a few paid options. Next, we began testing different systems. This portion of the process took much longer than we anticipated, as it was difficult to get a hands-on test of some of the different systems. The free options were easy to test; we downloaded the program and tried to use it. We tried to have multiple people use each free option, although it did not always work due to time constraints. However, the free options also tended to have inferior documentation, and so testing them to see if they would work for our purposes was time consuming. We had to learn enough about each notebook program to see if it would work in our specific situation, and not all our questions were quickly answered.

As we explored the different free options, it became obvious that they would all require compromises that we were unwilling or unable to make. One free ELN was designed almost entirely for DNA-based molecular biology experiments, which was not suitable for us. Another was effectively a blank word-processing document, which would only digitally replicate the problems we had with physical notebooks. A few had no method to lock experiments, to prevent future modifications. In the end, after learning how to use several free ELN options, we decided that none of them fit our needs.

As we moved into testing paid ELN options, we ran into another unexpected roadblock. Many of the paid ELNs did not really have effective demonstration options. Vendors would gladly run through a list of features and demonstrate basic usage, but few of them had hands-on demonstration options. We strongly felt the scientists and end users at our company needed time with the software to determine if they could work with it every day. Because of this, we did not move forward with any type of software that did not have a hands-on demonstration.

Advice: Test, test, test. Get input from multiple users, if possible, especially users who may use the software in different manner, managers versus *technicians, engineers* versus *biologists, the more input the better. Also, if you are not comfortable with the amount of help and support you get while shopping for a product, do not buy that product. The support will not get better once you have paid for it.*

14.5 A Decision

After all our research and testing, we decided to move forward using the SciNote. The decision was remarkably easy, given the long testing and research process; all the stakeholders were unanimous in selecting SciNote. SciNote had the flexibility that we needed for our varied experiments. It also was flexible enough to allow us to record our experiments ranging from less stringent R&D experiments to tightly controlled development projects. The web interface was simple enough that all our scientists felt comfortable using it, but it rarely felt like it was lacking features because of that simplicity.

Even that simplicity, however, required a lot of changes to our methods of recording experiments and the processes that we used to record and sign off on experiments. As a small company, we did not have a dedicated person to design the new processes for the ELN, and we were not sure how to approach the design and structure of moving our experiments from analog to digital. We wanted to identify and discard inefficiencies left over from using paper notebooks while not losing anything that may be important in the future.

14.6 How to Structure the ELN

To approach the structure building in a democratic way, we had every scientist who was going to use the ELN spend several weeks using it. During those weeks they entered in their experiments, created template protocols, managed inventories, prepared reports, and took exhaustive notes on what worked and what did not work. After that discovery time period, we reconvened and discussed how we structured our experiments, what benefits we found for using new features that were not available in an analog notebook (such as tagging experiments and commenting at collaborators), and what seemed extraneous. Much of our work is done using immunoassays; however, we also have worked on generating microfluidic devices, and our experiments run the range from answering basic questions about the physical design of a product to troubleshooting existing immunoassays. So, we knew that each scientist was going to have to determine what worked and what did not work for their niche. Over the course of several hours of meetings, we adopted a standard format for how we capture the experimental workflow, what critical pieces of information need to be captured, and how to capture them. In the end, we determined that some features of the ELN did not make sense for us to use, and some that were critical.

We have only made a few adjustments over the several years since adopting the ELN, but the general workflow has stood the test of time and changing personnel. The one drawback is that the first several weeks of captured experiments are not well organized. This is a minor problem though, given the ability to comment on and search for experiments in the ELN. Since adopting an ELN, our company has expanded into some more tightly regulated directions. Having a regulatorily compatible ELN already in place, such as SciNote, has been a great help in moving forward. Change tracking is already in place, and we have just needed to institute a few changes and set up sign-off protocols.

Advice: Of the advice here, this is probably the most critical. Once you have your ELN, regardless of which one you have chosen, you will need to figure out what features to use and what features are superfluous. You will also need to determine a consistent structure for how to record experiments, and answer questions such as what constitutes an experiment, or a task, or a subtask, or a project? We feel very strongly that the reason our ELN adoption worked so well is that we did not rush finding those answers, and we gave every user a voice in how they felt we could best utilize the ELN. Remember more voices are always better, everyone finds different ways to use the software.

14.7 Conclusion

Overall, our switch to an ELN system was a resoundingly positive experience. Had we known how smooth the switch was going to be, we would have done it years earlier. We save an inordinate amount of time while looking for historical data, more than we anticipated. Having a search function included in a notebook is also a benefit that we did not consider initially. If one is ever unsure of where something may have been filed, finding it is as easy as coming up with a few search terms. It literally saves hours of paging through old notebooks. We have also saved time in the recording of experiments and data; the ability to prepare templates for our few rote experiments and being able to cut and paste from previous, similar experiments saves, again, more time than we anticipated. One final unexpected benefit is that having an online ELN gives lab personnel some level of workplace flexibility. Typically, bench scientists need to be present in the lab to get any work done. Last winter was so cold that the office closed for two sequential days, and lab personnel were able to work from home, if they wanted, spending their time updating notebooks from the warmth of their homes. This is a small benefit, but it was greatly appreciated by the scientists who were able to get some work done without leaving their homes.

There are a few issues that we had to find work-arounds for, but they are minor compared to the ELN ease of use. As mentioned above, our method of implementing an ELN structure, while it worked well in the end, left the recording of some early experiments a little chaotic. One other con is specific to experiments that generate large data files, or a large volume of smaller data files. If we were to upload all the data that we generate, we would very rapidly use our data allotment, and buying enough additional data would become a very pricey proposition. To address that problem, we only upload a representative selection of images for each experiment. However, because of the flexibility of the ELN system, it is very easy to include a file path in the results that creates a link to the remainder of the data files on our in-house server, should someone need to view all the data. Even this drawback, which, on the surface seems relatively clunky, has morphed into a boon, because it allows us to quickly look at the important data for a historical experiment rather than flip through hundreds of data files to make sense of it. Then, if the synopsis is not detailed enough, one can go and find the remaining data files. Plus, we were never able to print out all the data files to paste into an analog notebook anyways.

While changing any system in a lab can be a daunting process, changing the method of recording experiments and data is likely the biggest single change most labs can make. Every scientist, at least currently, has been trained on how to record methods and data in a physical lab notebook, and very few have been trained on how to correctly use an ELN. Lack of training coupled with a resistance to change and a resistance to spending money where it is not needed causes a lot of inertia to overcome on the path to digitizing a lab. Our experience, as a small biotechnology company, was as positive as it could have been. Best of all, the reasons for the smooth transition have very little to do with our size and everything to do with how our personnel approached the transition. We exhaustively researched our options and spent time gauging our wants and our needs. Then, once we decided on an ELN, we gave all our end users time to generate structures and methods that they found useful and then semidemocratically synthesized those structures and methods into the accepted process that we currently use. Personally, I still smile when I search for and find a historical experiment, done by another scientist, in 30 seconds, and can move on with planning my own current experiments.

15

Identifying and Overcoming Digitalization Challenges in a Fast-growing Research Laboratory

Dorothea Höpfner[1, 2]

[1] Department of Biochemistry and Signaltransduction, University Medical Center Hamburg-Eppendorf (UKE), Martinistr. 52, 20246, Hamburg, Germany
[2] Center for Integrated Protein Science Munich (CIPSM), Department Chemistry, Technical University of Munich, Lichtenbergstrasse 4, 85747, Garching, Germany

15.1 Why Going Digital?

My first encounter with electronic lab notebooks (ELNs) took place very early in my career. At that time, I was interning in the pharmaceutical industry and was introduced to an ELN as part of normal data documentation and an important part of data integrity, reproducibility of experiments, and eventually compliance to good laboratory practice. While many colleagues – myself included – were often complaining about the tiresome and unintuitive interface of this particular ELN, I could not help but notice the many advantages of a mandatory standardized digital data documentation system: When I left the company a year later, there was no question about what experiments I had conducted, exactly how, and when. There were no loose papers or barely legible hard copy notebooks left behind. For the handover, my data was complete, nicely sorted, transparently documented, and in one place.

Coming back to academia to finish my master's program, the situation I encountered in this regard was quite different. Up until then I had never really thought much about data documentation and its effectiveness. After my previous experience in industry (and first major research activity), I may have subconsciously expected that data documentation requirements would be something normed and clearly defined. And while there were theoretic lectures and guidelines about good laboratory practice and proper scientific conduct, the question how data documentation should take place in practice remained only vaguely answered. The usual standard in all labs at university I encountered so far is a paper notebook and the requirement to document experiments in a way that they can be reproduced. While that seems to be a universally accepted consensus of proper data documentation etiquette, it is also very open to interpretation, especially for a beginner. I quickly learned that there was no such thing as a standardized style and very little control mechanisms. What amount of documentation was required to be able to reproduce data was mostly left to my own evaluation. To be fair, one learns and adapts quickly, and most

Digital Transformation of the Laboratory: A Practical Guide to the Connected Lab, First Edition.
Edited by Klemen Zupancic, Tea Pavlek and Jana Erjavec.

researchers I have met seem to manage quite well. While the order and legibility of paper notebook documentation varies greatly from person to person, all scientists I personally know faithfully document each day in the lab. The universally prevalent manner of documentation used is pen and paper, which – if properly executed (e.g. no loose paper collection, no ripped-out pages, and no referral to "as usual") – is a valid and accepted method of data documentation.

However, there are a few downsides to that style, which cannot be overcome.

The biggest one – in my opinion – is that hard copy lab notebooks cannot be searched. Unfortunately, most people have the tendency to only store data if the experiment had a successful outcome (regarding the method itself, not whether the scientific theory was proven right), while failed or unsuccessful experiments tend to be neglected. But if, for example, one wanted to know whether a predecessor ever tried a specific method or protocol and it failed, this method might never be mentioned in any paper or progress report. Having then to search a nondigital lab notebook is tedious and will take significant time. The same is true for successful experiments when dealing with unpublished data. Heaps of different projects and abandoned lines of work, most likely all documented within the same paper notebook, will create further difficulties in finding relevant experiments and protocols.

Another disadvantage of paper-based documentation lies in the fact that there is only one copy of it. Sharing extracts of a hardcopy paper notebook is only possible – if you do not want to give away your one original edition – by transferring it into another medium, e.g. by taking a picture or type it into a word file; in other words, digitizing it. Instead of this tedious two-step process, it would be much more sensible and time effective to directly generate and maintain one's documentation digitally right from the start. Everything that relies on sharing data will be facilitated: Handing over projects while still being able to access your own documentation or collaborating on the same project all become so much easier and quicker. What is even more, authorized members of the lab team could access ongoing documentation from anywhere and not just from within the lab. Instead of being limited by the necessity to leave all notebooks in the lab, documentation could be transferred and accessed flexibly.

In my experience, most labs nowadays encounter a staggering discrepancy between their digital experimental datasets and a still paper-based data documentation. Modern techniques and methods and their accompanying highly sophisticated hardware produce ever-growing amounts of data, which is almost exclusively digital. And while some of them – such as images of SDS-PAGE and Western blots – are easily transferred to the lab notebook by printing them out, most raw data is so complex and rich that nobody would ever try to print it on paper. In consequence, the description of the experiment's procedure and protocol is documented in a different place from where the output data is stored. This half-digital, half-analogous style makes it hard to connect data to documentation.

When my father wrote his doctoral thesis over 30 years ago on one of the (back then) most popular home computers, a C64, this neat piece of technology already enabled him to program a software, which could translate a three-letter codon sequence of DNA into its corresponding one-letter amino acid sequence of proteins.

However, most experimental data back then were still analogous, and the lab notebooks anyway were paper based. When writing his thesis, he had to draw figures by hand on millimeter paper, glue it into a blank spot of a printout, and then copy the page to make it more presentable. Since then, huge changes have taken place regarding data availability, quality, and amount. Today, my work in the lab completely relies on the use of computers and the generation of gigabytes of digitalized data. Despite this technological progress, however, my father's old paper-based lab notebooks look pretty much the same as mine: regarding data documentation, we are still stuck in the past.

When I look at the data documentation style of my coworkers, they – just as myself – tend to document their data in their (hardcopy) lab notebooks chronologically but not sorted according to project or topic. However, data storage on their PCs tends to be organized by project and/or by experiment type. While very intuitive to the scientist who conducted the experiments, this discrepancy between data documentation and data storage style could cause problems down the road when someone else, e.g. a project successor, is trying to match a certain experiment description with its related data. Ideally an ELN would fill this gap by being able to sort data according to topic, date, or method and directly link data with a topic, a method, and a specific procedure.

ELNs are not the magical solution to all data documentation problems of the modern scientist. They offer, however, enormous possibilities by being more coherent, accessible, searchable, sorted, and ultimately normed. Incorporating them into one's scientific routine is a logical step forward toward the ever-progressing digitalization of the lab.

15.2 Steps to Introduce ELNs in Lab Practice

In the very beginnings of my PhD I was asked by my supervisor to investigate ELNs as a potential topic for a workshop during our scientific retreat. He had recognized the drawbacks in the way data is traditionally documented in academia. I was initially thinking along the lines of a training session in digital data documentation and started searching for companies who might offer training. The workshop never came to pass, since not all participating research groups saw an added value in ELNs and alternatively opted for a practical illustration in Adobe Illustrator – no doubt also very useful. But the preparation for the proposed workshop gave me a good overview of what was to be had, simply by googling it. And it looked promising enough, so I suggested to my supervisor to pursue ELNs anyway and volunteered to come up with a strategy.

15.2.1 Step 1: Getting to Know the Market or What We Can Expect of an ELN

A good advice with anything one wants to start from scratch is to ask for recommendations and rely on experience from others. Unfortunately, when I asked around at university and adjoining institutes, ELNs were not something anybody could comment on much. There was no previous experience with any software or system to be found. So, I did what any proper scientist knows best: I researched. I searched for ELNs and companies that provide such software. In the beginning my ideas of ELNs and of what they could provide were rather vague, so I started very broad, looking into any software that deals with data documentation. I then quickly narrowed it down to those focusing on scientific research, and even further considering their ability to be applicable to our lab in terms of documentation style, and the kind of data that can be stored. Along my research I also reviewed some articles and blog entries about ELNs for biochemical research labs and some of the companies providing them.

Around the same time, I started to discuss with my supervisor how much money we would be able and willing to spend on data documentation, as it became very apparent that we had to expect to spend significantly more money than previously on hardcopy lab notebooks – which seems such an obvious realization, but coming from an environment where the amount of money you spend on data documentation is usually limited to the costs of pen and paper, the decision to spend a lot more for a software solution is a first obstacle that labs like ours with tight budgets might not want (or be able) to take. The good news is: it does not have to be hugely expensive! Of course, there are premium subscription plans that provide every comfort in data documentation and cost their price. But there are also quite a lot of "pay-by-data-amount" and monthly plans that are a lot easier to incorporate into one's budget. With these kinds of plans, one has to plan ahead to ensure that one is able to pay a constant monthly fee with options to upgrade, but it is a very good start for smaller labs who want to try out first whether the concept works for them.

With such considerations in mind, the field of software/software providers narrowed down to three to four candidates, which I took into serious consideration.

The truth is that up until then I was never really interested in – or kept track of – what alternatives to the traditional hard copy lab notebook were on the market. As mentioned before, my personal version of digital data documentation consisted of Word documents and Excel sheets, and I felt quite *en vogue* with my PowerPoint presentations. Searching for software that provides an ELN proved to be an eye opener in that regard: the capability of the software on offer and the beauty of user interfaces gave me an unprecedented surge of anticipation toward digital data documentation.

15.2.2 Step 2: Defining the Needs of Our Lab and Our Requirements for an ELN

The lab I work in used to be a small and young research group of around six PhDs, two postdocs, one technical assistant, and a constant stream of students and trainees. There were many reasons why I decided to conduct my PhD here: An exciting research topic; a dependable, approachable, and brilliant professor; and a wonderful team atmosphere, to name a few. The most important ones for this account, however, were the lab's overall characteristics to be very organized and to strive for constant improvement and state-of-the-art research.

As we are a biochemical research lab with strong emphasis on protein chemistry, recombinant protein production is our workhorse. Anything related to that is in consequence highly streamlined, automated, and digitalized in order to achieve highest throughput and product quality. Long before I joined the lab and the implementation of an ELN, we already had a lot of digitalized solutions in the lab. We maintain a digital plasmid and primer database that holds our ever-growing collection of over 6000 available plasmids including plasmid maps, sequencing results, and cloning strategies. In addition, we have access to a lab-intern wiki that is constantly updated and provides standardized protocols, protein purification strategies, protein and antibody lists, quality control elements, and a chemical inventory. However, the scientists' personal workflow and data documentation were not part of these solutions, as the first is limited to cloning, and the second is serving as a public summary of the lab's status quo and best working protocols. A platform was needed, where every scientist could access a personal space in order to daily document experiments in detail and link the corresponding raw data on our in-house server. Consequently, I was running in open doors with my professor when it came to digitalization of data documentation.

To my mind digitalizing data documentation using an ELN software only makes sense if it adds significant benefit in comparison to paper documentation. And as mentioned before, most software actually does a lot more than simply overcoming the disadvantages of analogous documentation. When deciding on an ELN it is very helpful to define research- and laboratory-specific needs and prioritize them beforehand, as it narrows down the selection of potential software and helps ranking one option over the other.

There are many topics to take into consideration:

15.2.2.1 Data Structure

A first point to discuss is the amount and structure of data one's lab generates. In our case, we were mostly interested in feeding in text-based data documentation, so I put an emphasis on software that was compatible with MS Word or was coming with an incorporated word editor. Due to the amount and size of generated experimental data as well as the availability of an in-house server, the data itself did not necessarily have to be incorporated but could also be linked.

15.2.2.2 Compatibility with Databases

We also made the possibility to integrate already existing databases a decision criterion. For coherence and reproducibility, it would be very beneficial to be able to link information about plasmids, proteins, or antibody batches and other samples directly from our preexisting databases to the actual application.

15.2.2.3 Flexibility of Documentation Style

Another point to consider is how flexible you can be, or need to be, with regard to documentation requirements. For example, highly streamlined processes with a lot of standard operating procedures might need (and even benefit from) reduced flexibility in documentation style in order to ensure uniformity, speed, and reproducibility. In contrast to that, research labs running individual projects with individual executing scientists and their unique documentation style might require very flexible interfaces, where each scientist can feel at home quickly. In our case, it became clear very quickly that no ELN is perfect for every single lab member considering their very own documentation style. Therefore, we required the ELN to offer a high degree of flexibility when feeding in data.

15.2.2.4 Report Options

In addition, a convenient and adaptable report structure to provide for fast project overviews or specific data summaries was on the top of our priority list. We wanted the ELN to be able to quickly and easily report specific subsections of projects in changing detail, whether as printouts for a spontaneous meeting or a broader overview for the next progress report.

15.2.2.5 Speed

In order to save time by applying an ELN and not make documentation more time consuming and frustrating, appropriate speed of data upload is also an important point. In addition, the interface of specific experiments should be quickly accessible without the need to click through convoluted folders or scroll through pages of previous work.

Additional topics to take into consideration might be the software's ability to admit multiple users and to apply different roles/user profiles (e.g. students should not be able to change protocols) or the availability of extra features such as registries for samples, inventories or protocols, features for project planning, or a citation- and library plugin. All these considerations depend highly on the needs of the specific laboratory and cannot be generalized.

15.2.3 Step 3: Matching Steps 1 and 2 and Testing Our Best Options

One major topic we discussed was the question of where to install the ELN. In the beginning we invested a lot of time and thought into the feasibility of setting up an ELN on our own server. While that option appealed to us on many levels, the main one being the (apparent) higher security of our data, running an ELN on our own server would have come with a number of significant challenges. As with most labs in academia, there is no IT infrastructure or IT support to speak of. Our "IT experts" are usually PhD students themselves, mostly self-trained, and hard to come by as they are doing all problem solving in addition to their actual PhD thesis work. Nevertheless, installing the program itself and managing access would have been feasible (some ELNs are available on GitHub) and comparatively easy. However, maintaining the program over the years with changing personnel and with no trained IT expert would have been a different story. In the end we decided to go with running the ELN on a foreign server (e.g. the one that comes with a subscription plan for most ELNs). The main reasons were that we did not think ourselves capable of properly maintaining the infrastructure and the program, the security of a foreign server and its protection against all kinds of data loss (not only due to cyber criminality but also fire, etc.) was in the end higher than anything we could provide, and that the full scope of a modern ELN including updates and incorporation of brand new features only comes with a subscription plan.

Before we signed up with a particular software provider, we decided to test out all remaining candidates. Most companies offer free 30-day trials or free accounts with limited data input, which I highly recommend making use of. After setting up the accounts I started at first to feed in random data from past projects to see how intuitive the interface was and whether all our predefined needs for data documentation and storage would be met. Unsurprisingly, random data input and software that is designed to not allow data deletion do not mix well.

Therefore, I subsequently tested every potential software painstakingly for convenience of the interface and its ability to realize our data documentation style. I firmly believe that a software program, which is unintuitive or time consuming to use and master, and which forces the user to significantly change their documentation style, will at the end of the day be a lot harder to implement and establish. The easier and more convenient it is to use, however, the more people will eventually voluntarily start using it. I therefore spent a significant amount of time on this step to ensure convenience and flexibility. I also started to play around to find a coherent input style and work-arounds for gaps in the software when it came to link in our already preexisting applications.

I quickly noticed significant differences between the different ELNs concerning the speed of data upload, an issue that, coming from paper documentation, never crossed my mind before. But there is little which is more inconvenient and annoying than being in a hurry, trying to quickly update some protocols and the program freezes while saving the changes. To best test for any forced and unwanted breaks, I used realistically sized files.

Whenever I encountered difficulties or found a certain ELN software solution very inconvenient, I contacted their customer service to check for their response. It is important to note that this response time might be crucial for one's future experiences with the ELN. There is a big opportunity in a close interface between scientists incorporating ELNs into their lab routine and the companies developing them. A constant feedback culture and a shared vision for the future of data documentation offers significant benefit to both sides: for the software company, by providing important insights into the real life struggle with digitalization and software application, thereby opening areas for further development and for the scientists, by offering the opportunity to directly report back bugs and gaps in the application and thus being able to influence software development around their needs. Therefore, I recommend the quality of communication with customer service to be a part of the decision-making process.

15.2.4 Step 4: Getting Started in Implementing the ELN

As soon as the decision for a specific software is made, there comes the crucial part of setting it up properly and of introducing it to the lab members. This step clearly deserves some thought and time and should not be rushed. Proper preparation and a plan of action will significantly influence the desired positive outcome and the rate of success in incorporating the ELN into the lab's day-to-day routine.

Because of my eye-opening data input experience while testing out the ELN of choice, I had a good guess as to what could be expected to happen if multiple lab members now started to use the software without further guidance. I also realized that early frustration would make the establishment of the ELN that much more difficult.

The first step before opening the ELN to the public should therefore include a definition of certain rules of usage. As data cannot be deleted from an ELN once uploaded, there should be a strict set of "dos and don'ts" in place, which helps to avoid the biggest mistakes for a beginner. The generation of an example project and training examples on the most common and popular techniques in the lab will also help to ease the initial hurdles in the transition from paper to digital. Only after that did I grant the team members access to the new software, while conducting a training session at the same time.

I then made the big mistake to think my job would be finished with that. I had researched the field of ELNs, defined the needs of our lab, figured out all the intrinsic details that come with advanced data documentation and software implementation, decided on a software program, presented the result to my lab members, and trained them in its use. I thought, finally my part was done. It would now be up to each individual to feed their data into the ELN and that would be it with a successful transition to digital documentation.

Unfortunately, I completely underestimated the power of habit and the need for a clear "extrinsic motivation" to make this transition happen for a lab crew of scientists. And while breaking habits and offering motivation is a very difficult and potentially quite vague topic, here are a few things I could and should have done additionally.

- One is the definition of milestones for users: Setting goals and communicating expectations, not just in the very beginning but especially all along the way, will help the individuum to measure progress and develop motivation to meet them.
- Another measure should be to prevent the mistake of allowing redundant documentation options: Being able and allowed to still use paper documentation while at the same time being required to use an ELN will actually delay successful establishment of routine digitalized documentation. There is a strong likelihood that feeding the ELN will, at least by some more traditional lab team members, be perceived as an "administrative add-on task." Define a time frame, in which the transition should take place, and after which only the ELN will be an acceptable documentation choice.
- It might also be helpful to design special roles and responsibilities when it comes to maintaining the software and its associated databases. Distinct responsibilities, such as maintaining the input repositories, looking after sample management, or making sure that the protocols are up to date, could be distributed among the group. Becoming involved and taking responsibility for part of the process will help the single user to start to identify him- or herself with the overall goal.

- Last, but not least, the development of a feedback culture related to challenges and issues with the transition to an ELN will facilitate the process. Along those lines the team could, and should, e.g. agree on control mechanisms to regularly check progress.

15.3 Creating the Mindset of a Digital Scientist

To date, it is approximately two years now since we established an ELN for our team. Since then, we have moved our labs from southern Germany to the north, and the team has grown significantly from a small research group to a whole institute, encompassing three separate research areas and more than 30 employees.

Our experience with digitalization of data documentation is still somewhat mixed: While all lab members have and maintain digitalized data, most – at least initial – documentation remains on paper. Many scientists tend to use a double documentation style, starting on paper, later digitalizing documentation of successful experiments (e.g. via Word, PowerPoint, Excel, Origin, and Adobe Illustrator), before ultimately feeding it into the ELN. Unfortunately, this style results in a significant time penalty and involves the risk of losing track and falling back into old patterns. In my experience, slow progress of digitalizing data documentation is mostly not due to a lack of motivation at the side of an individual scientist, instead, there may be specific reasons why establishing a digital data documentation routine does not seem to work for this individual, reasons for which solutions can be found. For some, a daily mandatory input scheme might help to set up a routine and to drive the initially necessary discipline. Other strategies might include a regular time slot in the calendar, solely dedicated to digital documentation. Sometimes the cause might also lie in lack of computer availability not only in the office but also directly in the lab. Here, the use of handheld devices might help to access the ELN directly from the site where experiments are conducted. Actually, data entry into ELNs is often feasible from smartphones or tablets via browser interfaces.

Despite a mixed progress after two years of using an ELN, I do not feel that the introduction of an ELN in our lab routine has failed. I rather consider our current situation an interim point in a longer journey. There are small successes along the way which we celebrate. Recently, for example, a new colleague took over one of my projects, and instead of pouring over my hardcopy lab notebooks and sending PowerPoint presentations, all my data were shared with one click. No copying pages, no transmission of multiple data files, no search; it was all there. I also savored the significant time savings when creating reports, setting up presentations, or writing the "Material and Methods" part of a paper. And, we are constantly working on new options to ease the documentation burden. Recently, we have been thinking about voice input for our ELN: I have the dream of one day being able to document my data by simply dictating the details of my experiments while I conduct them.

Digitalizing science, and in particular data documentation, is certainly a long-term process that needs to overcome long ingrained habits as well as it needs to master the many challenges that come with a young and fast developing field.

The transition from paper to digital documentation will not be terminated within a few short months or years, and it will start again with every new addition to the team. Once the decision is made to make the jump, it will actually not be the end, but rather the start of a long journey with many fallbacks in between. But it is a step that – in my opinion – every lab that wants to compete in an ever-digitalizing world must take sooner rather than later. With the implementation of an ELN, we took a logical next step to stay on top of the current state of art as scientists.

15.4 The Dilemma of Digitalization in Academia

The ideal modern scientist is supposed to be a jack-of-all-trades and master of all: They have broad expertise in a wide range of methods, successfully operate and maintain the accompanying hard- and software, have extensive literature knowledge in their research area as well as in adjoining fields, find creative and innovative solutions to scientific challenges, and have – mostly self-trained – skills in graphic design, software application, and other IT-related areas.

For most scientists this ideal is nearly impossible to meet, not due to lack of motivation or aspiration, but to a significant degree based on shortcomings in the scientific system, especially in academia. In any scientific education I have encountered so far, what I found severely lacking is the transmission of skills in software application. There are very few courses offered on script language, database systems, or even on practical use of vital software. Furthermore, while there are sophisticated software solutions on the market that ease lab routines and create intuitive interfaces, very little of what is available is known to and can be afforded by most labs. With budgets tight and money often perceived to be better spent elsewhere, many labs work with either very little or outdated software. There is also a generation challenge: A significant number of professors belong to a generation which is not "digitally native" and has learned science in the very beginnings of digitalization. So, the added value which such systems could offer may often not be seen. Consequently, when it comes to digitalization, it is a challenge to keep up with the latest changes in the field, especially when previous, well-known solutions (e.g. pen and paper) still work. However, future generations of scientists will have to operate in increasingly digitalized labs and an increasingly digital environment, where less and less of these old solutions might efficiently work. Many universities and laboratories still have to come up with adequate concepts on how to teach that: How can we prepare (future) scientists for digitalization in the lab?

Digitalization of data documentation offers many opportunities, not only in industry but also in academia. In this regard, the two are not so different as often perceived: Governmental regulations and research grants alike demand and require proper data documentation, and funds can be reclaimed if documentation is unsatisfactory. In this context, ELNs establish new standards and facilitate compliance with official requirements. It also stands to reason that academia could be significantly more productive, if proper and efficient data documentation would lead to less data loss or reproducibility issues.

Unfortunately, changing traditional data documentation style is hard work. While digitization of lab notebooks might help the group long term, it is often not perceived as an individual gain. In combination with scientists often being only employed on short-term contracts, data digitalization is difficult to sell. In addition, project funding is mostly approved for short terms of only two to three years, and therefore does not easily allow financing of solutions to long-term changes in documentation. The reaction from official sides such as universities and research foundations concerning data digitalization lags behind. Guidelines are often outdated, and most universities still do not offer digital solutions for data documentation nor IT-infrastructure or -support.

How we do science has changed quickly in the past decades, and that change will most likely not slow down in the future. Digitalization of data documentation is one crucial element in it. The sooner we get on track and take part in that process, the easier the transition will be.

16

Turning Paper Habits into Digital Proficiency

Tessa Grabinski

Michigan State University – Grand Rapids Research Center, Translation Neuroscience, 400 Monroe Avenue NE, Grand Rapids, MI 40503, USA

16.1 Five Main Reasons for the Implementation of a Digital System to Manage the Research Data

I have always been one of those people who love the feel of paper in my hands. I must have the paperback in my hands and smell the paper in order to immerse myself into the story. This peculiarity has always worked for me.

The same has been true for my laboratory notebook for the past 15 years. Whether it was the carbon copy notebooks in college or the hard cover computational notebooks used in my various places of employment, my trusty, old, paper laboratory notebooks were there to provide me with vital information of the numerous experiments done over the years.

However, as my lab notebooks continued to accumulate data and as our lab grew in size, I was stuck with the dilemma of searching through paper notebooks to try and find particular data sets or switching to an electronic laboratory notebook where searching would take less than half the time. On average, I became aware that at least two to three hours a week were spent sifting through old notebooks trying to locate particular information. This took even more time when it was a notebook from a previous employee. It became clear that in order to grow as a lab, we needed to make a change in how we stored our data.

Our paper laboratory notebooks always followed the standard good laboratory practice quality system to ensure consistency, reproducibility, and integrity of experimental data. Each entry was required to include the date, title of the experiment, hypothesis or purpose, protocol, data, analysis, and finally, future directions. Within any of these sections, additional notes would be taken if certain things deviated from the normal protocol or if case numbers changed. For example, during my numerous ELISAs, the samples and development times changed regularly, so these were noted (usually in a different color ink to draw my attention to a slight change in the regular protocol). Each notebook was also supposed to include a table of contents. In my experience, this rarely happened, mostly because I did not leave enough empty pages in the front of my notebook to account for the number of experiments in the

Digital Transformation of the Laboratory: A Practical Guide to the Connected Lab, First Edition.
Edited by Klemen Zupancic, Tea Pavlek and Jana Erjavec.

rest of the book. I did notice that notebooks that did have a table of contents did not help me search for information any faster than just flipping through the entire notebook.

Reasons that helped our lab reach the decision to go digital were the following:

16.1.1 Scale-up of the Laboratory

We started to think about integrating data into an electronic version because our lab grew in size very rapidly, and with several different people working on the same project, being able to look in one place for results instead of having to go to several different people would be extremely beneficial.

We also considered how easy the transition would likely be because of how much data was currently being generated in electronic form. For example, all our plate readers, western blot scanners, gel imagers, and NanoDrop systems generated electronic data in the form of TIFFs and Excel files, so it seemed logical to convert our paper notebooks to electronic based on the fact that including the electronic data would be easier than printing out pictures or spreadsheets and taping them to paper notebooks. Our institution also charges for printing, so anything we printed to paste into our notebook cost our lab money, especially when printing in color.

16.1.2 Protocol Management Issues

When technicians were troubleshooting a new reagent or modifying a protocol, the results were not always being effectively communicated to the rest of the lab, so mistakes were being made. We also had problems with protocols being modified and then never being fully incorporated into every lab members' repertoire, which led to some variability in our results. This was extremely problematic, especially when different technicians would do the same experiment for other personnel, only to find out that they did it completely different from each other because one of them was using an outdated protocol. There was no way to compare results from two different test runs because the protocols used were different. This became more of an issue as our lab received more funding and more personnel were hired.

16.1.3 Environmental and Financial Factors

One of the biggest factors for me, personally, for trying to convince the lab to make the switch from paper to electronic was the environmental consideration. Scientific research is a messy business, and we generate a lot of waste, most of which is unfortunately, not recyclable due to contamination with potentially hazardous materials. I wanted to be proud of trying to minimize our waste, and when the ability to use digital instead of paper was available, why not?

We also had to consider physical space. We already had two full bookshelves of old notebooks. Considering the constant work and the increase in personnel that inevitably happens, we were using a lot of notebooks. I started to take note of how much money we were spending on paper laboratory notebooks. We had eight lab members plus some rotating students who, on average, would use four notebooks a year. Accounting for the personal requests of hardcover notebooks versus cheaper softcover notebooks and factoring in the cost of colored printing so you could put data pictures in your notebook, we were spending approximately $1600 in paper laboratory notebooks a year. Some people might find that amount of money trivial, but when you consider what you could do with that money at the lab bench, it seems significant to me.

16.1.4 Introducing the Benefits of Technology to Younger Employees

The biggest pushback I had when I initially presented this idea was the mentality that "this is the way we've always done it." That comment is just plain silly when you think about it. If that was the case, we would still be mouth pipetting and smoking in the lab. Also, being one of the senior employees in the lab, I was not willing to accept the fact that the younger generation would not be able to grasp the concept or think the electronic notebook would be too difficult to learn and use.

16.1.5 Remote Access to Data by Authorized Supervisors

There were also a few other labs in our research unit that were using some sort of electronic notebook system. After conversing with them, my boss felt it would be beneficial for us to start using electronic notebooks as well. One of the biggest benefits for him was that he would be able to access our data and results anytime he was traveling. Previously, lab members would have to scan their paper notebook pages and send the PDF files to him or send TIFFs of pictures and describe what we did in length in our e-mail correspondence if he was not in the lab. This was tedious and time consuming, which is why converting to an online system where everything was easily accessible was an alluring concept and one that helped cement his decision in making the switch from paper to an electronic laboratory notebook.

16.2 The Six-step Process of Going from Paper to Digital

Once we came to terms with making the switch to an electronic lab notebook, we started searching online for options that would best suit our needs. My boss found several different variations of electronic lab notebooks and decided to present them to our lab during a full staff meeting. The entire lab liked the idea of moving to a digital format, but we needed to discuss the logistics of how we were going to convert from paper to digital and identify what features of the electronic notebook were necessary for us before proceeding.

16.2.1 Defining the Specific Needs of the Laboratory

When trying to decide which software to use, we had to think about what our laboratory's specific needs were. Some of the criteria that had to be met were that the software had to allow for multiple users at the same time, an unlimited number of users to account for anticipated growth of personnel, and the ability to be used on a laptop or smartphone. Our lab has several employees, and we could not be limited by only being able to have one person using the electronic laboratory notebook at one time. Therefore, we needed a multiuser interface that was still private to our users but was easily accessible to everyone in the lab at the same time. We also needed a program that would not limit us by the number of total users. Most laboratories are subject to expansion and contraction of personnel based on funding availability, so we needed to anticipate these sorts of changes in our staff and not be restricted to a set number of users. The biggest deciding factor for which electronic laboratory notebook to use came down to us was a system that could be used on tablet PCs or smartphones. Not only did the interface have to work on these platforms, but it needed to be user friendly and have a nice mobile site.

When we decided to switch to the electronic laboratory notebook, we knew we had to have something portable that users could take with them throughout our laboratory space and easily input protocol notes and data. Our lab decided to treat the electronic notebook exactly like our paper notebooks, meaning it would be located on our laboratory bench beside us while we worked and carried to various procedure rooms when work was being done elsewhere. We elected to get tablet PCs for all laboratory members and immediately establish guidelines, so the tablets were only to be used for the electronic notebook and therefore were subject to our institution's laboratory notebook regulations (such as not leaving the premises). Given all the criteria we needed the electronic lab notebook to meet, we ultimately chose to move forward with SciNote because we felt the software best met all of our specific needs.

16.2.2 Testing the Software and Defining the Standard Way to Use It

The logistics of converting from paper to digital was fairly easy. Initially, we created one account for all members of the lab to try for two weeks. At the time, our lab consisted of one postdoc, two graduate students, two technicians, one summer student, one manager, and one principal investigator. During this two-week trial period, we each tried the electronic notebook in a way that we saw fit to our personal needs. Everyone had their own idea about how we should set up the projects and experiments.

At our next staff meeting we each presented how we had been using SciNote for our data by simply logging in to the webpage and showing our initial trial period data and setup. During this time, my boss took notes and commented on what he liked, what he did not like, and what he needed to see in order to be an accurate duplicate of our paper notebooks. At our next staff meeting, he presented how he wanted us using the electronic lab notebook and came up with the standard way everyone in the lab would be required to use it. As far as how each of us were required to keep track of experiments, my boss was lenient and did not mind how people used the notebook to input their experiments as long as the nomenclature matched the file format we previously used and that the data was accurately labeled. This has never been a problem when looking up experiments or protocols because of the search feature, which will pull up all instances of whatever you are searching for, regardless of whether it is an experiment, task, step, or result. That being said, my boss did have one very specific rule of being up-to-date every day just like a regular paper notebook should be. At any time, he wanted to be able to look at our electronic notebook and expect to see all our protocols and data present.

16.2.3 Organizing the Collaboration Between Lab Members and Supervisors

Our approach to using the electronic lab notebook consists of individuals making projects specific to what they are working on. We then add technicians or other personnel as users to those projects so they can add their protocols and data as they become available. Our lab functions as a collaborative unit where our technicians provide support to both postdocs and graduate students, while maintaining their own personal projects. Therefore, we always have numerous people working on the same project, so it was imperative to have a system where several different users could add their work to that one particular project. This is where the electronic laboratory notebook has had the biggest impact on our work. We no longer have technicians searching through several paper notebooks to find one project among many. We simply look up the specific project of interest in the electronic laboratory notebook to find all information regarding that particular project. Also, with several users assigned to one project, we have a lot of data and results being generated daily so being able to either verbally tell the main user or to simply mark a protocol step as being completed alerts them that there is new data for them to look at when they have a chance.

Previously, the technician would be responsible for notifying the main user that an experiment was done so they could show them their results from their paper notebook. This was sometimes difficult considering classes and meetings requiring attendance of the postdocs and graduate students. Now, the main project owner can simply access their project results from anywhere an Internet connection is available. Not only does this save time for everyone involved, it also allows the boss and project owner time to critically analyze the results when they best see fit.

We do still have projects where only one person is the sole user, so the ability to manage who can access and edit the project is a vital part of our SciNote usage. Part of our initial decision when selecting how to set up projects and users in the electronic notebook was to ensure our boss was always included as a "user" for all projects. This ensures his ability to edit and insert comments into any of our projects as he sees fit. One example of this situation is my "Mouse Colonies" project. I am the only member of our team who manages our animal colonies; therefore, I am the sole "owner," and my boss is listed as the only "user" of that project. This prevents any unwanted or accidental editing of projects that are meant for certain users only. Deciding to include or exclude certain users is something that should be discussed as a team when establishing your electronic notebook organization.

16.2.4 Managing Projects and Setting Up Work Processes

As far as setting up a systematic flow, I found it easier to create one project and then have several different experiments pertaining to that project. Under each experiment, I have one protocol and then several different tasks and steps. For example, under my profile you will find a project "Antibody Production" (Figure 16.1.) Within that project, you would find several different experiments, each named for the particular antibody being made. Under one antibody, I will have several different tasks named such as "Purification, Western Blots, and ELISA Screenings." Under each of these tasks, I have every purification, western blot, or ELISA screening I have ever done for that antibody as a new step. The only difference between the various steps is the date and results with an occasional note about varying incubation times or other differences that might occur. This format makes sense to me and seems most logical as I routinely retest our antibodies and did not want to create new tasks of the same antibody.

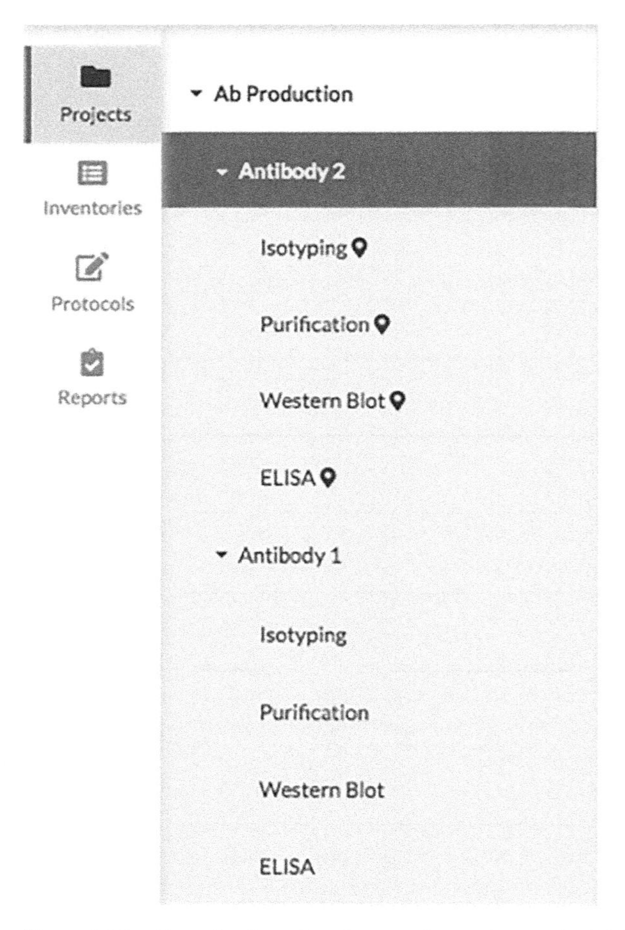

Figure 16.1 Organization of projects and experiments in SciNote. Source: Figure created by Tessa Grabinski.

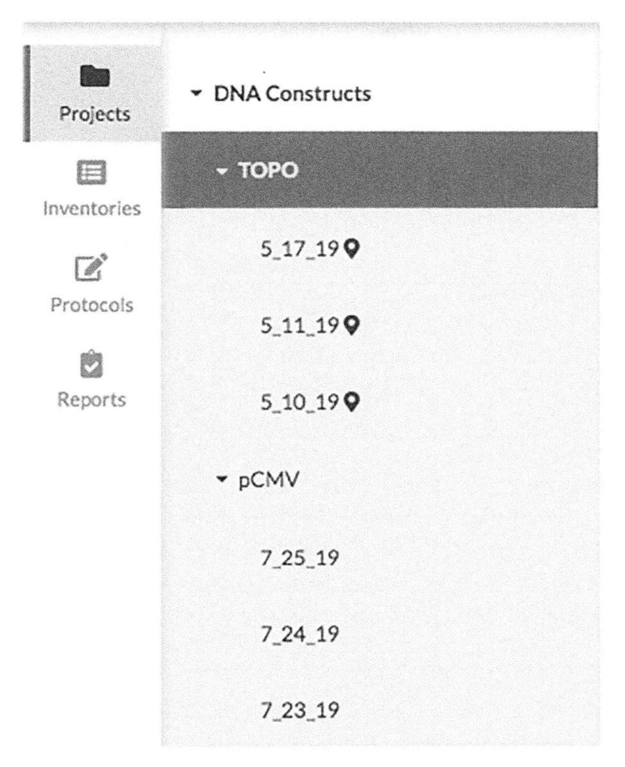

Figure 16.2 Example of a different approach to organization of projects and experiments in SciNote. Source: Figure created by Tessa Grabinski.

However, this is not how other users input their experiments. One of our technicians prefers to create one project and then have new tasks named only by the date of whatever protocol or technique they were doing for that day (Figure 16.2.). Within that daily task, you would find all kinds of experimental procedures listed as different steps, even if they are completely different assays. For example, under one date, you might find a western blot, DNA isolation, and gel extraction, all under the same task. This format works for her, and as I mentioned previously it does not impede the ability to find specific information because the search feature within SciNote is all inclusive.

Our users also vary in how they prefer to input images. I like to take screenshots of my gels, westerns, and ELISA data after I have modified them in Adobe Photoshop by labeling bands, samples, etc. I then take a screenshot and upload the .png file into the results section of the electronic notebook. Other users will upload the .jpeg image after they modify it in Photoshop. Again, it does not matter how the data gets there, only that it is accurately labeled and uploaded on time.

Users also use the integrated Microsoft features differently. Several people will still use Excel and Word on their desktop and then copy and paste into a table in their steps or results section of the electronic notebook. Personally, I have run into a problem with the copy and paste method when I had to repeat another technician's transfection experiment. I needed to see and modify the Excel formulas in their spreadsheet, but was unable to do so because they had simply copied and pasted the numbers into a table in their step section. Not only did this cause extra time on my part to try and reproduce their math, it made me question whether or not I was actually doing the experiment the exact same way they did.

I am a huge advocate of using the "New Office file" feature of SciNote. For example, I have a project titled "Mouse Colonies," where I keep track of all my animals, including matings, litters, and genotype. By creating an Office file, I simply click on the file and select "Edit in Excel Online" to modify my data. This has been extremely helpful for me because animal numbers are constantly changing, and the integration with Microsoft Office allows me to make changes daily (if necessary) and simply download and save the file at the end of the month when my census is due. My boss has also told me how useful the Office File integration has made it for him, especially with taking a look at our animal numbers. He used to ask me verbally throughout the month how we were looking as far as having the correct number of animals for an experiment, and I could only give him a rough estimate off the top of my head, but now, he can simply look at the Office File that shows the most accurate numbers. He can also reassign an animal for a particular project by modifying my file. Giving my boss the ability to access this data at any time has been a huge time-saver for me. It does not take long to pull up a file, but I usually have to stop what I am doing and change gears, so I can get him the information he is requesting, and that takes time and also distracts me from my previous engagement. You can see an example of my integrated Office File for one set of my animals pictured (Figure 16.3).

Using the Office File also allows you to see any formulas in Excel sheets which would eliminate any questions regarding how a number was attained and make it easier for other users to make changes.

One key feature that I appreciate and will mention about the Office file integration is that anytime you type either in the Word or Excel file, it is automatically saved, so there is no more accidental loss of data because you forgot to hit the save button.

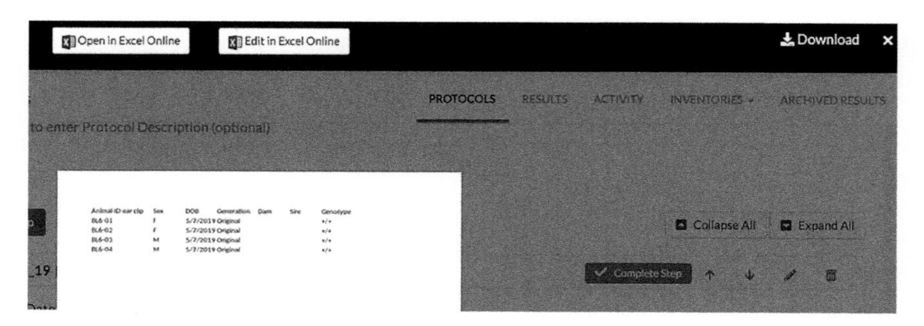

Figure 16.3 Example of using Microsoft Office online files in SciNote. Source: Figure created by Tessa Grabinski.

16.2.5 Versioning of Protocols and Keeping the Protocol Repository Up to Date

One of the key features of SciNote that our lab uses are the Team Protocols. These get used often and are the most up-to-date protocol in the lab that everyone should be using. When users incorporate these protocols into their steps, they simply edit them if changes were made during their experiment. For example, I use the Indirect ELISA protocol several times weekly, but the proteins, dilutions, and development times change every time, so I simply edit the protocol and highlight my changes using boldface type and sometimes changing the font color. This keeps the original protocol intact in our Team Protocols but allows me to make the necessary changes that arise when actually doing the experiment (Figures 16.4 and 16.5).

When you create a new step in a project, you can select the "Load Protocol" button at the top of the screen, where it will pull up your entire protocol repository. From here, you simply click the protocol you wish to load. I can make changes to the uploaded protocol (such as different sample names, volumes used, etc) in my project without compromising the original protocol in the repository. You can select the "Load Protocol" button at the top of the screen, And SciNote pull up your entire protocol repository. From here, you simply select the protocol you wish to load.

I will be honest and say that our lab does not utilize all the features SciNote has to offer. One of these items is the "Inventory" section. We initially talked about inputting all of our DNA constructs and tissue samples into the Inventory, but decided it was too great of a task to devote the amount of time it was going to take. However, the example below could make use of inventory management functionalities.

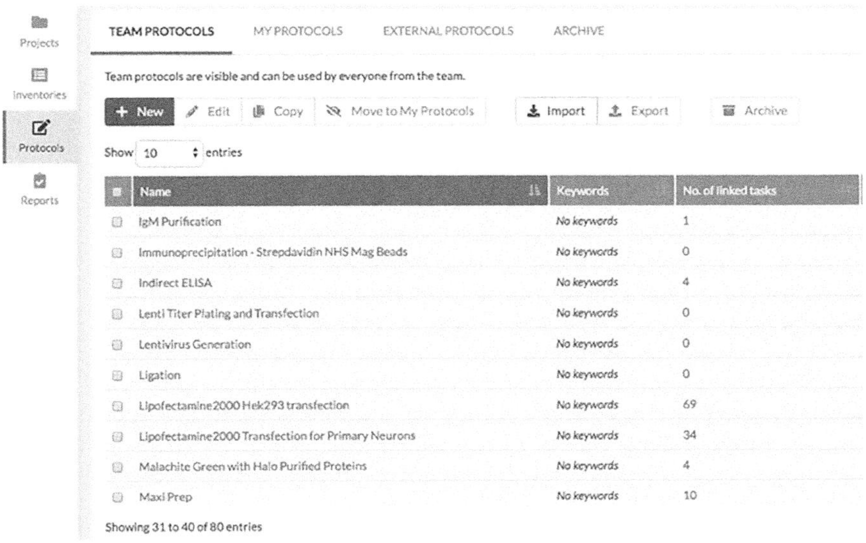

Name	Keywords	No. of linked tasks
IgM Purification	No keywords	1
Immunoprecipitation - Strepdavidin NHS Mag Beads	No keywords	0
Indirect ELISA	No keywords	4
Lenti Titer Plating and Transfection	No keywords	0
Lentivirus Generation	No keywords	0
Ligation	No keywords	0
Lipofectamine2000 Hek293 transfection	No keywords	69
Lipofectamine2000 Transfection for Primary Neurons	No keywords	34
Malachite Green with Halo Purified Proteins	No keywords	4
Maxi Prep	No keywords	10

Figure 16.4 Team protocols repository in SciNote. Source: Figure created by Tessa Grabinski.

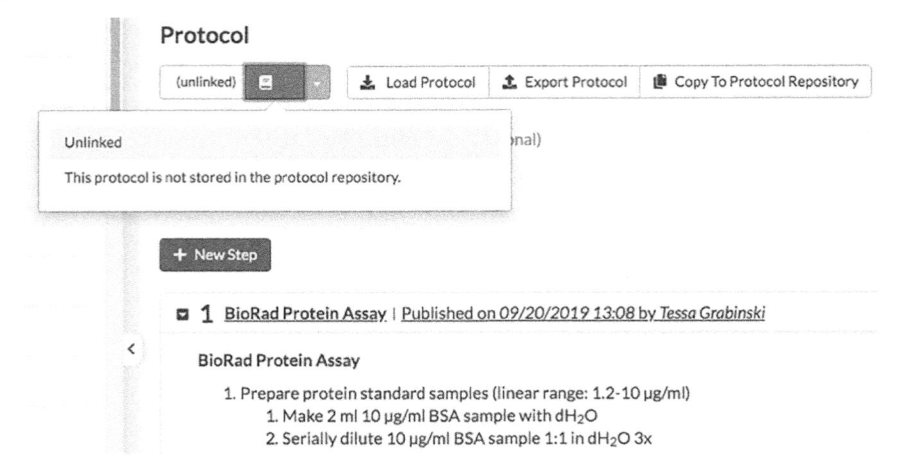

Figure 16.5 Choosing to link (or not) the protocol with the version saved in the repository. Source: Figure created by Tessa Grabinski.

16.2.6 Choosing to Digitize Only New Projects

Initially, our lab members went into great discussion about whether or not we should retroactively digitize data. Some of us had ongoing projects that would be input going forward but consisted of years of previous work still on paper. Ultimately, we chose to not input any old data from our paper notebooks, simply because it would be too time consuming and we needed to focus on our current and future research.

16.3 Onboarding All Team Members and Enhancing the Adoption of the New Technology in the Lab

Discussing and creating an acceptable model to all parties involved early on for how our lab would be using the notebook helped us avoid any discrepancies between our lab members' notebooks. This approach also allowed us to maintain a similar nomenclature that corresponded to our previous data stored on our servers as well as our paper laboratory notebooks. As a manager, it is my job to ensure that all users are familiar with the electronic laboratory notebook and are using it in the manner our lab has established. To introduce new users to SciNote, I simply have them sit down at a computer and watch the video tutorials on the SciNote Support webpage. Once they complete the tutorials, we sit down together while I personally show them how our lab uses the program. Since our users structure their profiles differently, I show them a few different users' experiments and work with them to decide which makes the most sense to them.

The only complaints I hear about the electronic laboratory notebook tend to come from one person, and it is usually right after an update is pushed through. People do not like change, and certain personalities find it more frustrating than others and are usually more hesitant to embrace new ideas. That being said, I have never had a problem with updates to the software. Communication is always a top priority for SciNote, and they make sure to include e-mails and links to what features were updated and how to utilize them.

When it comes to making sure all users are inputting their experimental data into the electronic notebook, the responsibility lies solely on the end users. As the laboratory manager, I can take note of data that is not being input and ask people to update their experiments, but my boss is ultimately the one who makes sure people are using the electronic notebook properly and according to his lab standards.

One of our biggest challenges early on was determining how to keep track of the simple day-to-day tasks that technicians do routinely in the lab. For example, how should we track different stock reagents that were being made for the entire lab to use? This became a topic of discussion between us because some of us thought it was necessary to keep track of these seemingly simple tasks, where others did not think it was important. We ultimately decided it was necessary to include these things so if everyone started having problems with a 10× stock solution, we could track down which technician made it and how it was made. We chose to include a single project titled "Lab Reagents" where we could make individual experiments for all common shared reagents. For example, in our "Lab Reagents" project, you will find several different experiments with titles from "Protease Inhibitors" to "10× Tris Buffered Saline," as shown in the screenshot (Figure 16.6).

Anytime one of our technicians makes these stock solutions, they add a new step to the protocol that shows exactly what they did, when they did it, who made it, and any additional notes about issues that came up during the protocol.

For example, you can see in the screenshot (Figure 16.7) that ELISA wash buffer was made according to our standard recipe, which is for a 2-l amount. However, 4 l

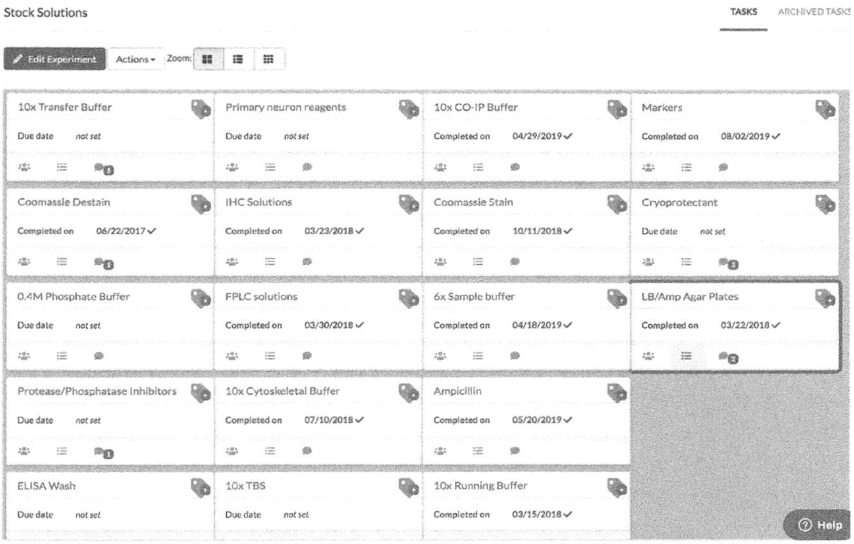

Figure 16.6 Tracking the day-to-day tasks and their completion progress. Source: Figure created by Tessa Grabinski.

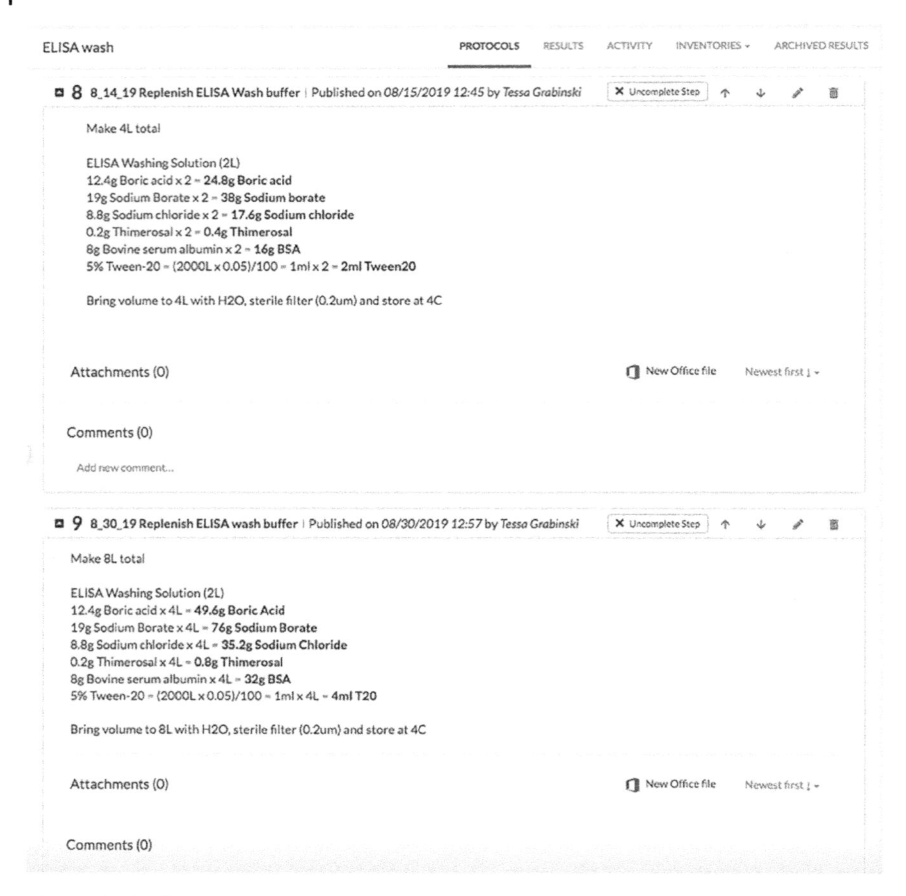

Figure 16.7 Records of lab technicians' work progress in SciNote. Source: Figure created by Tessa Grabinski.

was made the first time and 8 l the second time, so the protocol was modified in **BOLD** type to highlight the math changes that account for those differences.

Not only does this method help us track down who made something last, but it also helps account for a technician's time and productivity.

Our laboratory is also structured where we have a full staff meeting every Monday morning and are required to give a short PowerPoint presentation of everything we did the previous week and discuss what we plan to do in the coming week. Since the presentation is required to include all data, we already have pictures and graphs made, so using those same pictures and data in the results section of SciNote is tremendously easy. I simply copy and paste my results from my presentation directly into the SciNote Results section as a screenshot or vice versa from SciNote to my PowerPoint. I do this daily at the same time I am updating my electronic notebook, so my PowerPoint presentation is being put together throughout the week. That way, there is not any scrambling on my part on Friday afternoon to make sure my presentation matches what my electronic notebook shows I accomplished for the week. Here, you

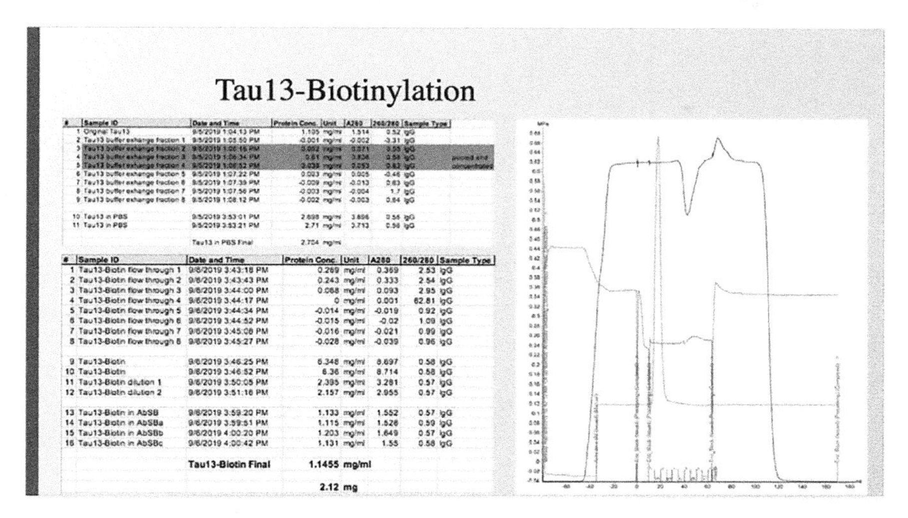

Tau13-Biotinylation

#	Sample ID	Date and Time	Protein Conc.	Unit	A280	260/280	Sample Type	
1	Original Tau13	9/5/2019 1:04:13 PM	1.105	mg/ml	1.314	0.52	IgG	
2	Tau13 buffer exchange fraction 1	9/5/2019 1:05:50 PM	-0.001	mg/ml	-0.002	-3.31	IgG	
3	Tau13 buffer exchange fraction 2	9/5/2019 1:06:16 PM	0.052	mg/ml	0.071	[illegible]	IgG	
4	Tau13 buffer exchange fraction 3	9/5/2019 1:06:34 PM	0.61	mg/ml	[illegible]	[illegible]	IgG	pooled and
5	Tau13 buffer exchange fraction 4	9/5/2019 1:06:52 PM	[illegible]	mg/ml	0.053	[illegible]	IgG	concentrated
6	Tau13 buffer exchange fraction 5	9/5/2019 1:07:22 PM	0.003	mg/ml	0.005	-0.46	IgG	
7	Tau13 buffer exchange fraction 6	9/5/2019 1:07:39 PM	-0.009	mg/ml	-0.013	0.83	IgG	
8	Tau13 buffer exchange fraction 7	9/5/2019 1:07:56 PM	-0.003	mg/ml	-0.004	1.7	IgG	
9	Tau13 buffer exchange fraction 8	9/5/2019 1:08:12 PM	-0.002	mg/ml	-0.003	0.84	IgG	
10	Tau13 in PBS	9/5/2019 3:53:01 PM	2.698	mg/ml	3.696	0.56	IgG	
11	Tau13 in PBS	9/5/2019 3:53:21 PM	2.71	mg/ml	3.713	0.56	IgG	
		Tau13 in PBS Final	2.704	mg/ml				

#	Sample ID	Date and Time	Protein Conc.	Unit	A280	260/280	Sample Type
1	Tau13-Biotin flow through 1	9/6/2019 3:43:18 PM	0.269	mg/ml	0.369	2.53	IgG
2	Tau13-Biotin flow through 2	9/6/2019 3:43:43 PM	0.243	mg/ml	0.333	2.54	IgG
3	Tau13-Biotin flow through 3	9/6/2019 3:44:00 PM	0.068	mg/ml	0.093	2.95	IgG
4	Tau13-Biotin flow through 4	9/6/2019 3:44:17 PM	0	mg/ml	0.001	62.81	IgG
5	Tau13-Biotin flow through 5	9/6/2019 3:44:34 PM	-0.014	mg/ml	-0.019	0.92	IgG
6	Tau13-Biotin flow through 6	9/6/2019 3:44:52 PM	-0.015	mg/ml	-0.02	1.09	IgG
7	Tau13-Biotin flow through 7	9/6/2019 3:45:08 PM	-0.016	mg/ml	-0.021	0.99	IgG
8	Tau13-Biotin flow through 8	9/6/2019 3:45:27 PM	-0.028	mg/ml	-0.039	0.96	IgG
9	Tau13-Biotin	9/6/2019 3:46:25 PM	6.348	mg/ml	8.697	0.58	IgG
10	Tau13-Biotin	9/6/2019 3:46:52 PM	6.36	mg/ml	8.714	0.58	IgG
11	Tau13-Biotin dilution 1	9/6/2019 3:50:05 PM	2.395	mg/ml	3.281	0.57	IgG
12	Tau13-Biotin dilution 2	9/6/2019 3:51:16 PM	2.157	mg/ml	2.955	0.57	IgG
13	Tau13-Biotin in AbSB	9/6/2019 3:59:20 PM	1.133	mg/ml	1.552	0.57	IgG
14	Tau13-Biotin in AbSBa	9/6/2019 3:59:51 PM	1.115	mg/ml	1.526	0.59	IgG
15	Tau13-Biotin in AbSBb	9/6/2019 4:00:20 PM	1.203	mg/ml	1.649	0.57	IgG
16	Tau13-Biotin in AbSBc	9/6/2019 4:00:42 PM	1.131	mg/ml	1.55	0.58	IgG
		Tau13-Biotin Final	**1.1455**	**mg/ml**			
			2.12	**mg**			

Figure 16.8 Organizing results in SciNote. Source: Figure created by Tessa Grabinski.

Figure 16.9 Organizing results in SciNote, under the dedicated Results section. Source: Figure created by Tessa Grabinski.

can see an example of one of my PowerPoint slides alongside the same images in my SciNote Results section (Figures 16.8 and 16.9).

16.4 Benefits of Switching from Paper to Digital

When thinking about the benefits of the electronic laboratory notebook over the paper notebooks, the element that has had the biggest impact for me personally is being able to search for a simple phrase or keyword. The ability to search every part of the electronic notebook for a word or phrase is vital when my boss comes in and asks me if I remember a particular stain I did a few years ago. I can easily type in a search phrase and pull up every instance of what the boss is asking for. This cuts my search time down significantly by eliminating any sifting through paper notebooks, using only my memory and approximate dates of when we think I may have done that particular experiment.

Another benefit to using the electronic laboratory notebook is that we are no longer using paper. Considering how much nonrecyclable waste we already generate in the lab, the fact that we are no longer using paper has a significant impact on the environment. Given the current data on climate change, I feel it is our duty to do as much as we can to help save the planet. Being able to see how many tree branches our lab has saved in the digital forest using SciNote makes that process even more enjoyable. Anytime my benchmate and I are filing out our electronic notebook and we get a popup saying we have saved another tree branch; we announce it to each other with pride. We enjoy knowing we are helping the earth, and it adds a bit of friendly competition between members of the lab.

Every member of our lab is required to use the electronic laboratory notebook as part of their job and is subject to disciplinary action for not complying. Using the electronic notebook has become part of our daily tasks, and it helps us maintain accurate, concise protocols and results from every member of our team. As our lab continues to grow and change, I look forward to implementing even more features that the SciNote has to offer. Personally, I believe our lab has benefited immensely by making the switch to electronic, and I encourage those of you who are still thinking about it to take the plunge!

17

Going from Paper to Digital: Stepwise Approach by the National Institute of Chemistry (Contract Research)

Samo Andrensek and Simona L. Hartl

Center for Validation Technologies and Analytics, National Institute of Chemistry, Ljubljana, Slovenia

17.1 Presentation of our CVTA Laboratory

CVTA or Center for Validation Technologies and Analytics is a department that performs qualitative and quantitative analysis of different samples, optimizes and develops analytical procedures (APs), and validates APs for different pharmaceutical companies [1].

Analyses in CVTA are performed within requirements of good manufacturing practice (GMP) and good development practice (GDevP) in accordance with requirements of ISO/IEC 17025:2017 standard and also in agreement with ISO 9001.

CVTA is registered at FDA (Food and Drug Administration) as GMP analytical testing site. CVTA is GMP certificated by JAZMP (Public Agency of Republic of Slovenia for Medicines and Medical Devices) and SA (Slovenian Accreditation) and a holder of national etalon (NNE) for mole (amount of substance).

Different regulatory authorities or medicine agencies have very similar requirements for data integrity assurance with appropriate management of documentation, change control management, complaints and product recall, quality risk management, self-inspection system, personnel organization and education, quality control (QC), outsourcing activities, handling samples and reference standards, premise and equipment, computerized systems, and others.

17.2 Data Management Requirements Explained in Detail

Documentation may exist in a variety of forms, including paper based, electronic, or photographic media, as described in *EudraLex, The Rules Governing Medicinal Products in the European Union, Volume 4, Good Manufacturing Practice, Medicinal Products for Human and Veterinary Use, Chapter 4: Documentation* [2].

While avoiding unacceptable risk to product quality, patient safety, and public health, different GMP-regulated authorities, such as MHRA (Medicines and

Digital Transformation of the Laboratory: A Practical Guide to the Connected Lab, First Edition.
Edited by Klemen Zupancic, Tea Pavlek and Jana Erjavec.

Healthcare Products Agency) or PIC'S (pharmaceutical inspection convention), have accepted that all data and metadata should be complete, consistent, and accurate, throughout the data life cycle. In other words, data should be attributable, legible, contemporaneous, original, and accurate (ALCOA).

17.2.1 Meaning of ALCOA

Attributable: It should be possible to identify the individual who performed the recorded task. The need to document who performed the task/function is in part to demonstrate that the function was performed by trained and qualified personnel. This applies to changes made to records as well: corrections, deletions, changes, etc.

Legible: All records must be legible – the information must be readable in order for it to be of any use. This applies to all information that would be required to be considered complete, including all original records or entries. Where the "dynamic" nature of electronic data (the ability to search, query, trend, etc.) is important to the content and meaning of the record, the ability to interact with the data using a suitable application is important for the availability of the record.

Contemporaneous: The evidence of actions, events, or decisions should be recorded as they take place. This documentation should serve as an accurate attestation of what was done, or what was decided and why, i.e. what influenced the decision at that time.

Original: The original record can be described as the first capture of information, whether recorded on paper (static) or electronically (usually dynamic, depending on the complexity of the system). Information that is originally captured in a dynamic state should remain available in that state.

Accurate: Ensuring results and records are accurate is achieved through many elements of a robust pharmaceutical quality management system.

This can comprise:

a. equipment-related factors such as qualification, calibration, maintenance, and computer validation;
b. policies and procedures to control actions and behaviors, including data review procedures to verify adherence to procedural requirements;
c. deviation management including root cause analysis, impact assessments, and Corrective and Preventive Actions (CAPA);
d. trained and qualified personnel who understand the importance of following established procedures and documenting their actions and decisions.

Together, these elements aim to ensure the accuracy of information, including scientific data, that is used to make critical decisions about the quality of products.

Complete: All information that would be critical to recreating an event is important when trying to understand the event. The level of details required for an information set to be considered complete would depend on the criticality of the information. A complete record of data generated electronically includes relevant metadata.

Consistent: Good documentation practices should be applied throughout any process, without exception, including deviations that may occur during the process. This includes capturing all changes made to data.

Enduring: Part of ensuring that records are available is making sure they exist for the entire period during which they might be needed. This means they need to remain intact and accessible as an indelible/durable record.

Available: Records must be available for review at any time during the required retention period and accessible in a readable format to all applicable personnel who are responsible for their review whether for routine release decisions, investigations, trending, annual reports, audits, or inspections.

(This resource is recommended on the subject: PIC/S guidance, GMPs for data management and integrity, August 2016) [3].

17.2.2 FDA and CFR 21 Part 11

For the purpose of public health protection, FDA in the United States has published Guidance for Industry, Part 11 of Title 21 of the Code of Federal Regulations; Electronic Records; Electronic Signatures – Scope and Application (21 CFR Part 11) [4].

The guide was issued in August 2003 (in 1997 for the first time). It describes requirements for electronic records and electronic signatures on computerized systems where industrial production and analyses (food or drug samples) under GMP regulations are performed.

Analytical software where GMP analyses of pharmaceutical samples or validations of APs are performed must be FDA 21 CFR Part 11 consistent and must be validated to ensure data accuracy, reliability, and consistency. Data must be stored in a secure way, i.e. backed up and archived, and these processes must be periodically checked. Change, deletion, or moving original data should be avoided, but if so it must be done in a controlled way and with the noted reason. Employees must have appropriate education and training before using analytical software.

In our lab, we chose to use the electronic laboratory notebook (ELN) SciNote which is FDA 21 CFR Part 11 consistent software, suitable for use in GMP-regulated environments (as well as in research laboratories with no regulation).

17.2.3 MHRA and GxP Data Integrity Guidance and Definitions

In addition, the document MHRA and GxP Data Integrity Guidance and Definitions; Revision 1: March 2018 provides guidance for UK industry and public bodies regulated by the UK MHRA including the Good Laboratory Practice Monitoring Authority (GLPMA) [5].

This guidance has been developed by the MHRA inspectorate and partners and has undergone public consultation; where possible the guidance has been harmonized with other published guidance and is a UK companion document to PIC/S, WHO, OECD (guidance and advisory documents on good laboratory practice [GLP]), and EMA guidelines and regulations.

Among others this guide indicates the following:

This guidance primarily addresses data integrity and not data quality.

The organization needs to take responsibility for the systems used and the data they generate. The organizational culture should ensure that data is complete, consistent, and accurate in all its forms, i.e. paper and electronic.

Arrangements within an organization with respect to people, systems, and facilities should be designed, operated, and, where appropriate, adapted to support a suitable working environment, i.e. creating the right environment to enable data integrity controls to be effective.

Organizations should be aware that reverting from automated or computerized systems to paper-based manual systems or vice versa will not in itself remove the need for appropriate data integrity controls.

Where data integrity weaknesses are identified, companies should ensure that appropriate corrective and preventive actions are implemented across all relevant activities and systems and not in isolation.

The guidance refers to the acronym ALCOA rather than "ALCOA+,". ALCOA being attributable, legible, contemporaneous, original, and accurate and the '+' referring to complete, consistent, enduring, and available.

Data has varying importance to quality, safety, and efficacy decisions. Data criticality may be determined by considering how the data is used to influence the decisions made.

The risks to data are determined by the potential to be deleted, amended, or excluded without authorization and the opportunity for detection of those activities and events.

Data may be generated by:

- Recording on paper, a paper-based record of a manual observation or of an activity; or electronically, using equipment that range from simple machines through to complex highly configurable computerized systems; or using a hybrid system where both paper-based and electronic records constitute the original record; or by other means such as photography, imagery, and chromatography plates.
- The use of available technology, suitably configured to reduce data integrity risk, should be considered. Complex systems require "validation for intended purpose." Validation effort increases with complexity and risk (determined by software functionality, configuration, the opportunity for user intervention, and data life-cycle considerations).
- Where the data generated is captured by a photograph or imagery (or other media), the requirements for storage of that format throughout its life cycle should follow the same considerations as for the other formats, considering any additional controls required for that format.

Automation or the use of a "validated system" (e.g. e-CRF; analytical equipment) may lower but not eliminate data integrity risk. Where there is human intervention, particularly influencing how or what data is recorded, reported, or retained, an increased risk may exist from poor organizational controls or data verification due to an overreliance on the system's validated state.

Systems and processes should be designed in a way that facilitates compliance with the principles of data integrity.

Enablers of the desired behavior include but are not limited to:

- At the point of use, having access to appropriately controlled/synchronized clocks for recording timed events to ensure reconstruction and traceability, knowing and specifying the time zone where this data is used across multiple sites.

- Accessibility of records at locations where activities take place so that informal data recording and later transcription to official records does not occur.
- Access to blank paper proformas for raw/source data recording should be appropriately controlled. Reconciliation, or the use of controlled books with numbered pages, may be necessary to prevent recreation of a record. There may be exceptions such as medical records (GCP) where this is not practical.
- User access rights that prevent (or audit trail, if prevention is not possible) unauthorized data amendments. Use of external devices or system interfacing methods that eliminate manual data entries and human interaction with the computerized system, such as barcode scanners, ID card readers, or printers.
- Inclusion of subject matter experts in the risk assessment process.

The use of scribes to record activity on behalf of another operator: the recording by the second person should be contemporaneous with the task being performed, and the records should identify both the person performing the task and the person completing the record. The process for supervisory (scribe) documentation completion should be described in an approved procedure that specifies the activities to which the process applies.

17.2.4 Definition of Terms and Interpretation of Requirements

Data – facts, figures, and statistics collected together for reference or analysis. All original records and true copies of original records, including source data and metadata and all subsequent transformations and reports of these data, that are generated or recorded at the time of the GxP activity and allow full and complete reconstruction and evaluation of the GxP activity.

Raw data (synonymous with "source data" which is defined in ICH GCP (ICH harmonised guideline for Good Clinical Pratice)) – the original record (data) which can be described as the first capture of information, whether recorded on paper or electronically. Information that is originally captured in a dynamic state should remain available in that state.

Raw data must permit full reconstruction of the activities. Where this has been captured in a dynamic state and generated electronically, paper copies cannot be considered as "raw data."

In the case of basic electronic equipment that do not store electronic data, or provide only a printed data output (e.g. balances or pH meters), the printout constitutes the raw data. Where the basic electronic equipment do store electronic data permanently and only hold a certain volume before overwriting, this data should be periodically reviewed and where necessary reconciled against paper records and extracted as electronic data where this is supported by the equipment itself.

Metadata – data that describe the attributes of other data and provide context and meaning. Typically, these are data that describe the structure, data elements, interrelationships, and other characteristics of data, e.g. audit trails. Metadata also permits data to be attributable to an individual (or if automatically generated, to the original data source).

Data integrity – the degree to which data are complete, consistent, accurate, trustworthy, and reliable, and that these characteristics of the data are maintained throughout the data life cycle. The data should be collected and maintained in a secure manner, so that they are attributable, legible, contemporaneously recorded, original (or a true copy), and accurate. Assuring data integrity requires appropriate quality and risk management systems, including adherence to sound scientific principles and good documentation practices.

Data life cycle – all phases in the life of the data from generation and recording through processing (including analysis, transformation, or migration), use, data retention, archive/retrieval, and destruction. Data governance, as described in the previous section, must be applied across the whole data life cycle to provide assurance of data integrity. Data can be retained either in the original system, subject to suitable controls, or in an appropriate archive.

Recording and collection of data – organizations should have an appropriate level of process understanding and technical knowledge of systems used for data collection and recording, including their capabilities, limitations, and vulnerabilities. When used, blank forms (including, but not limited to, worksheets, laboratory notebooks, and master production and control records) should be controlled. For example, numbered sets of blank forms may be issued and reconciled upon completion. Similarly, bound paginated notebooks, stamped or formally issued by a document control group, allow detection of unofficial notebooks and any gaps in notebook pages.

Data transfer/migration – the process of transferring data between different data storage types, formats, or computerized systems.

Data migration is the process of moving stored data from one durable storage location to another. This may include changing the format of data but not the content or meaning.

Data transfer should be validated. The data should not be altered during or after it is transferred to the worksheet or other application. There should be an audit trail for this process.

Data processing – a sequence of operations performed on data to extract, present, or obtain information in a defined format. There should be adequate traceability of any user-defined parameters used within data-processing activities to the raw data, including attribution to who performed the activity. Audit trails and retained records should allow reconstruction of all data-processing activities regardless of whether the output of that processing is subsequently reported or otherwise used for regulatory or business purposes. If data processing has been repeated with progressive modification of processing parameters this should be visible to ensure that the processing parameters are not being manipulated to achieve a more desirable result.

Original record – the source capture of data or information, e.g. original paper record of manual observation or electronic raw data file from a computerized system and all subsequent data required to fully reconstruct the conduct of the GxP activity. Original records can be static or dynamic.

A static record format, such as a paper or electronic record, is one that is fixed and allows little or no interaction between the user and the record content. For example, once printed or converted to static electronic format, chromatography records lose the capability of being reprocessed or enabling more detailed viewing of baselines.

Records in dynamic format, such as electronic records, allow an interactive relationship between the user and the record content. For example, electronic records in database formats allow the user to track, trend, and query data; chromatography records maintained as electronic records allow the user or reviewer (with appropriate access permissions) to reprocess the data and expand the baseline to view the integration more clearly.

Where it is not practical or feasible to retain the original copy of source data (e.g. MRI scans, where the source machine is not under the study sponsor's control and the operator can only provide summary statistics), the risks and mitigation should be documented.

True copy – a copy (irrespective of the type of media used) of the original record that has been verified (i.e. by a dated signature or by generation through a validated process) to have the same information, including data that describe the context, content, and structure, as the original.

Audit Trail – a form of metadata containing information associated with actions that relate to the creation, modification, or deletion of GxP records. An audit trail provides for secure recording of life-cycle details such as creation, additions, deletions, or alterations of information in a record, either paper or electronic, without obscuring or overwriting the original record. An audit trail facilitates the reconstruction of the history of such events relating to the record regardless of its medium, including the "who, what, when, and why" of the action.

Where computerized systems are used to capture, process, report, store, or archive raw data electronically, system design should always provide for the retention of audit trails to show all changes to, or deletion of, data while retaining previous and original data. It should be possible to associate all data and changes to data with the persons making those changes, and changes should be dated and time stamped (time and time zone where applicable).

The reason for any change should also be recorded. The items included in the audit trail should be those of relevance to permit reconstruction of the process or activity.

Audit trails (identified by risk assessment as required) should be switched on. Users should not be able to amend or switch off the audit trail. Where a system administrator amends or switches off the audit trail, a record of that action should be retained.

The relevance of data retained in audit trails should be considered by the organization to permit robust data review/verification. It is not necessary for audit trail review to include every system activity (e.g. user log on/off and keystrokes).

Routine data review should include a documented audit trail review where this is determined by a risk assessment. When designing a system for review of audit trails, this may be limited to those with GxP relevance. Audit trails may be reviewed as a list of relevant data or by an "exception reporting" process. An exception report is a validated search tool that identifies and documents predetermined "abnormal" data or actions that require further attention or investigation by the data reviewer.

Reviewers should have sufficient knowledge and system access to review relevant audit trails, raw data, and metadata (see also "data governance").

Where systems do not meet the audit trail and individual user account expectations, demonstrated progress should be available to address these shortcomings. This should either be through add-on software that provides these additional functions or by an upgrade to a compliant system. Where remediation has not been identified or subsequently implemented in a timely manner, a deficiency may be cited.

Electronic signature – a signature in digital form (biometric or nonbiometric) that represents the signatory. This should be equivalent in legal terms to the handwritten signature of the signatory.

The use of electronic signatures should be appropriately controlled with consideration given to:

- How the signature is attributable to an individual.
- How the act of "signing" is recorded within the system so that it cannot be altered or manipulated without invalidating the signature or status of the entry.
- How the record of the signature will be associated with the entry made, and how this can be verified.
- The security of the electronic signature, i.e. it can only be applied by the "owner" of that signature.

It is expected that appropriate validation of the signature process associated with a system is undertaken to demonstrate suitability and that control over signed records is maintained.

Where a paper or pdf copy of an electronically signed document is produced, the metadata associated with an electronic signature should be maintained with the associated document.

The use of electronic signatures should be compliant with the requirements of international standards. The use of advanced electronic signatures should be considered where this method of authentication is required by the risk assessment. Electronic signature or E-signature systems must provide for "signature manifestations," i.e. a display within the viewable record that defines who signed it, their title, and the date (and time, if significant) and the meaning of the signature (e.g. verified or approved).

An inserted image of a signature or a footnote indicating that the document has been electronically signed (where this has been entered by a means other than the validated electronic signature process) is not adequate. Where a document is electronically signed the metadata associated with the signature should be retained.

Data review and approval – the approach to reviewing specific record content, such as critical data and metadata, cross-outs (paper records), and audit trails (electronic records), should meet all applicable regulatory requirements and be risk based. There should be a procedure that describes the process for review and approval of data. Data review should also include a risk-based review of relevant metadata, including relevant audit trail records. Data review should be documented, and the record should include a positive statement regarding whether issues were found or not, the date that review was performed, and the signature of the reviewer.

Computerized system user access/system administrator roles – full use should be made of access controls to ensure that people have access only to functionality that is appropriate for their job role, and that actions are attributable to a specific individual.

For systems generating, amending, or storing GxP, data-shared logins or generic user access should not be used. Where the computerized system design supports individual user access, this function must be used.

System administrator rights (permitting activities such as data deletion, database amendment, or system configuration changes) should not be assigned to individuals with a direct interest in the data (data generation, data review, or approval).

Individuals may require changes in their access rights, depending on the status of clinical trial data. For example, once data management processes are complete, the data is "locked" by removing editing access rights. This should be able to be demonstrated within the system.

Data retention – may be for archiving (protected data for long-term storage) or backup (data for the purposes of disaster recovery). Data and document retention arrangements should ensure the protection of records from deliberate or inadvertent alteration or loss. Secure controls must be in place to ensure the data integrity of the record throughout the retention period and should be validated where appropriate (see also data transfer/migration).

Archive – a designated secure area or facility (e.g. cabinet, room, building, or computerized system) for the long-term retention of data and metadata for the purposes of verification of the process or activity.

Archived records may be the original record or a "true copy" and should be protected so they cannot be altered or deleted without detection and protected against any accidental damage such as fire or pest.

Archive arrangements must be designed to permit recovery and readability of the data and metadata throughout the required retention period. In the case of archiving of electronic data, this process should be validated, and in the case of legacy systems the ability to review data periodically verified (i.e. to confirm the continued support of legacy computerized systems). Where hybrid records are stored, references between physical and electronic records must be maintained such that full verification of events is possible throughout the retention period.

Backup – a copy of the current (editable) data, metadata, and system configuration settings maintained for recovery including disaster recovery.

Backup and recovery processes should be validated and periodically tested. Each backup should be verified to ensure that it has functioned correctly, e.g. by confirming that the data size transferred matches that of the original record.

The backup strategies for the data owners should be documented.

Backups for recovery purposes do not replace the need for the long-term retention of data and metadata in its final form for the purposes of verification of the process or activity.

Validation – for intended purpose (GMP; See also Eudralex; Volume 4; Annex 11, 15) – Computerized systems should comply with regulatory requirements and associated guidance. These should be validated for their intended purpose which requires an understanding of the computerized system's function within a process. For this reason, the acceptance of vendor-supplied validation data in isolation of system configuration and user's intended use is not acceptable. In isolation from the intended process or end-user IT infrastructure, vendor testing is likely to be limited to functional verification only and may not fulfill the requirements for performance qualification.

Functional verification demonstrates that the required information is consistently and completely presented. Validation for intended purpose ensures that the steps for generating the custom report accurately reflect those described in the data checking standard operation procedure (SOP) and that the report output is consistent with the procedural steps for performing the subsequent review.

IT Suppliers and Service Providers (including Cloud providers and virtual service/platforms (also referred to as software as a service (SaaS)/platform as a service (PaaS)/infrastructure as a service (IaaS) – Where "cloud" or "virtual" services are used, attention should be paid to understanding the service provided, ownership, retrieval, retention, and security of data.

The physical location where the data is held, including the impact of any laws applicable to that geographic location, should be considered.

The responsibilities of the contract giver and acceptor should be defined in a technical agreement or contract. This should ensure timely access to data (including metadata and audit trails) to the data owner and national competent authorities upon request. Contracts with providers should define responsibilities for archiving and continued readability of the data throughout the retention period.

17.3 Going from Paper to Digital

Recording notes on paper is a very slow process, writing can be unreadable, data cannot be filtered or compared, and there is no audit trail of critical data such as changes, deletions, or moving of data.

While instruments were equipped according to the requirements of FDA 21 CFR Part 11 from the end of the year 2017, many other processes in the laboratories were still recorded manually in a paper form or in a form of controlled laboratory notebooks, books of events, or notebook lists of standards and samples.

As a first step that CVTA made to digitalize the recording of the abovementioned other processes was the introduction of electronic signatures to sign the PDF documents (e.g. change control, deviation reports, Excel spreadsheets, APs, and reports) with the state certificate Sigen-ca.

Before electronic signatures, everything was processed in a paper form with manual signatures. During the same year, electronic signatures were implemented on CVTA analytical systems for signing analytical raw data (e.g. chromatograms and spectrums). CVTA introduced preprinted laboratory notebooks for routine analyses and lists of standards and samples (for receipt of samples, reference standards, and chemicals).

Preprinted notebooks are still in use, they are approved by Quality Assurance unit: peer-reviewed, tested before use, and the number of issued notebooks is controlled. When introducing preprinted notebooks we got very positive responses from our analysts because it facilitated their work. Laboratory notes are recorded faster, up to date, readability improved, and the rate of mistakes is lower (or zero). Notes are being reviewed faster, labor productivity has increased, and pressure on analysts and quality control dropped.

It was all fine at the start, but regarding the data integrity we identified some weaknesses of our documentation system.

First of them was an audit trail, i.e. who, when, what, and why something was created or changed.

Second weakness was that manual inspection and statistical evaluation of different data in paper form demands long and tedious work as the data cannot be filtered by project, person, client, reason, error, technique, instrument, etc.

That requires an appropriate software and electronic (digital) documentation system.

These are the main reasons why CVTA started to search for the solution for management of documentation within a solid electronic documentation system, while at the same time we hoped to even further diminish the labor load, automatize it, and make it less stressful and less open for human errors.

We found ELN SciNote consistent with FDA 21 CFR part 11 requirements.

Advantages of this software are:

- Automatic saving of data on the server system at the time of its creation, assuring valid versions of APs and protocols for our analyst.
- Data can be quickly found, filtered, and compared.
- Because SciNote is an ELN software consistent with FDA 21 CFR part 11 requirements, it enables us to assign and manage user roles and permissions.
- There is also a time-stamped audit trail of creating, changing, or deletion of data.
- When a document is electronically signed, the following data is recorded: who (identity of a signer), what (description of activities made), and when (date, time, and geographic latitude) an activity was done.
- Entries in this ELN software are controlled with a unique user's name (ID code) and password. Passwords can be periodically changed.
- Administrators can have different levels of accessibility.

Even though ELN SciNote was designed as an electronic notebook, in CVTA, during close collaboration with ELN SciNote development team, we decided to use that software also as documentation system for other activities which are still paper based.

17.4 Implementation of SciNote (ELN) to CVTA System

It took us about six months to implement SciNote to the CVTA system.

Implementation included the following activities by two new employees:

- two weeks of studying of CVTA SOPs and workflows in CVTA,
- three weeks of studying SciNote features and functionalities (video tutorials, etc.),
- three months: testing functionalities and validation of SciNote due to CVTA user's requirements (URS) (including QA, QC, and Head of CVTA)
- two weeks: preparing validation report (including QA),
- one month: creating a manual for users (including QA),
- one week: education of other CVTA employees.

In the GMP environment it is required to document URS and change control when purchasing new equipment, software, or hardware.

17.4.1 Some of CVTA user's Requirements (URS)

- URS 1: It is possible to edit inventories by a certain person even if a document is e-signed and locked.
- URS 5: Items will stay assigned even if a task is unlocked.
- URS 6: When a document is signed and locked once by an author, only another person with the same or higher rights can unlock or cosign a document.
- URS 3: Within an inventory, it is possible to filter by keywords.
- URS 4: When documents are revised, links within a document remain.
- URS 2: Organization admin should have the right to unlock a retired user and right to transfer projects of a retired user to a new person, not to himself.
- URS 7: It is possible to see the detailed audit trail of a single experiment.
- URS 8: It is possible to filter by experiments in the audit trail.

17.4.2 From Documentation Review and Approval to ELN Implementation

We have reviewed and approved Installation Qualification (IQ) and operation qualification (OQ) protocols for installation and qualification of ELN SciNote. Our IT service installed software, and we had intention to test it first.

SciNote team organized educational courses that we could start, first with setting user's roles, creating teams and finally creating protocols, reports, and inventories. After the courses we created a CVTA team and invited employees to this group.

Nothing happened over a period of one year for the very obvious reason: every employee was fully occupied with their own work. Because of that CVTA hired two people with priority tasks to test the software.

First step was to educate them about GMP, standard operating procedures of CVTA and protocols that are followed in CVTA, and how all prescribed procedures reflect in CVTA environment (workflows in CVTA). We decided to start testing ELN SciNote with something easy. First challenge was testing software with the introduction of a form for change control management. While testing the process of change control in ELN SciNote to meet our specific needs, in a paper environment in

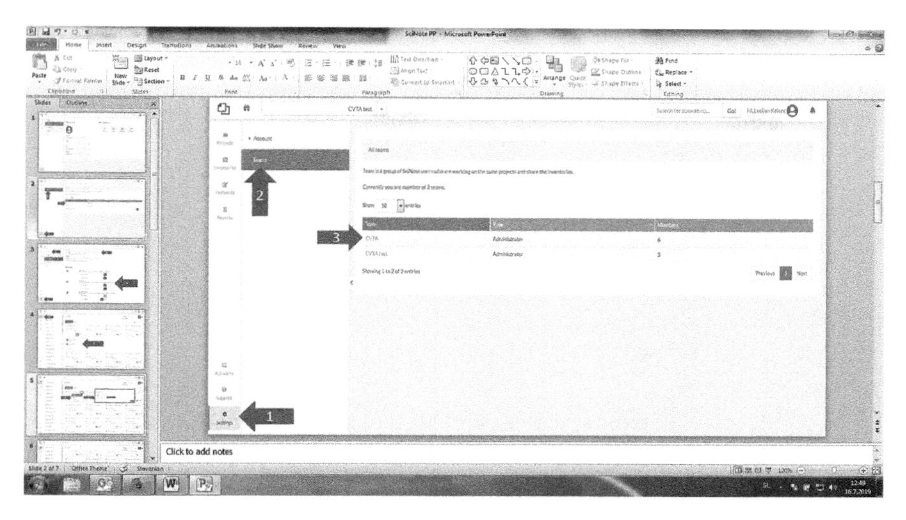

Figure 17.1 Creating a team in SciNote electronic lab notebook. Soure:Figure created by Samo Andrensek, and Simona Lojevec Hartl.

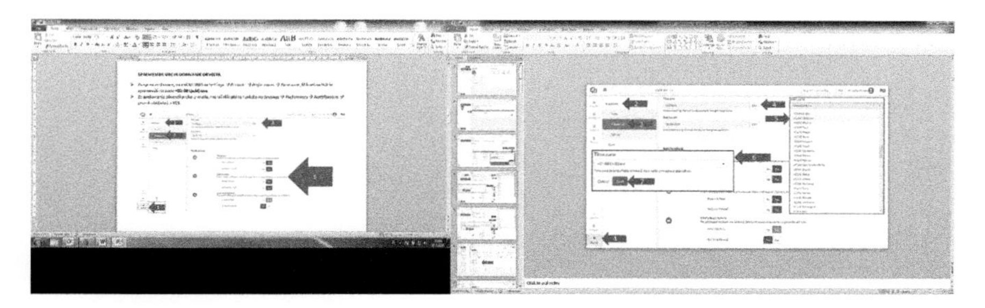

Figure 17.2 Setting time and date records in SciNote electronic lab notebook. Soure:Figure created by Samo Andrensek, and Simona Lojevec Hartl.

CVTA they had to write user manuals for SciNote as well and check functionalities of software in the perspective of GMP and data integrity at the same time.

Organization Admin (IT of CVTA) had to send invitations to dedicated persons first to create their own account. Dedicated persons started with studying SciNote features and functionalities watching video tutorials.

The first team created was the testing team "CVTA test" (Figure 17.1; please note that the interface of SciNote has been updated since the screenshots were taken, 2019).

They set time zone, date record, and time record (Figure 17.2).

They started with creating protocols and inventories due to CVTA form No. 49 for change control and physical list of changes. They checked the user roles and permissions (Figure 17.3). Meanwhile, they also tested the different functionalities listed in OQ protocol of SciNote and formed OQ report supported by screenshots.

		TEAM ROLES DEFINITIONS			PROJECT ROLES DEFINITIONS							
ACTIONS	Organization admin	Team admin	Normal user	Guest	Owner	User	Technician	Viewer		Creator	Public project	Public protocol
ORGANIZATION												
view organization members	X											
invite new users	X											
view organization members	X											
invite new users to organization	X											
create new team	X											
promote other members to the level of Organization administrator	X											
change password policy	X											
access system logs	X											
invite users to team	X											
change team member's permissions	X											
invite members to teams	X											
PROJECTS												
view projects		X	X	X								
create project		X	X									
view project					X	X	X	X	OR		X	
view project activities					X	X	X	X				
view project users					X	X	X	X	OR		X	
view project notifications					X	X	X	X	OR		X	
view project comments					X	X	X	X	OR		X	
edit project					X							
archive project					X							
restore project					X							
add user to project					X							
remove user from project					X							
edit users on project					X							
add comment to project					X	X	X					
edit project comment					X				AND	X		
delete project comment					X				AN	X		

Figure 17.3 Levels of user roles and permissions in SciNote.

17.4.3 Step-by-Step Implementation of Change Control Management in SciNote

First, CVTA form no. 49 (Module 1, Module 2) for change control management was usually recorded as a physical list of changes, then completed, converted to pdf, signed with state certificate Sigen-ca, and manually stored to server.

Module 1 represents a proposal for change and consists of ID number of change, signature of person who suggested a change, table of suggested changes for quick orientation, description of change, activities to do if the change is accepted, signature

National Institute of Chemistry
Center for validational technologies and analytics
CHANGE CONTROL
Form 49
MODULE 1
Page 1/1

ID number of change control	Record ID number as "KSserial number from the list of changes/YY"; e.g.: KS123/19
Date:	Record a date (format: DD.MM.YYYY) of change control proposal
Proponent (name, surname, sign, date)	

Change: (mark with x)

SOP		Software	
Form		Instrument	
Validated Excel spreadsheet		Person	
Analytical procedure		Module of instrument	
Validation protocol		Space of measurement equipment	
Validation report		Measurement equipment out of use	
Validation protocol of producer		Other	

Description of change and a reason for change:
Describe change and a reason for change.
Make risk analysis.

Activities to do connected with proposed change:
Record all changes and activities connected with a proposed change.

Head of CVTA: (name, surname, sign, date)	Put e-sign with Sigen-ca (Proposal of change control is signed from Head of CVTA)
Approved by QA CVTA: (name, surname, sign, date)	Put e-sign with Sigen-ca (Proposal of change control is approved from QA of CVTA)

National Institute of Chemistry
Center for validational technologies and analytics
CHANGE CONTROL
Form 49
MODULE 2
Page 1/1

ID number of change control	Record ID number as "KSserial number from the list of changes/YY"; e.g.: KS123/19
Date of conclusion of change	Record a date (format: DD.MM.YYYY) when change and all activities connected with a change were concluded.
Notes:	Record what changes and activities connected with those changes were made.
Head of CVTA: (name, surname, sign, date)	Put e-sign with Sigen-ca (Conclusion of change control is signed from Head of CVTA)
Approved by QA CVTA: (name, surname, sign, date)	Put e-sign with Sigen-ca (Conclusion of change control is approved from QA of CVTA)

Figure 17.4 CVTA form no. 49 for change control management (consists of Module 1 – M1 and Module 2 – M2). Soure: Figure created by Samo Andrensek, and Simona Lojevec Hartl.

of Head of CVTA, and signature of QA manager who approves the change control form.

Module 2 represents finalization of change and consists of ID number of change, confirmation of changes and activities made, signature of head of CVTA, and signature of QA manager who approves the finalization of the change control form (Figure 17.4).

17.4.3.1 Creating Projects in SciNote

- APs, protocols, SOPs, change control, and out of specification (OOS) investigation, due to a name of CVTA client,
- CVTA SOPs,
- Templates, where we created different workflows, e.g. change control management and investigation of OOS results.

17.4.3.2 Creating a Workflow

For creating the protocol of change control in SciNote, we created the workflow with tasks M1 and M2; M1 means proposal of change, and M2 conclusion of change (Figure 17.5).

17.4.3.3 Creating the Tasks and Protocol Steps

1. Proposal for change control with task M1 (Figure 17.6).
   Conclusion of the change with task M2 (Figure 17.7).

Figure 17.5 Workflow for change control management in SciNote. Soure: Figure created by Samo Andrensek, and Simona Lojevec Hartl.

17.4.3.4 Filtering, Overview of Data and Inventory for Change Control Management

In SciNote, it is possible to filter the data, e.g. connection of change control form and type of change (e.g. SOP, new person, AP, validation protocol or report, new instrument, and new software), or even to see the connection with the name of the subscriber. What is important as well is that it is possible to see if a change has been finished or not (Figure 17.8).

Workflow and inventory were created analogous for investigations of OOS results also. In the inventory we can filter, e.g. person, substance, and instrument, if the corrective and preventive actions were issued, finished and efficient (yes/no), and was OOS investigation finished (yes/no) (Figure 17.9).

17.4.3.5 Audit Trail of Changes

What is very important from a GMP or research standpoint is that ELN SciNote includes a detailed **audit trail of changes** (Figure 17.10), e.g. history of e-signatures and text changes within a single task – once a task is signed and locked.

To make sure that only authorized personnel can access these details in SciNote, if you want to see an audit trail you must be the team administrator and/or organization admin.

17.4.3.6 Overview of all Activities

In SciNote, we can also see all activities that can be filtered to meet our needs. We would usually filter them by:

- Team
- Date when the activity was performed
- Activity type
- Activity groups (Projects/Task results/Task/Task protocol/Task inventory/Experiment/Reports/Inventories/Protocol repository/Team)
- Activity types within activity groups (e.g. within activity group task result you can see the following global activities: Result added, Result edited, Result archived, Result deleted, Result comment added, Result comment edited, Result comment

M1 (managing changes) preview

Created at: 06/26/2019 12:35
Last modified at: 08/19/2019 10:31
Added by: L
Keywords: No keywords
Authors: No authors

No protocol description

Protocol Steps

1 1. Change suggestion mark | Published on 07/18/2019 09:52 by L

Attachments (0) Newest first

2 2. The date of registration of the change | Published on 07/18/2019 09:52 by L

Attachments (0) Newest first

3 3. Change | Published on 07/18/2019 09:52 by L

Attachments (0) Newest first

Change
- SOP
- Form
- Validated Excel calculation
- Analysis procedure
- Validation protocol
- Validation report
- Manufacturer's validation protocol
- Software
- Instrument
- Person
- The instrument module
- The space of measuring equipment
- The elimination of measuring equipment
- Other

4 4. The description of change, reason | Published on 07/18/2019 09:52 by L

Attachments (0) Newest first

5 5. Activities related to the proposed change: | Published on 07/18/2019 09:52 by L

Attachments (0) Newest first

6 6. Proposer | Published on 07/18/2019 09:52 by L

Attachments (0) Newest first

7 7. Head of the CVTA | Published on 07/18/2019 09:52 by L

Dr. Samo Andrenšek

Attachments (0) Newest first

8 8. Approved QA CVTA | Published on 07/18/2019 09:52 by L

Simona Lojevec Hartl

Attachments (0) Newest first

9 9. Valid version | Published on 07/18/2019 09:52 by L

Form 49, MODUL 1, Managing change, version: ____, valid from: ____

Figure 17.6 Example from SciNote – change control, task M1. Soure: Figure created by Samo Andrensek, and Simona Lojevec Hartl.

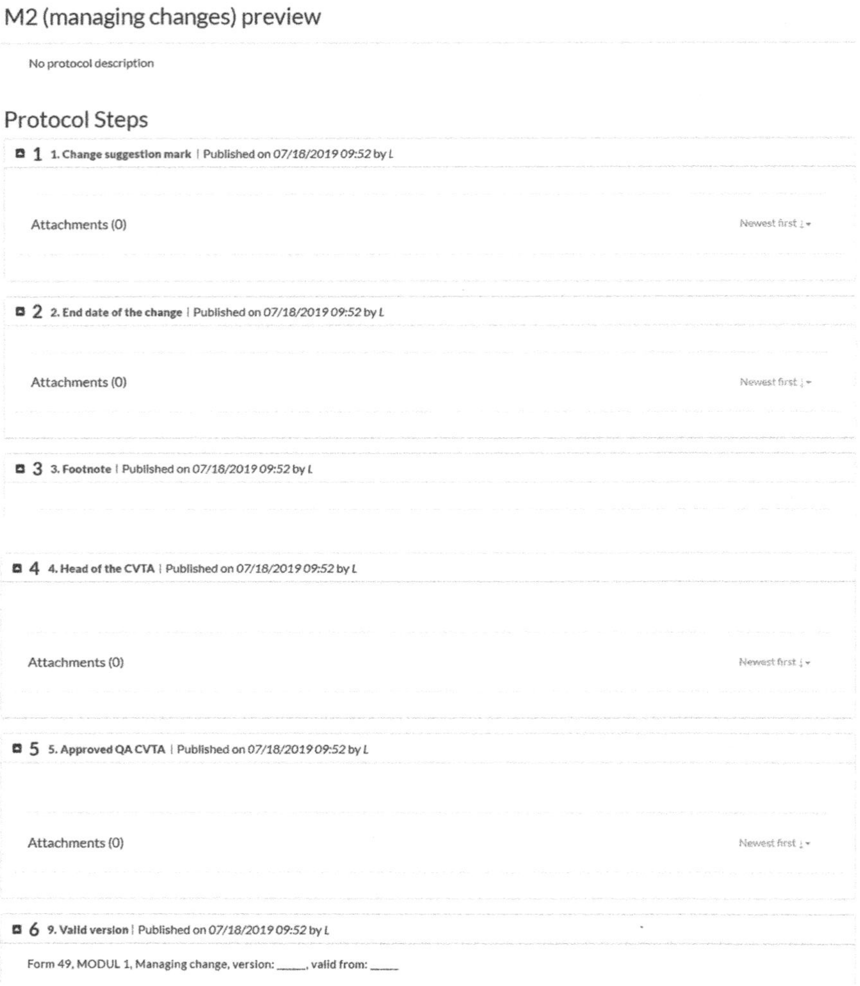

M2 (managing changes) preview

No protocol description

Protocol Steps

1 1. Change suggestion mark | Published on 07/18/2019 09:52 by L

Attachments (0) Newest first

2 2. End date of the change | Published on 07/18/2019 09:52 by L

Attachments (0) Newest first

3 3. Footnote | Published on 07/18/2019 09:52 by L

4 4. Head of the CVTA | Published on 07/18/2019 09:52 by L

Attachments (0) Newest first

5 5. Approved QA CVTA | Published on 07/18/2019 09:52 by L

Attachments (0) Newest first

6 9. Valid version | Published on 07/18/2019 09:52 by L

Form 49, MODUL 1, Managing change, version: _____, valid from: _____

Figure 17.7 Conclusion of the change – task M2. Soure: Figure created by Samo Andrensek, and Simona Lojevec Hartl.

deleted, Office online file on result edited, Image on result edited, Chemical structure on result crested, Chemical structure on result edited, and Chemical structure on result deleted)

- Users
- Projects, experiments, and tasks
- Inventories
- Protocols
- Reports

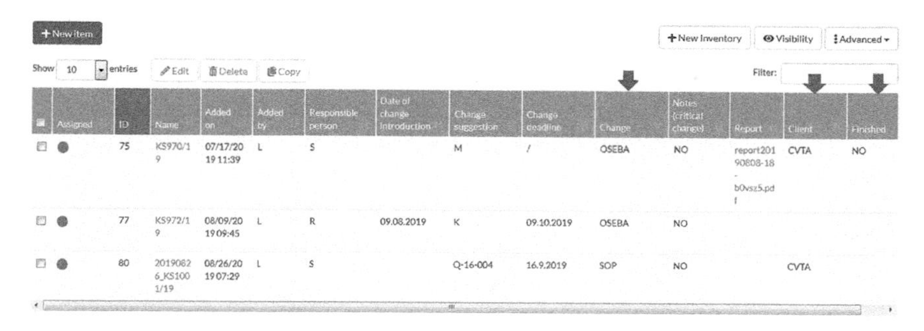

Assigned	ID	Name	Added on	Added by	Responsible person	Date of change introduction	Change suggestion	Change deadline	Change	Notes (critical change)	Report	Client	Finished
	75	KS970/19	07/17/2019 11:39	L	S		M	/	OSEBA	NO	report20190808-18-b0vsz5.pdf	CVTA	NO
	77	KS972/19	08/09/2019 09:45	L	R	09.08.2019	K	09.10.2019	OSEBA	NO			
	80	20190826_KS1001/19	08/26/2019 07:29	L	S		Q-16-004	16.9.2019	SOP	NO		CVTA	

Figure 17.8 Inventory for change control management. Soure: Figure created by Samo Andrensek, and Simona Lojevec Hartl.

Figure 17.9 Workflow for investigations of OOS results. Soure: Figure created by Samo Andrensek, and Simona Lojevec Hartl.

Figure 17.10 Detailed audit trail within a task, where change in text (old value/new value) is visible. Soure: Figure created by Samo Andrensek, and Simona Lojevec Hartl.

17.4.4 Organization and Signing of CVTA Documentation in ELN SciNote Due to User Roles and Permissions

17.4.4.1 Managing the Team Roles and Responsibilities within SciNote

Organization Admins:

- IT personnel (they have all rights; special rights: add/remove users from organization or team, they can create a new team).

  Teams:

- Administrators: QA, QC, head, deputy of head (special rights: user's management in a Team: add/remove users and determination of user's roles)
- Normal users: analysts and technicians
- Guests: role dedicated to personnel in case of inspections

  Projects:

- Owners: QA, QC, head, deputy of head (special rights: archiving projects and user's management in a project)
- Users: analysts of CVTA
- Technicians: n.a. (too less rights)

  Viewers:

- Personnel in the phase of education – Inventories for CVTA are lists (e.g. of change control and investigations of OOS results).

17.4.4.2 Managing Projects for Efficient Work with Clients

Names of projects are names of CVTA clients and CVTA SOPs also. As CVTA wants that all different clients' data is separated, every project is dedicated to a different client.

Every client has a separate protocol for APs, protocols, change control documents, and OOS investigations, e.g. Client 1-Change control; Client 1-Analytical procedures, Client 1-Protocols, and Client 1-OOS investigations.

Tasks within each project have their own history of changes, e.g. "CVTA SOPs" is a protocol that includes tasks with all different CVTA SOPs and a task with a list of CVTA SOPs. "Client 1_Analytical Procedures" includes tasks with all different Client 1 APs. Within each task a document can be revised including audit trail of changes and link to change control document.

New versions of documents (e.g. CVTA SOPs, APs, protocols, and reports) can be linked to change control documents. While issuing a new version of a document, the old version is manually archived. Analyst uses only a valid electronic version of CVTA SOP, AP, or validation protocol.

In change control forms, the number of signatures is minimized to two dedicated persons; forms are authorized by CVTA personnel. A form is signed and locked by the head of CVTA and cosigned by the QA manager to approve the change. When a document is signed and locked once, only people with the same or higher rights can unlock the document.

All activities within a single document are automatically recorded in detailed audit trails.

17.5 Suggestions for Improvements and Vision for the Future

From the point of data integrity, analytical equipment that has no configurable software available is a very critical part of equipment in GMP or research laboratories right now.

Examples of that kind of equipment are balances, pH meters, pipettes, temperature meters, and hygrometers. It is important not to overlook systems of apparent lower complexity. (A good resource on this matter and further recommended reading: MHRA, GxP Data Integrity Guidance and Definitions; Revision 1: March 2018.)

There is no physical evidence for recording activities executed on equipment that could replace software that records all these activities, e.g. daily calibration, measurements, date, time, user's data, and correcting data.

ELN SciNote enables integration with Gilson electronic pipettes already. We are planning to test these pipettes and implement them in our system together with ELN SciNote.

In CVTA we still use validated Excel spreadsheets for calculating analytical results. It is one step further from calculating results on unvalidated spreadsheets, but it is still a very slow process (area and weights from balances must be retyped and data of samples, reference substances, and specifications must be typed down by hand), and after e-signing Excel spreadsheets must be manually saved to the server. Human mistakes can be done anytime.

All these deficiencies of using validated spreadsheets require digitalization as soon as possible. The situation is very similar when creating analytical reports. When there is a human error possibility, mistakes can be done despite very strict quality control procedures.

In order to maximize automation of calculating and reporting results we are looking forward to the possibility of further integrations.

References

1 Kemijski inštitut, Širimo meje znanosti, *Ljubljana*, p. 219–224, 2016

2 EudraLex, The Rules Governing Medicinal Products in the European Union, Volume 4, Good Manufacturing Practice, Medicinal Products for Human and Veterinary Use, Chapter 4: Documentation. https://ec.europa.eu/health/sites/health/files/files/eudralex/vol-4/chapter4_01-2011_en.pdf

3 PIC/S guidance, Good practices for data management and integrity in regulated GMP/GDP environments, PI 041-1 (Draft 2), August 2016. https://picscheme.org/users_uploads/news_news_documents/PI_041_1_Draft_2_Guidance_on_Data_Integrity_2.pdf.

4 FDA, Part 11 of Title 21 of the Code of Federal Regulations; Electronic Records; Electronic Signatures – Scope and Application (21 CFR Part 11). https://www.fda.gov/regulatory-information/search-fda-guidance-documents/part-11-electronic-records-electronic-signatures-scope-and-application.
5 Document MHRA, GxP Data Integrity Guidance and Definitions; Revision 1: March 2018 https://assets.publishing.service.gov.uk/government/uploads/system/uploads/attachment_data/file/687246/MHRA_GxP_data_integrity_guide_March_edited_Final.pdf.

18

Wet Lab Goes Virtual: In Silico Tools, ELNs, and Big Data Help Scientists Generate and Analyze Wet-lab Data

Jungjoon Lee[1] and Yoonjoo Choi[2]

[1] *Toolgen, Gasan-dong, Geumcheon-gu, Seoul, 08501, Republic of Korea (South)*
[2] *Combinatorial Tumor Immunotherapy MRC, Chonnam National University Medical School, Hwasun, Jeonnam 58128, Republic of Korea (South)*

> "We have now entered what I call 'the digital age of biology,' in which the once distinct domains of computer codes and those that program life are beginning to merge, where new synergies are emerging that will drive evolution in radical directions."
>
> Source: From Craig Venter (Life at the speed of light, © 2013, Hachette).

Life in its essence is information composed of digital codes. For example, human genome contains 2.9 billion base pairs equivalent to 2.9 gigabytes (GB). Today, advances of information technology (IT) enable scientists to deal with data of this size.

IT advances are revolutionizing how life scientists keep records of their daily wet-lab results and processes. They are using cloud-based electronic lab notebooks (ELNs) which enable data and protocols to be shared instantly with international collaborators. These protocols include not only the wet-lab experiment design but also "big data" collection and interpretation. As a result, the productivity of the early adopters among scientists has increased dramatically. The size of the data that one scientist can work with is rapidly expanding.

In this chapter, we provide cases for the use of in silico tools that can improve productivity of the scientists. For example, our laboratory is specialized in high-throughput screening. With less than eight lab members, we routinely screen from thousands to millions of mutants generated by various experimental methods.

We "engineer" biology via two fundamentally different approaches. First, the genomic DNA is edited randomly with various tools and the resulting phenotype is observed. The process resembles the natural evolution process by which beneficial random mutations are selected and accumulated. Second, we take a "rational design" approach to engineer proteins. Structural information of the target protein is needed in order to make intelligent changes. The insight often comes from structural information of other proteins. In both cases, scientists utilize and generate

* These authors contributed equally.

Digital Transformation of the Laboratory: A Practical Guide to the Connected Lab, First Edition.
Edited by Klemen Zupancic, Tea Pavlek and Jana Erjavec.

"big data" in the format of genetic codes and image data. In the following chapter, we demonstrate the size of data generated from genome-editing and protein design works, and how the use of ELN can record the data efficiently to keep everything in one place.

For decades, scientists have been developing tools to edit the genome of living cells. The ideal tool would be able to target any combination of nucleotides in the genome. In 2012, several scientists reported that they have developed such a tool that works in mammalian cells [1, 5, 8, 9]. They have used the system called CRISPR (clustered regularly interspaced short palindromic repeat)-associated protein 9 (Cas9) – (CRISPR-Cas9) – from Nature that was developed by bacteria to fight against invading virus. Since then, the tool has been optimized so that genome editing has become more popular and it became a daily routine to edit the genome of various cells.

18.1 CRISPR-Cas9 Explained

CRISPR-Cas9 has two components: Cas9 protein and single guide RNA (sgRNA). sgRNA sequence starts with spacer sequence at the 5′ end of RNA which is 19 mer sequence. The sequence should be the same sequence with the one in the target sequence which should also contain the "PAM sequence," is the NGG codon sequence (AGG, CGG, UGG and GGG) at the 3′ end. The rest of the sgRNA sequence is called transcrRNA which is conserved. To change the target, only the 19mer spacer sequence needs to be changed.

For knockout studies, the first step is therefore listing candidate sgRNA spacer sequence in the exon region of the target gene. The process is done by searching all the exon region sequences which contain NGG sequence at the 3′ end. As the number of candidate sequences is more than thousands, scientists use in silico tools to perform the job.

In addition, each candidate sgRNA can also edit another genomic sequence that has sequence similarity to it. This is called an "off-target" sequence. Due to the large size of the genomic DNA, many candidate sgRNAs have hundreds of candidate off-target sites. In silico tools exist that enlist a number of off-targets to each candidate sgRNA so that scientists can select the one with fewer mismatches.

After treating the Cas9 to the cell, checking whether CRISPR-Cas9 has edited the intended target sequence is very challenging. Previously, the success of genome editing was checked by polymerase chain reaction (PCR) amplification of the target site which was cloned and sequenced to check the editing patterns. The development of Next-Generation Sequencing (NGS) techniques has enabled scientists to check the editing pattern of thousands of single cells at the same time.

As it became easy to generate genome-edited cells, high-throughput screening of these cells is now feasible. The most popular phenotype of the edited cells that is measured is image data. However, the difference between images of the wild type and that of the edited cell is often too subtle to be quantitatively differentiated by the human eye. In addition, the number of images is hundreds to thousands for high-throughput studies. Various in silico tools have been developed to help scientists accumulate and handle these "big data."

18.2 Introduction of the Digital Solutions and ELN into the Laboratory

Research is all about communication. Therefore, one of the major challenges to become a successful scientist or laboratory is writing lab notes well. As explained earlier, the tasks for scientists in the genome-editing field are becoming constantly more complex, and large size data is being generated. Therefore, recording all details of lab work is becoming cumbersome.

ELNs have been developed in time to meet the unmet needs. Using ELN can resolve many problems for scientists faced with big data. However, it is an implementation step that prevents and often hinders scientists from changing their daily routine to a new, unfamiliar one. It is therefore helpful to observe how early adopters among scientists use ELN and implement the ones that are helpful.

In the following chapter, the process of genome editing is divided into three steps. In the first step, scientists design a single guide RNA (sgRNA) with minimum off-target effect. Next, the sgRNA is synthesized and delivered to the cell, and the efficiency of genome editing is measured by NGS. Finally, the phenotype of the edited cell is measured. In each step, use of ELN is essential in keeping records and sharing of the data among authorized lab members. Examples for the use of ELN are demonstrated in each sector.

18.3 The Role of the ELN and In Silico Tools in the Genome-editing Process

18.3.1 Designing sgRNA

Genome editing of mammalian cells with CRISPR Cas9 involves designing 20 mer sgRNA that targets the gene to be edited. However, there are many potential off-target sites in the rest of the target genome. Many computational tools are available which predict potential off-target sites and their mismatch number to the target sgRNA, to name a few: Cas – OFFinder, CHOPCHOP, CRISPR Design, CRISPR Design Tool, CRISPR/Cas9 gDNA finder, CRISPRfinder, E-CRISP, and ZiFiT.

In many cases, scientists use CRISPR-Cas9 to knock out specific gene in mind. In this case, it will be cumbersome to enlist possible sgRNA sequences and find off-targets of each sgRNA. In silico tool, Cas-Designer, a web-based tool for choice of CRISPR-Cas9 target sites, that enlists sgRNAs with potential off-target sequences can help scientists with that task. Using this tool, scientists enter just the name of the gene to the tool and select the sgRNA with minimum off-target sequence.

18.3.2 Issues with Paper-based Processes and the Use of ELN

In most cases, the number of candidate sgRNAs for a single gene is hundreds to thousands. If the number of candidate off-targets for each sgRNAs is included, the number is multiplied by factor of 10 at least. It is therefore impossible to keep these records in paper-based lab notebooks.

Using ELNs, it is possible to keep thousands of candidate sgRNAs and off-target sequence data (that can be linked, added to inventories, or organized in Excel files) and explain rationale for selecting a few sites with a minimum number of candidate off-target sites.

For example, our company is developing CRISPR-Cas9 as possible therapeutic modalities. Therapeutics work by knocking out a specific gene. There are thousands of possible sgRNAs that can knock out the gene. To obtain food and drug administration (FDA) approval, we need to justify which sgRNA we selected. One of the most important criteria for the selection is the number of possible off-target sites. To obtain FDA approval, we should be able to explain why certain sgRNA sequence is selected. The number of mismatches for all the candidate sgRNAs is tabulated in Excel file. The file is saved in SciNote ELN, and the top 10 candidates are highlighted. These candidates are experimentally tested, and the one with highest efficacy with minimum off-target effect is selected. All results are saved in ELN in the Excel file so that other people do not have to repeat the same process.

After the genome editing, the editing pattern and efficiency should be checked. Decades ago, this was done by PCR amplification of the target sequence, which is cloned, and hundreds of clones are sequenced. Recent development of NGS techniques enabled scientists to sequence thousands of copies of PCR products at the same time. By analyzing PCR products that contain target locus using NGS methods and analyzing the result with in silico tool Cas analyzer, scientists can check thousands of insertion, deletion (indel) patterns at the same time.

Since hundreds of indel patterns are generated for each sample, scientists can record only a few representative indel patterns into paper-based lab notebooks with overall indel percentage. Instead, complete data can be stored in the ELN.

18.3.3 High-content Imaging for the Target Discovery

Before the discovery of CRISPR-Cas9, most of the genome-wide screens involved small interfering RNA (siRNA). While siRNA knocks down the expression level of the gene, CRISPR-Cas9 can completely "knock out" the gene. As a result, target discovery via genome-wide screen with CRISPR-Cas9 has become very popular. Since there are more than 20 000 genes in the human genome, knocking out each gene and screening its phenotype of the KO (knockout) cell involves a lot of labor.

Development of automated screening platforms increased the throughput of the screening process dramatically. Several vendors now manufacture these automated microscopes. Using high-content imaging platform, mammalian cells treated with siRNA or CRISPR-Cas9 are incubated inside the microscope machine, and the images of cells in hundreds of different wells are captured automatically. As the

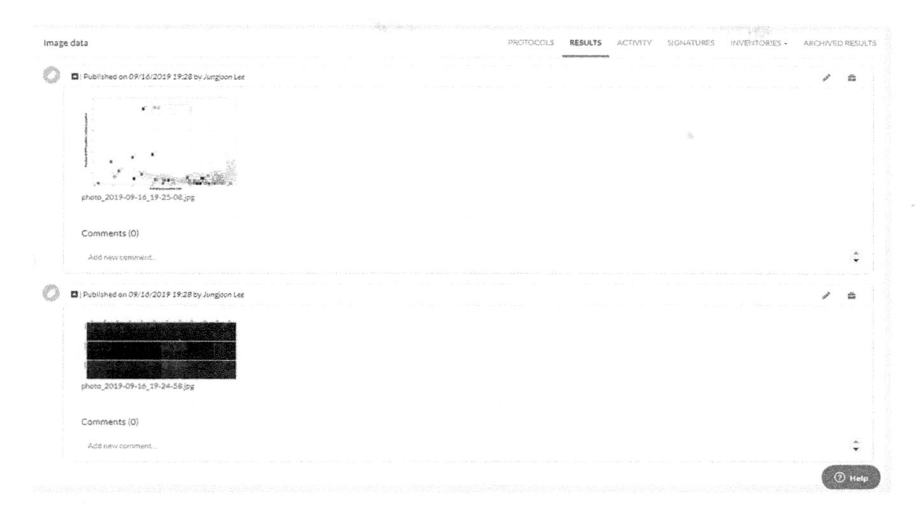

Figure 18.1 Examples of the saved results comprising hundreds of image files with quantitative results. Source: Figure created by Jungjoon Lee.

number of samples is large (hundreds of image files), it is often beyond human capacity to qualitatively extract image features of each file for comparison. Instead, the image feature can be automatically analyzed by image analysis software. Using these in silico tools, scientists can design pipelines which computers follow repetitively.

Storing thousands of images in the paper-based lab notebook is cumbersome, and scientists usually store multiple GB-sized image files on hard drives. However, the link between the lab notebook and the hundreds of image files in the hard drive is not obvious in most cases. Using ELN, the image files can be stored in cloud data storage and linked via hyperlink located at various results sections of the ELN. Example from our laboratory of images saved in SciNote ELN and hyperlinked is below (Figure 18.1):

The pipeline for the analysis can also be stored online, which can be reused when scientists want to reanalyze the previous data with different parameters.

18.3.4 Plant Virtual Laboratory

CRISPR-Cas9 can be used to edit plant genomes. In most cases, scientists modify the plant genome to obtain breeds with the desired phenotype. Seeds of various genotypes were generated and planted in the soil. The images of the plants in various stages were taken and compared to their genotypes which were obtained by genotype analysis of the plants. The whole process is automated. The seed is planted into a small flower pot which is placed in the conveyor belt. The conveyor belt is connected to the NGS analyzer, which captures the leaf of the plant and analyzes its genotype by NGS. The conveyor belt is also connected to a high-resolution camera which takes images of the plants at various stages.

Since the number of plant seeds to be tested is often hundreds to thousands, the scientist using a paper lab notebook would repetitively record genotypes and images in the lab notebook in chronological order. Using ELN, the genotype and image data are stored in a computer, which is easily programmed so that scientists can click the sample number to watch images of the sample at various stages. The quantitative features of the images of all the samples can also be automatically measured and compared to each other to generate graphs. This is called a virtual tool, and many advanced labs have constructed their own virtual tools to increase the productivity of their research, e.g. Key gene's Virtual Reality Breeding Tool.

18.4 The Role of the ELN and In Silico Tools in the Protein Design Process

Though it is difficult to answer what life is, we know how life works. Living organisms use biomolecules to maintain their functions, and in the core, there are proteins. Due to the endless effort of scientists, we now understand what they could look like. The development of X-ray crystallography, nuclear magnetic resonance (NMR) spectroscopy, and cryogenic electron microscopy (Cryo-EM) technologies now allows us to grasp the image of their chemical structures.

German chemist August Wilhelm von Hofmann developed molecule models for public lectures at the Royal Institution, representing atoms as balls and chemical bonds as sticks. Apparently, atoms are not solid spheres, and bonds are not hard rods. However, such models have been useful representations of reality. Modern scientists have brought the atom balls and bond sticks to the virtual world: Computer. Simulation and manipulation of virtual biomolecules in the artificial biological environment have become the essence of modern biology.

18.4.1 Protein Modeling

Proteins are composed of special building blocks, amino acids. The central atom of an amino acid is carbon which has four electrons that can establish chemical covalent bonds (which can be represented as "rods"). There are 20 proteinogenic amino acids, and all the amino acids look alike except for one group of atoms which we call "side chain" (or residue). Each side chain has characteristic features such as size, charge, and hydrophobicity. Scientists often use letters for the side chains, and the primary structure is the sequence of letters.

The primary structure is critical in protein folding. Christian Anfinsen performed a series of pioneering experiments with bovine pancreatic RNase in the 1960s. He discovered that a protein structure is in thermodynamic equilibrium with its environment. In other words, atomic interactions (the letter sequence) determine its three-dimensional structure. This hypothesis, also called the thermodynamic hypothesis, is the fundamental basis of computer-based protein modeling. Imagine you construct a virtual environment filled with two hydrogen balls connected to an oxygen ball (water) and you put complex atom balls (protein) based on its

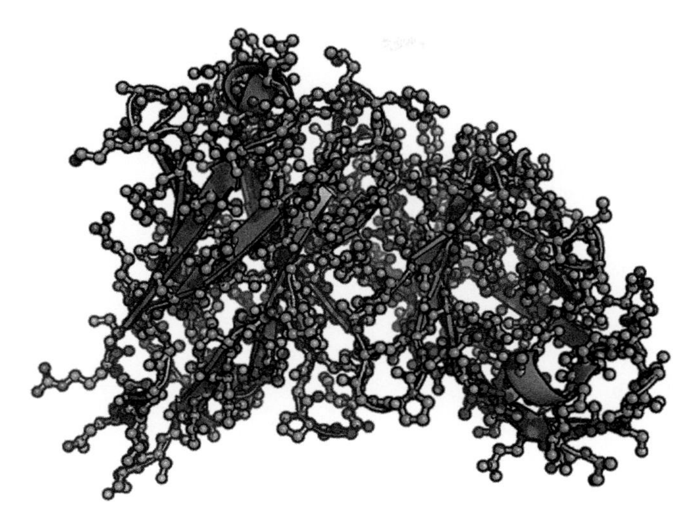

Figure 18.2 Folded protein structure in stick and rod representation. Image is created by the author of this text.

primary structure into the virtual water molecules. If you have precise rules for interactions between the atom balls, your atom balls must be folded into the real protein structure and perform specific functions.

The rules are termed as the force field in molecular modeling. An all-atom force field is composed of sets of parameters for all atom types. While parameters may differ from force fields, the general form of force field energy function is substantially similar to each other.

$$E_{Total} = (E_{bond} + E_{angle} + E_{dihedral}) + (E_{vdw} + E_{elec.})$$

The first three terms are bonded terms (atom balls connected by rods), and the last two terms are nonbonded terms (interactions between balls without rods). Many studies have proven the powerful utility of molecular modeling.

The structure (Figure 18.2.) of adalimumab (trade name: Humira), which is one of the top selling therapeutic antibodies for rheumatoid arthritis, is represented in stick and rod (PDB ID: 4NYL). Atoms of the protein interact with each other and fold into a unique three-dimensional structure. Image is created by the author of this text.

18.4.2 Protein Redesign

The success of protein molecular modeling naturally leads us to exploring protein sequences Nature allows. Based on the existing protein structure, one may virtually change residues and find sequences. Computation and minimization of side-chain interactions give insight into sequences that can be folded to the same structure. Full Sequence Design 1 (FSD-1) is the first redesigned protein which was built based on the structure of the zinc finger (zif268) [6]. FSD-1 is 28 amino acid residues in length, but only shares 6 identical residues to zif268.

One of the first computationally redesigned proteins (Figure 18.3.), called FSD-1 (thick gray, PDB ID 1FSV), was completely redesigned from a zinc finger backbone

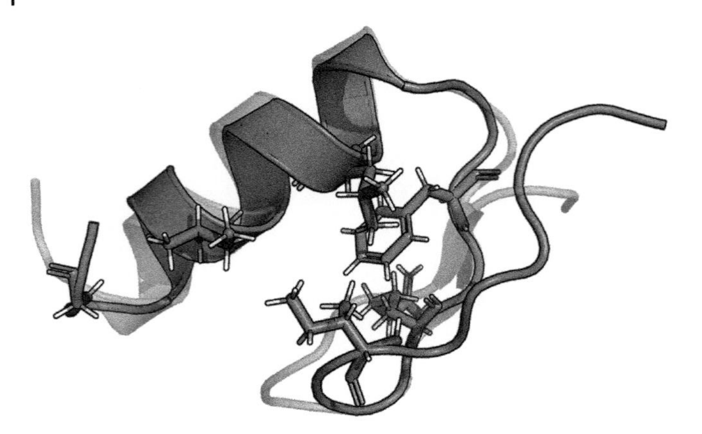

Figure 18.3 Redesign of zinc finger backbone structure.

structure (transparent cyan, PDB ID 1AAY). While having similar three-dimensional structures, they only share six amino acids (highlighted as stick). Image is created by the author of this text.

18.4.3 Importance of Keeping the Electronic Records

While zif268 is a very small protein, the possible sequence combination is not. The authors (10.1126/science.278.5335.82) calculated how immense the combination can be: Considering the physicochemical properties of the proteinogenic amino acids, they estimated that there could be 1.9×10^{27} possible protein sequences. Comparing the estimated age of the universe (approximately 4×10^{17} s), it is simply impossible to explore the entire protein sequence space. As most therapeutic proteins are bigger than the zinc finger (for example, antibodies are 40 times bigger than FSD-1), a typical and practical protein design strategy is not to directly find all possible sequences but to generate hundreds and thousands of near-optimal protein sequence designs for selection. Such information is so large that electronic records are the only solutions. In the following section, we demonstrate how protein designers make use of ELNs in the development of therapeutic antibodies, which has been one of the major applications of the protein redesign.

18.4.4 Development of Therapeutic Antibodies

The current biologics drug market is driven by antibodies which are key agents in the immune system. Understanding the antibody structure is critically important for developing a novel therapeutic agent. The functional units of antibodies consist of four polypeptide chains, two identical heavy chains and light chains linked by disulfide bonds. The typical antibody structure looks like "Y," and each chain has three hypervariable binding sites called complementarity determining regions (CDRs). These CDRs are found in the two variable regions of the light and heavy chains and consist of three hypervariable loops in each of the light (CDR-L1, -L2, and -L3) and the heavy (CDR-H1, -H2, and -H3) chains.

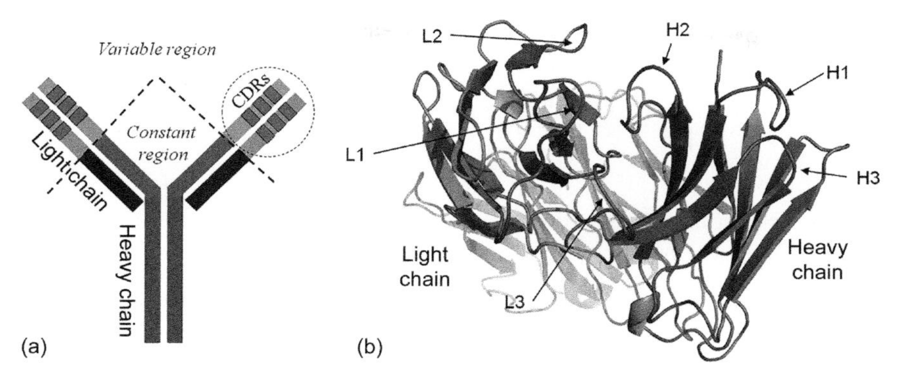

Figure 18.4 Antibody structure in (a) schematic view and (b) cartoon view. Image is created by the author of this text.

Figure 18.4(a) shows a schematic view of antibody structure. The circled region is magnified in (b), which is a cartoon view of the framework regions and six CDRs of a mouse antibody (PDB ID 1ACY). Image is created by the author of this text.

As experimental structure determination is costly and time consuming, prediction and modeling of CDR structures have become very important. CDR modeling has several important applications such as binding affinity enhancement, epitope/paratope prediction, and docking. Accurate CDR modeling has been of critical significance since the advent of NGS techniques. While most CDRs (CDR-L1 to H2) are relatively easy to predict, CDR-H3 is extremely diverse both in sequence and structure owing to the way antibodies are encoded in the genome. There are a number of prediction algorithms for CDR-H3 (reviewed in reference 10), and they generate structural information (three-dimensional coordinates) in the protein data bank (PDB) format. A structure file is typically a plain text file with long lines of information which can be rendered to a three-dimensional figure (Figure 18.5.).

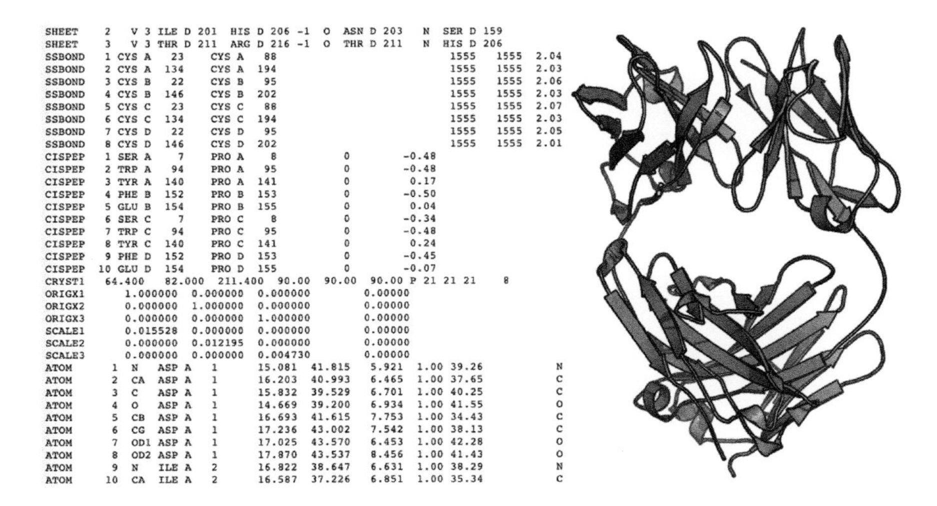

```
SHEET    2   V 3 ILE D 201  HIS D 206 -1  O  ASN D 203   N  SER D 159
SHEET    3   V 3 THR D 211  ARG D 216 -1  O  THR D 211   N  HIS D 206
SSBOND   1 CYS A   23    CYS A   88                          1555   1555  2.04
SSBOND   2 CYS A  134    CYS A  194                          1555   1555  2.03
SSBOND   3 CYS B   22    CYS B   95                          1555   1555  2.06
SSBOND   4 CYS B  146    CYS B  202                          1555   1555  2.03
SSBOND   5 CYS C   23    CYS C   88                          1555   1555  2.07
SSBOND   6 CYS C  134    CYS C  194                          1555   1555  2.03
SSBOND   7 CYS D   22    CYS D   95                          1555   1555  2.05
SSBOND   8 CYS D  146    CYS D  202                          1555   1555  2.01
CISPEP   1 SER A    7    PRO A    8          0        -0.48
CISPEP   2 TRP A   94    PRO A   95          0        -0.48
CISPEP   3 TYR A  140    PRO A  141          0         0.17
CISPEP   4 PHE B  152    PRO B  153          0        -0.50
CISPEP   5 GLU B  154    PRO B  155          0         0.04
CISPEP   6 SER C    7    PRO C    8          0        -0.34
CISPEP   7 TRP C   94    PRO C   95          0        -0.48
CISPEP   8 TYR C  140    PRO C  141          0         0.24
CISPEP   9 PHE D  152    PRO D  153          0        -0.45
CISPEP  10 GLU D  154    PRO D  155          0        -0.07
CRYST1   64.400   82.000  211.400  90.00  90.00  90.00 P 21 21 21     8
ORIGX1      1.000000  0.000000  0.000000        0.00000
ORIGX2      0.000000  1.000000  0.000000        0.00000
ORIGX3      0.000000  0.000000  1.000000        0.00000
SCALE1      0.015528  0.000000  0.000000        0.00000
SCALE2      0.000000  0.012195  0.000000        0.00000
SCALE3      0.000000  0.000000  0.004730        0.00000
ATOM      1  N   ASP A   1      15.081  41.815   5.921  1.00 39.26           N
ATOM      2  CA  ASP A   1      16.203  40.993   6.465  1.00 37.65           C
ATOM      3  C   ASP A   1      15.832  39.529   6.701  1.00 40.25           C
ATOM      4  O   ASP A   1      14.669  39.200   6.934  1.00 41.55           O
ATOM      5  CB  ASP A   1      16.693  41.615   7.753  1.00 34.43           C
ATOM      6  CG  ASP A   1      17.236  43.002   7.542  1.00 38.13           C
ATOM      7  OD1 ASP A   1      17.025  43.570   6.453  1.00 42.28           O
ATOM      8  OD2 ASP A   1      17.870  43.537   8.456  1.00 41.43           O
ATOM      9  N   ILE A   2      16.822  38.647   6.631  1.00 38.29           N
ATOM     10  CA  ILE A   2      16.587  37.226   6.851  1.00 35.34           C
```

Figure 18.5 PDB file format and its three-dimensional structure.

A protein structure is typically stored as a plain text file which contains three-dimensional coordinate information. The coordinate can be depicted using a protein file viewer such as PyMol. The structure shown here is a monoclonal antibody, cetuximab (trade name: Erbitux, PDB ID 1YY8). Image is created by the author of this text.

18.4.5 Importance of Electronic Lab Notebook for Communication Between Team Members

As mentioned earlier, a protein's function is specified by its structure. In order to investigate how CDR structures may impact on antibody function, sharing three-dimensional structure information is critical. Thanks to the advent of digital media such as ELNs, researchers now can easily share structural information encoded in thousands of lines with the authorized collaborators.

Nowadays, antibody engineering is carried out in close collaborations between computational structural biologists and wet-lab researchers. Computational techniques are applied to not only modeling of antibody structure but also antibody redesign for improved properties such as high binding affinity and stability. Antibody redesign typically requires several rounds of mutual feedback between computational and experimental efforts. After modeling a target antibody structure, the computational structural biologist generates a large number of variants according to certain criteria including the force field energy. Researchers share information about the antibody designs to select a handful number of variants for further processes such as protein expression and characterization. It is quite common that tested designs may not have desirable properties. The wet-lab researcher reports back to the computational antibody designer to feed the results into the next round of antibody redesign. Each round produces a vast amount of information, which may not be efficiently shared in the paper format of lab notes. Figure 18.6

Figure 18.6 An example of antibody redesign information saved in the results section in SciNote. Source: Figure created by Jungjoon Lee.

shows an example of humanized cetuximab [3] which were redesigned using a computational protein redesign method [2, 4].

The computational antibody engineer may produce a large number of variants for further testing. Typically, the design process goes through multiple rounds, and each round yields additional information about each variant. ELNs are effective means of sharing information. SciNote was used to electronically store and share mutual information.

As nonantibody proteins such as enzymes, hormones, receptors, toxins, and other classes of proteins also constitute a huge reservoir of prospective therapeutic candidates, in silico drug development will be more available than ever in the near future. The recent advances in creating artificial therapeutic proteins will also enrich the pool of potential therapeutics and may soon enable tailored medicine [7, 11]. While all are promising, the development of nonantibody therapeutics is known to be more challenging than that of therapeutic antibodies, and thus close collaborations between computational biologists and wet-lab researchers would be vital.

Sharing scientific knowledge and information electronically will boost efficiency and availability of future therapeutics.

References

1 Cho, Seung Woo, Kim, Sojung, Kim, Jong Min, and Kim, Jin-Soo (2013). Targeted genome engineering in human cells with the Cas9 RNA-guided endonuclease. *Nature Biotechnology* 31(3): 230–232.
2 Choi, Yoonjoo, Hua, Casey, Sentman, Charles L., Ackerman, Margaret E., and Bailey-Kellogg, Chris (2015). Antibody humanization by structure-based computational protein design. Paper presented at the MAbs.
3 Choi, Yoonjoo, Ndong, Christian, Griswold, Karl E., and Bailey-Kellogg, Chris (2016). Computationally driven antibody engineering enables simultaneous humanization and thermostabilization. *Protein Engineering, Design and Selection* 29 (10): 419–426.
4 Choi, Yoonjoo, Verma, Deeptak, Griswold, Karl E., and Bailey-Kellogg, Chris (2017). EpiSweep: Computationally driven reengineering of therapeutic proteins to reduce immunogenicity while maintaining function. In: *Computational Protein Design*, pp. 375–398. Springer.
5 Cong, Le, Ran, F Ann, Cox, David, Lin, Shuailiang, Barretto, Robert, Habib, Naomi, and Marraffini, Luciano A. (2013). Multiplex genome engineering using CRISPR/Cas systems. *Science* 339 (6121): 819–823.
6 Dahiyat, Bassil I., and Mayo, Stephen L. (1997). De novo protein design: fully automated sequence selection. *Science* 278 (5335): 82–87.
7 Fleishman, Sarel J., Whitehead, Timothy A., Ekiert, Damian C., Dreyfus, Cyrille, Corn, Jacob E., Strauch, Eva-Maria, and Baker, David (2011). Computational design of proteins targeting the conserved stem region of influenza hemagglutinin. *Science* 332 (6031): 816–821.

8 Hwang, Woong Y., Fu, Yanfang, Reyon, Deepak, Maeder, Morgan L., Tsai, Shengdar Q., Sander, Jeffry D., and Joung, J. Keith. (2013). Efficient genome editing in zebrafish using a CRISPR-Cas system. *Nature Biotechnology* 31(3): 227–229.

9 Mali, Prashant, Yang, Luhan, Esvelt, Kevin M., Aach, John, Guell, Marc, DiCarlo, James E., and Church, George M. (2013). RNA-guided human genome engineering via Cas9. *Science* 339 (6121): 823–826.

10 Marks, C., and Deane, C.M. (2017). Antibody H3 structure prediction. *Computational and Structural Biotechnology Journal* 15: 222–231.

11 Silva, Daniel-Adriano, Yu, Shawn, Ulge, Umut Y., Spangler, Jamie B., Jude, Kevin M., Labão-Almeida, Carlos, and Leung, Isabel. (2019). De novo design of potent and selective mimics of IL-2 and IL-15. *Nature* 565 (7738): 186–191.

19

Digital Lab Strategy: Enterprise Approach

Cesar Tavares

Universidade Federal do Rio de Janeiro, Rio de Janeiro, RJ – Rio de Janeiro, Brazil

19.1 Motivation

> "With your gaze on the horizon and your mind up to the sky, never stop. Walking brings us happiness" [1].

When we are working to solve a complex problem, we need more than ingenuity; we need purpose and focus. We have to do what is correct and work outside of our comfort zone.

19.1.1 Which Problem Do We Want to Solve?

The first thing we must do is define the question or the problem we are solving. This is not as obvious as it seems.

One of the most frequent errors in the realm of digital innovation is to provide the answers before forming the questions, or to demand results before defining a proper strategy. This approach frequently leads to poor technological choices, such as the selection of a product solely because it is used by a competitor or is the market leader.

For example, we might evaluate Amazon's success, conclude that we should copy this model by selling our products on the web, and build a team to lead this innovation path. But wait – we need to understand the problem first. Are we a B2B, B2C, or C2C business? What is our volume? Are we a retail or key account sales business, or both? Are we a small business trying to scale up, or are we already a giant? If we are small, we can easily adapt our internal process to enable online business. If we are a long-lasting company adapting to new technologies, we will probably need to launch several initiatives: one to enable online to offline sales and another to modernize and simplify internal processes (through an investment in robotic process automation, for example). In a nutshell, we must make the company scalable and sustainable.

Digital Transformation of the Laboratory: A Practical Guide to the Connected Lab, First Edition.
Edited by Klemen Zupancic, Tea Pavlek and Jana Erjavec.

With this in mind, which problem do I want to solve regarding laboratory digitalization at enterprise level? The scenario I would like to explore in this chapter is:

1. Hundreds of laboratories worldwide
2. Presence in dozens of countries
3. A wide range of lab types: chemical, material, mineral, environmental, safety, life science, food, seed, soil, toys, cosmetics, transportation, etc.
4. Constant growth (new labs are acquired or organically grown)
5. A weak matrix organizational structure (although we are managed globally, each region and each lab are responsible for their own profit and loss)

These business constraints can then be translated into needs:

- Due to the size of the company (items i and ii), we need a scalable solution that can be quickly deployed.
- Due to the company's variety and growth (items iii and iv), we need a flexible and adaptable architecture.
- Due to the company's organizational structure (item v), we need a strong global digital program strategy to manage stakeholder expectations.

The business constraints and the needs that they imply effectively define the problem that must be addressed as the company pursues a digital laboratory framework. The complex and diverse environment created by a network of hundreds of labs, several business lines with different (and sometimes conflicting) requirements, and growth through acquisition require fast and inexpensive solutions that motivate the labs to push for digitalization and innovation and keep the company up to date with technology. These solutions must succeed in a scenario where only 25% of digital projects achieve their goals and bring the expected benefits to the organization.

Before searching for solutions, a final caveat must be considered. According to The C-Suite Outlook, a report by Forbes Insights and the Project Management Institute, "Nearly 80% of organizations have undergone a significant transformation over the past year using disruptive technologies. But only 25% of those transformation projects yielded tangible benefits that were realized against their original goals" [2]. Digital transformation will only succeed if it brings business benefits. The business lines have a full-time job to keep the company profitable and growing, so they will only support and push the necessary digital transformation when they identify how the 4.0 revolution will help them meet their goals.

19.1.2 New Problems Require New Answers

In a business environment defined by disruptive technologies, there are no easy answers. A common first action is to try to maintain control and stay in the comfort zone, but that does not work. We must brace for impact.

The traditional approach, in which the company defines a single tool to be deployed globally while all labs change their processes to meet the solution constraints, is not effective. The company will lose business by not delivering customized products. The perceived inferior solution will cause the team to disengage from the vision, and the knowledge already captured in existing solutions will be lost.

The more effective approach is to create a Lab Digitalization Program with the support of company executives and to build a solid professional network within company resources. The Lab Digitalization Program will have a broad scope within the organization, touching disciplines such as network, architecture, software development, product life cycle, cybersecurity, lab automation, and innovation. The program will be structured in waves, starting with the foundation and enabling constant evolution to the digital environment in the lab. The next wave must start before the previous wave is completed in order to catch up with current technology and eventually thrive in this new reality.

The strategy pillars of the Lab Digitalization Program are:

- **Flexible solutions** – reuse existing infrastructure and adapt to new technologies.
- **Scalable tools** – easy to learn and smart on reusable content so that the program can be completed quickly.
- **Measurable goals** – the program must produce results for the organization and shareholders.

19.2 Designing a Flexible and Adaptable Architecture

The solution has two key requirements. It must be flexible enough to handle the existing diverse laboratory IT environment and operational processes, and it must be Digital Lab Strategy: Enterprise Approach adaptable enough to support new business strategies and technologies with minimal disruption. The architecture of the Lab Digitalization Program is outlined in Figure 19.1.

The architecture includes a service bus layer to orchestrate communication between all enterprise systems. It is not the objective of this chapter to describe all the benefits of a service bus layer, but there are several reasons I prefer to use such a tool. It assures the correct synchronization of data packages between systems, automatically queues the data packages, has all the most common and effective communication protocols and methods, and keeps updating when new ones are available to the market. Maintenance is reduced if it is needed to integrate N to M systems. Instead of N times M connections (using 1 : 1 integration), the service bus layer needs N plus M connections, which makes communication scalable when dealing with hundreds of systems.

In addition to a service bus layer, the architecture also includes a laboratory instrument middleware layer. This layer facilitates instrument integration, which is critical to our operations because almost all of the generated data start with the instruments. The data only become available for use if the instrument is integrated, or if the data are typed in our existing systems. The instrument middleware layer will also standardize communication between the instruments from different vendors and different laboratory systems. A scalable solution requires the reuse of automation from one lab to another for the same family of instruments. This topic needs

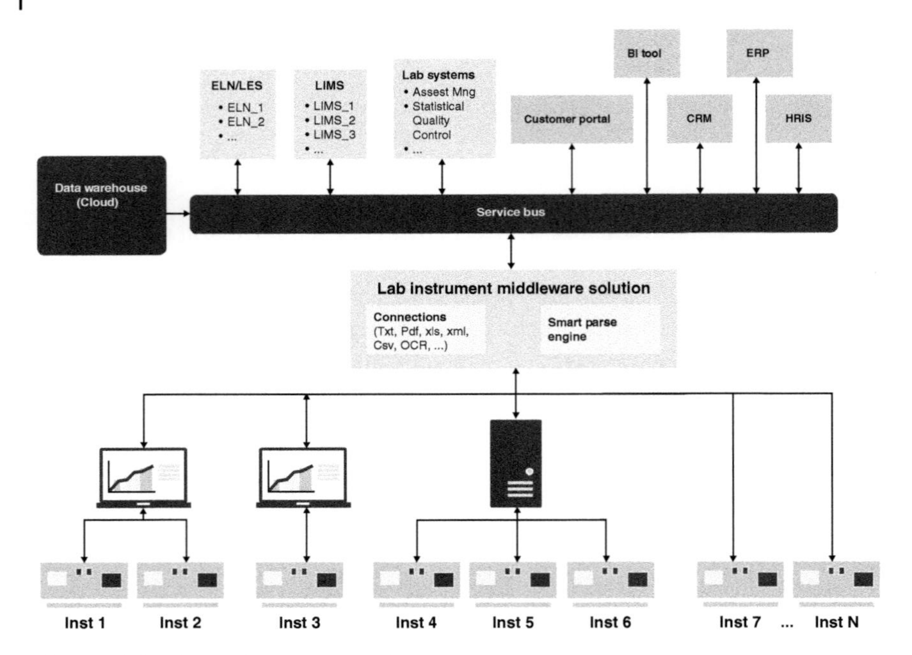

Figure 19.1 Enterprise system architecture diagram.

further attention, and it is an issue in the laboratory digitalization world that has never been tackled properly.

Connectivity within the physical network is just as important as instrument integration. There are recurrent issues of connectivity between laboratory instruments and the corporate network. Consider the following network use cases:

- Case 1: The instrument is connected to a computer in the corporate network, and it cannot receive automatic updates because it might affect the instrument integration to the laboratory information management system (LIMS). It is a point of vulnerability in the network and should be removed.
- Case 2: The instrument is connected directly to the network, and it is not a managed device by the corporate IT team. It is a point of vulnerability in the network and should be removed.
- Case 3: The supplier discontinues support for old versions of the OS. All computers using the old version should be upgraded to supported versions. When support is discontinued for the OS in computers connected to instruments, it is not always possible to update them because the instrument control software is not compatible with the new versions of the OS. The computer becomes a point of vulnerability in the network and should be removed.

The solution proposed is simple: to create a DMZ (demilitarized zone: a buffer between the uncontrolled Internet and your internal networks) in the network for all the instruments and all the computers to which they are connected. A simplified diagram is presented in Figure 19.2. The firewall will be configured to enable specific and unidirectional communication between the instrument and the LIMS server.

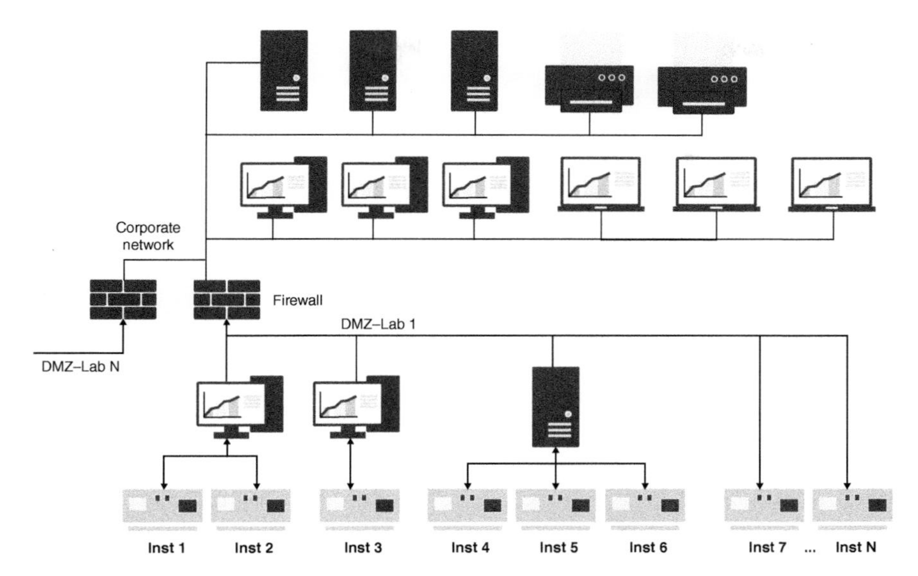

Figure 19.2 Simplified physical network diagram.

It might not be the ideal solution, but it is cost-effective, protects your assets, accommodates changes, and is scalable. When you are dealing with a single lab, it may seem like a cheap solution, but remember that we are digitalizing a few thousand labs.

19.3 There is Only One Rule: No Rules

Ironically, I will start this section with a rule: Moore's Law. The simplified version states that the overall processing power of computers doubles every two years. Thus, according to Moore's Law, technology progresses geometrically, in an ever-accelerating mode, rather than at a constant pace. It constantly leaps forward, bringing to the market new products, new ideas, and new possibilities. How can a company stay relevant in a market like this? How can a company be built to last thrive in this scenario?

One solution is to create a digital data environment independent of any single digital tool in order to avoid putting all of our eggs in one basket. We must diversify the data environment and create patterns that enable communication between systems. In other words, we must focus on the data, not the tool.

Thus, the Lab Digitalization Program has to develop and maintain data packages to communicate between systems. Of course, it will only work if the program is oriented toward continuous improvement. Otherwise, it will be obsolete in a few months or even weeks.

The service bus layer of the proposed architecture will orchestrate the flow of data packages and provide the flexibility to add, remove, or change any of the tools used in the lab with minimum impact on operations. To deploy a new tool, the deployment team must follow the data package specifications, and if needed, add new packages or new line items in existing packages. During go-live, we can switch lab by lab or, in big labs, department by department, making the technological changes more manageable and increasing the success rate for the project.

As an example, consider the data packages that communicate between the LIMS and the supply chain tool:

- Case 1: Purchase a reagent, which will expire in 30 days according to the inventory module at LIMS. The data package includes the product name, product spec, quantity, preferred vendor, expected delivery date, ship to, bill to, cost center, and notes to the vendor.
- Case 2: The supply chain tool confirms that the reagent purchase was executed. The data package includes PO number, vendor, delivery date, confirmation of product name, product spec, and quantities.
- Case 3: Confirm with the supply chain tool the receipt of the reagent as purchased, such that the quantities will be automatically received. The data package includes the PO number, quantity received, and receiving date and time.
- Conclusion: With the simple data package definition, we enable a centralized purchasing process and corporate inventory control. It enables "just-in-time" inventory, reduces waste, improves negotiation conditions, and allows for a modern supply chain process.

As technology changes, we must also have a strategy to "enforce" the use of new technology in the lab using common market practices that have proven their benefits and results. The goal is to keep the approach bottom-up; the drive must come from the lab itself. We must have a reward system based on the labs' adoption rate, deployment speed, thrive, and so on. It should take into consideration the financial benefits brought to the company and the size of the challenge.

A gamification strategy could be an effective way to increase the employee engagement rate within the Lab Digitalization Program. For example, the program decided to focus on instrument integration, so monthly, we will reward the lab with the highest number of automatic test results, as well as the lab with the highest increase over the previous month.

19.4 The Lab Digitalization Program Compass

The Lab Digitalization Program will have greater efficacy when it creates the correct environment for changes to occur, rather than forcing the adoption of those changes. But what does it mean to create the correct environment?

Companies with a clear focus on innovation and digitalization will not require a strong Lab Digitalization Program, because it is already part of the company culture. Unfortunately, this is not the case in all companies, and in those cases, the Lab Digitalization Program must play a more decisive role in creating a culture of innovation within the corporation. The Lab Digitalization Program has to be at the epicenter of these changes, serving as a compass and providing direction.

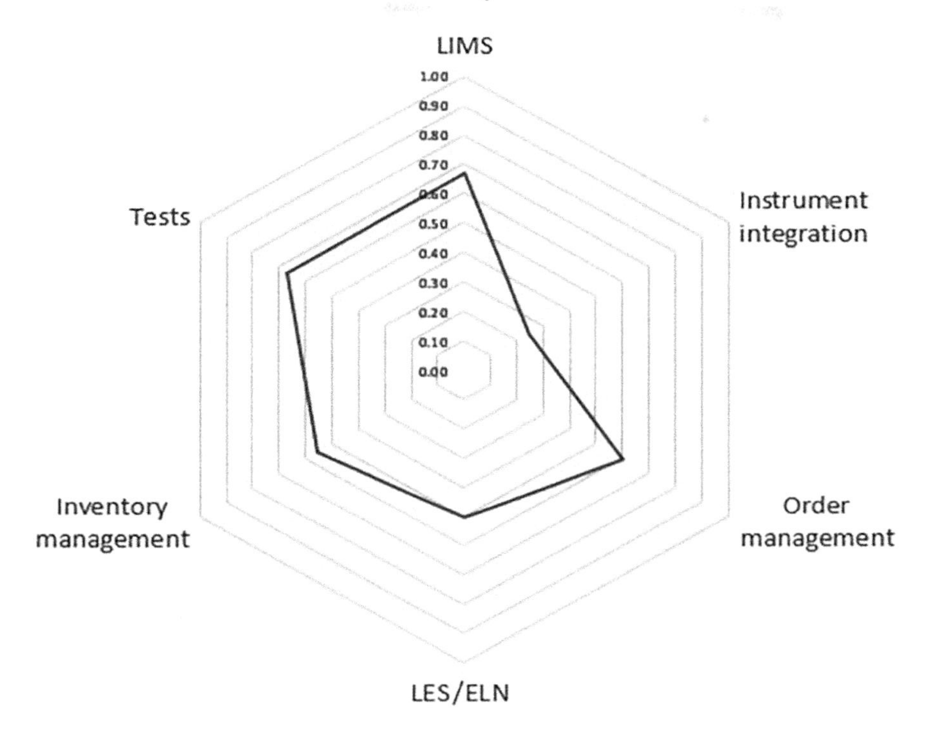

Figure 19.3 Example of a radar chart to measure the laboratory digital maturity level.

The first step is to assess the digital maturity of each lab, which will provide a clear picture of the investment needed. The Lab Digitalization Program team can then plan resources, infrastructure, and budget accordingly. Additionally, the team can rank the laboratories, and the ones in critical situations can cut the line, becoming the initial focus. This assessment needs to be made year after year to evaluate the progress and pace of the Lab Digitalization Program. It may confirm that we are doing the right thing or that we need to pivot to increase the program's effectiveness.

Each lab will need to answer a questionnaire (e.g. SharePoint or SurveyMonkey), and the answers will generate a radar chart (Figure 19.3) with the digital maturity level of the laboratory.

The goal of the Lab Digitalization Program is not to have all labs at 100% in all dimensions. A proper internal benchmark must be developed, which can then be sliced by type of laboratory, region, and line of business, based on the sample population.

The Lab Digitalization Program must supply the laboratories with an evaluation of current technologies and their fit within the organizational strategy, enabling the laboratories to make informed decisions. The goal is to advise the decision-makers, not to impose a solution.

Table 19.1 Example of a lab solutions portfolio.

	LIMS A	LIMS B	ELN X
Lab size	- Up to 50 users - 1000 tests per month - 150 methods	- From 50 users to 100 - 1000 tests per week - 300 methods	- Up to 30 users - 500 tests per month - 100 methods
Hosting options	Locally	Cloud	Cloud and private cloud
Time to deploy	Six months	Nine months	Three months
CAPEX	$$$	$$$$	$
OPEX	$	$$	$
Already integrated into corporate solutions	Y	N	N
Price negotiated by procurement	Y	Y	N
Already deployed in the company	Yes: labs 1, 2, 3, and 4	Yes: labs 11, 12, 13, and 14	No

Table 19.1 presents a simplified example of how the portfolio of tools can be organized. To be more effective, a questionnaire must guide the filtering process, in addition to the solutions portfolio.

How does it work?

1. The lab requests support from the Lab Digitalization Program.
2. The Lab Digitalization Program starts with the questionnaire:
   a. How many employees work in the laboratory and would need licenses?
   b. How many test methods are executed in the Lab?
   c. Does the lab have any certifications or goals to become certified?
   d. And so on.
3. Based on the reply, the Lab Digitalization Program will come up with a shortlist of options to start the conversation with the lab and set expectations about price and timeline.
4. A proper business justification must be made, and the Lab Digitalization Program should have it as a template as well.
5. The laboratory makes a go/no-go decision

When we have the go decision, the laboratory will decide between using internal resources and having an external consultant. The Lab Digitalization Program should present the costs, and the laboratory should fund all the investment needed.

The goal is to keep the program lean, fast, and highly skilled. The best way to achieve that goal is through competition. The Lab Digitalization Program should have the business acumen to offer its expertise outside of the corporation, either as a service provider or through the digital tools it has developed.

19.5 Conclusion

In order to develop a successful Lab Digitalization Program, flexibility is essential. The ability to adapt and pivot is the only way to survive in the current technological world. To achieve effective changes, you must measure your current state (baseline), identify roadblocks, develop the appropriate strategy, pivot, and measure again. Otherwise, the Lab Digitalization Program will not achieve the goals expected by the company.

Since I started working in lab digitalization, I have had the opportunity to learn from my mistakes, change course, and achieve some important milestones. I feel both energized and challenged by the future. I hope you can find useful insights based on the ideas shared in this chapter.

References

1 "Com o olhar para a frente, e a mente para o alto, o caminhar nos traz felicidade," MSc. Thaheciyl Tavares, Brazilian philosopher, sociologist, professor, and Orthodox priest.

2 Forbes Insights (2018). The C-suite outlook: how disruptive technologies are redefining the role of project management.

Part V

Continuous Improvement

After the insightful expert comments and case studies that provided the overview, advice, best practice, and ideas you can implement in your own lab's digitalization strategy, we will continue with the steps that follow after the digitalization. How to focus on the continuous improvement of your lab?

20

Next Steps – Continuity After Going Digital

Klemen Zupancic

CEO, Revenue department, SciNote LLC, 3000 Parmenter Street, 53562, Middleton, WI, USA

20.1 Are You Ready to Upgrade Further?

This chapter is about the digital evolution that follows the revolution. Every lab has some form of digitization/digital tools implemented already; if nothing else, everyone uses MS Excel and other software solutions for analyzing results, storing lists of samples and reagents, and communication tools such as e-mail and Slack. There are no labs that are completely paper based. The biggest first step toward digitalization is, however, the introduction of a centralized software that can handle even more: data, team, inventories, and/or project management.

Electronic lab notebooks serve that central role of laboratory digitalization strategy, because they connect all the different data sources: from when the data gets generated, which protocols were used, list of reagents used, samples analyzed, and instruments that analyzed them. Implementing one should definitely be the first bigger project in your road to digitalization as it starts the digital mindset of an organization, and it serves as the base to connect other pieces of software to work together.

So if your lab moved away from using paper notebooks and implemented such central software, first congratulate yourself, as you made a very difficult and very important step into the future. You are one of 10% of labs worldwide that have done so and are ready for the rest of this chapter.

20.2 Understanding the Big Picture

To achieve digital transformation, it is not enough to just implement as many software tools as possible, it is equally, if not more, important that these tools communicate with each other. For example, if you are using a tool for sample management where you store sample information and data analytics software, where you analyze raw data obtained from those samples, it is important that you can look at the analyzed results and have a direct connection to the data in the sample management

Digital Transformation of the Laboratory: A Practical Guide to the Connected Lab, First Edition.
Edited by Klemen Zupancic, Tea Pavlek and Jana Erjavec.

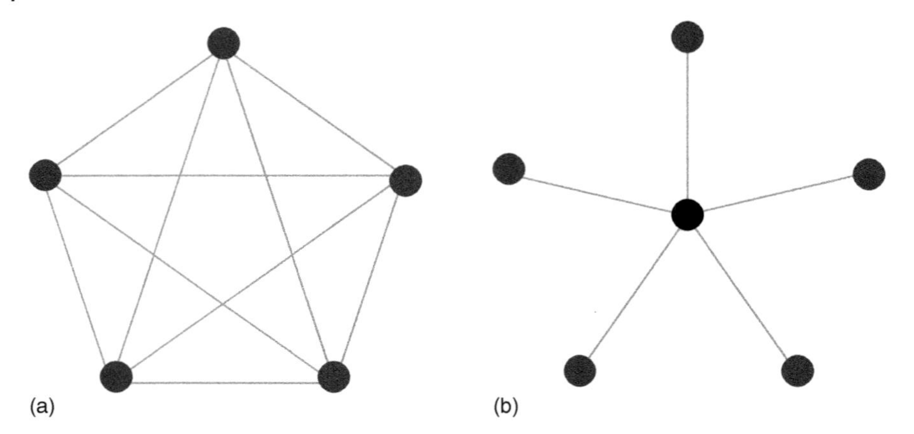

Figure 20.1 Schematic representation of connections if all software tools communicate with each other directly (a) or through a central source of truth such as ELN (b).

software. If this link has to be created manually, you will eventually make a mistake and therefore jeopardize the integrity of any conclusions you make.

Therefore, it is crucial that the software tools you are implementing can talk to each other. If you want each new tool to talk to all other tools, you will need to create exponentially more integrations (see Figure 20.1), which is not a sustainable way to go forward. Instead, you should have a centralized system to which you connect all other tools, and electronic laboratory notebooks (ELNs) serve as a great basis for that. Therefore, it is important that an ELN of your choice offers integrations and has an open application programming interface (API).

Figure 20.1 shows that, in the first case, the number of connections grows exponentially with every new piece of software added, whereas in the second case, the number grows linearly as each new type of software only needs one connection to the ELN.

The end goal of digitalization is removing human operators from recording the data and transferring it between different systems. If we completely automated data capture (both the experimental measurement and surrounding parameters) and workflow of data handling and processing, we would completely remove human error and assure complete traceability, the lack of which has resulted in the reproducibility crisis in science [1–3]. And where there is room for human error, chances are that software will be faster to perform the task, so we save a lot of time too. And what is the role of humans in all of this? What we do better than machines; creative thinking, problem solving, and articulating the right questions.

Commodity apps that everyone is using are provided by tech giants who offer everything from search engines, to shopping, content, and social platforms in one product. The biggest example of this is the Chinese WeChat, a single go-to app for everyday life. In contrast, science is not a commodity market. A typical scientist may be using as many as 70 different software tools to do their work. There is also a great variety of tools one is using, depending on their niche of research and needs. In science, variety is actually appreciated and needed, we will never have one product that fits all the needs of all scientists. Every piece of software is designed for a specific niche, and we need this variety to explore the unknown. Science also has a long-lasting culture of openness and sharing. Not only in academia but also in industry, which is leveraging the published knowledge, and also contributing in published papers, patents, and open source products.

It is important that we embrace that complexity and start thinking how to connect different pieces of the software together. These integrations, connection of different software apps, will play an integral (pun intended) part of your lab's digital evolution.

20.3 What to Integrate First?

Let us look at an example of a typical data flow in the lab to see how to start thinking about integrations. Every lab is, of course, different, but there are some basic principles that are the same, or at least similar in every lab.

Typically it all starts with a collection of samples or biological specimens. They might come from external sources, such as blood samples of patients or plant material from field experiments. Alternatively, they can also come from internal sources, such as cell lines. Either way the accompanying information is usually a spreadsheet with some information about them. Sometimes they also come with attached files (e.g. consent forms) or with relations to other samples. All of this information does not necessarily have to be passed over every step of experiments, but it is very important that you can access this information at any time in an accurate, unambiguous way.

Reagents, which are used in experiments, come with their own set of information, such as lot numbers, concentrations, and volumes. The same goes for the equipment used, from pipettes and centrifuges to scales and PH meters. Each of those devices also carries their own information from device model, equipment settings, software and firmware versions to calibration dates and other maintenance information.

Next is the protocol a lab worker will use to prepare samples, reagents, and equipment to perform the analysis. This is not just a list of steps to be performed but also calculations for solutions, equipment settings, and operators' decisions that are used to perform experiments. When the measurements are performed, not only raw data is generated but also data about environmental information, date and time, and other metadata, often as important as raw data for the purposes of traceability and reproducibility.

After the experiment is done, a set of software tools are used to analyze the raw data. It is important to know exactly which was used in the analysis, how the data was manipulated, and which version of software was used to do so. Sometimes we do not know how the data should be analyzed in advance until we have a chance to play with it and generate some intermediary results. It is important that we save that and record our reasoning for any further decisions we make on how the analysis should be done.

The last part of the research process is typically collecting the data from multiple experiments, making sense of it, and presenting it in a way that others will

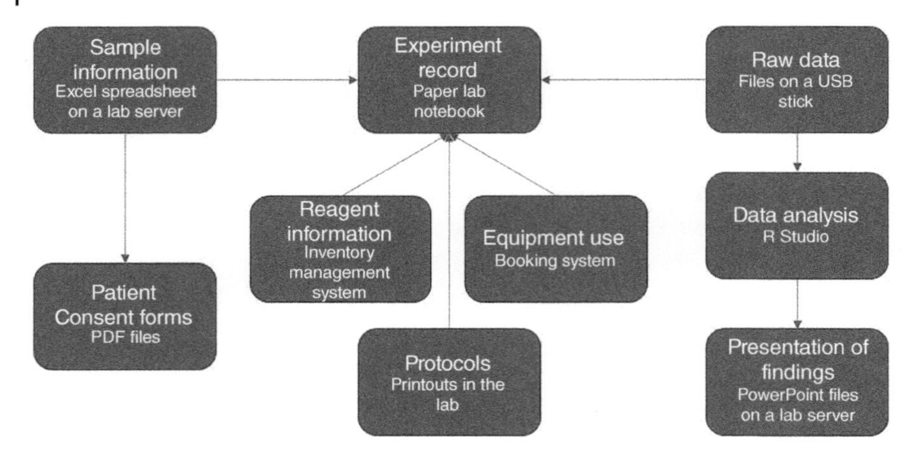

Figure 20.2 An example of a typical data workflow map in a "paper-based" lab. Arrows represent transitions where data has to be manually connected between different systems.

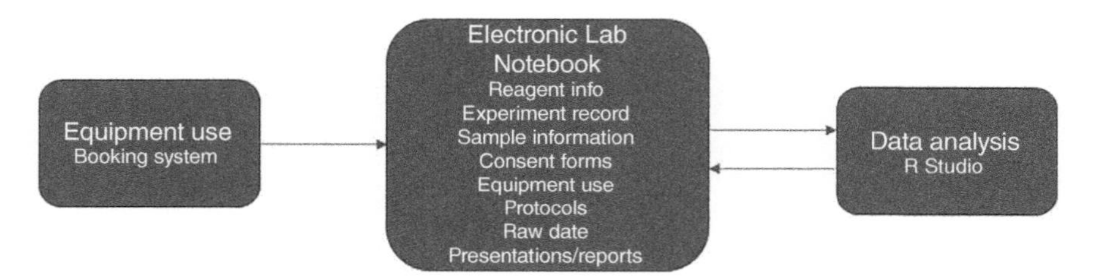

Figure 20.3 An example of a much simpler data workflow in a digital lab after implementing an electronic lab notebook and integrating it to existing digital systems. Since everything is digital, the data transfer can happen automatically.

understand. For academics, this typically means a paper, poster, or a lecture, but it is not much different for industrial scientists who have to present it to decision-makers and other colleagues.

As you are reading this, you must also have other ideas about data that is generated in your use cases. Do we really need all of it?

You cannot know in advance. Is it important to know if a patient had breakfast before their blood was drawn? Maybe. But unless we record it, we will not know; therefore, we should collect as much as we can.

It is a good idea to draw a map of your data flow in order to visualize it (Figure 20.2). This will help you understand it and see where bottlenecks and potential errors can occur. It makes sense to start integrating different pieces of software at these places (Figure 20.3).

20.3.1 Integrations

The three good places to start to think about integrations are laboratory equipment, data repositories, and data analytics platforms where you are probably already using software. There are also other integration possibilities that might not seem directly connected to the flow of information in a specific experiment, but their inclusion can be crucial for traceability and productivity of your lab. We explore those in the Section 20.3.5. There are also other integration possibilities of the workflow that you do manually and could benefit by software introduction (Example?). There are plenty of tools that you can choose from, but the most important characteristic is that the tool can be integrated with other tools now so that these different pieces of software can communicate to one another.

20.3.2 Laboratory Equipment – Concepts of IoT and Lab 4.0

The dream of having all laboratory equipment connected and being controlled in a centralized manner is not new, but it has been very rarely achieved. It is often described with buzzwords such as Internet of Things (IoTs) in the lab or Lab 4.0 (comes from Industry 4.0). In this dream, each piece of equipment is "smart," and it connects to the centralized "mission control" platform where it deposits all the data and is synced with all other equipment to work in perfect unison. In reality, this feat has been realized very seldomly. Achieving this dream requires a lot of custom software development and has been practically employed only in the pharmaceutical industry and only in processes that are very stable.

Although it is perhaps the most obvious of integrations, connecting laboratory equipment to the rest of the workflow is very challenging, primarily due to lack of standardization. Each piece of scientific equipment is different as it serves different purposes, so it is almost impossible to standardize. There are different settings for different types of machines, different user inputs, and different outputs. Adding to this the ever-changing R&D workflows and exponential growth of new advanced scientific instruments generating a ton of data makes this dream more of an asymptote than a reality. However, there is still a lot a lab can do.

Although there are products on the market that promise plug and play integration with lab equipment (TetraScience, QubusLab, and SciTera), chances are you will have to do some custom development. Although custom development of software should be avoided, especially if something exists off the shelf, it is more straightforward to do when it comes to integrations, especially if you know what you need. When considering which equipment to connect, you should ask yourself the following questions.

20.3.2.1 Does the Equipment Support Integrations?

Does the equipment come with an open API? Is the API built and documented well? Does the provider offer some examples of use of the API? Is perhaps the equipment already integrated with an ELN? If not, then it will be difficult to connect it to another system. One way to bypass this is to have the equipment deposit raw data files in either a folder or a database, which can then be imported to an ELN on a regular basis, so it is worth checking if the equipment software supports that. It is ok if the equipment you have scores poorly on this topic, but it is essential that any new equipment you consider buying is addressing the external connectivity.

20.3.2.2 How Often Is the Instrument Being Used?

If you are a molecular biology lab, chances are that you are running a lot of polymerase chain reactions (PCRs). Even though they might not be the primary source of raw experimental data, it might be important for you to follow the specific conditions of each PCR, especially if they vary significantly. Even if you are using very sophisticated instruments, sometimes a good place to start would be with a simple lab scale.

20.3.2.3 Is There a High Chance for Human Error?

Is a human operator responsible for collecting raw data, appropriately marking and storing it in an ELN? Even the most diligent people will make mistakes; especially when it comes to raw experiment data, you want to be sure that the data file corresponds to the samples used, that the right data was used for analysis and interpretation, and that the file was not accidentally copy-pasted to a wrong folder.

20.3.2.4 Do You Need One- or Two-way Sync?

Even if the instrument is just used to measure something simple, it typically needs an input (settings) and output (raw data). There are a few examples though when an instrument only needs input (e.g. PCR machine) or only produces output (e.g. lab scale). Such one way of transferring information will simplify the integration a lot. Even if you would require two-way communication in order to fully digitalize the process, it is a good idea to see the input and output as two separate pieces of that integration and approach them one at a time, rather than all at once.

20.3.2.5 Is the Equipment Using Any Standards?

Although there is a certain lack of standardization when it comes to scientific equipment, there are certain standards that are becoming mainstream, and having an instrument that is capable of communicating in them is definitely beneficial. The main standards to look out for are SiLA and AniML, and equipment supporting them should be easier to integrate. Standards are explained in more detail in the Chapter 08 of this book: Understanding Standards, Regulations, and Guidelines.

20.3.2.6 Is Equipment Cloud Connected?

If the equipment has an ability to connect to the Internet either via Local Area Network (LAN) cable or Wi-Fi, it will make integration and monitoring of it seamless. Having this possibility is definitely a more sustainable option for the future as well.

Although we have listed the questions in order of importance, it is difficult to "calculate" the priority of integrations. The more "no's" are used to describe the factors above, the more difficult and expensive it will be to integrate. However, even if these integrations require a degree of customization, the return should come very quickly, given how expensive scientific experiments can be.

20.3.3 Data Repositories

Data repositories are lists of things that you use in the experiments, from sample collections, to reagents, protocols, equipment, antibodies, etc. From the perspective of traceability, it is very important to have a very clear understanding of what was used in the experiments and what was the state of it when used; therefore, the link between the experiment and an item in a repository is a crucial piece of metadata.

Generally speaking, there are two types of repositories, internal and public. Typical internal repositories would include lists of lab reagents (inventory) and a list of standard protocols that laboratories and organizations use for their work. It is important that the internal cataloging system is compatible with the ELN of choice. Ideally ELN would already have its own repository system that allows for such annotations, and your existing lists should be easily importable. The end goal should be that whoever is looking at experimental results can track back the exact bottle of the reagent or vial of sample that was used to generate them.

Public or external repositories are databases of reagents, protocols, annotations, and vendor catalogs. These are more difficult to incorporate in an ELN. More and more ELNs are being integrated with such databases, but when they are not, these databases often offer exportable versions of their information, which should be transferred to an ELN. If both the public database and the ELN, for example offer API access, it is typically straightforward to do an integration where the information is automatically populated in the ELN.

Apart from traceability, there is another reason why you should integrate with public databases, particularly when it comes to reagents. Having the link between reagent inventory and experiments that consume them, you can simplify tracking the amount of use and automate (or at least simplify) ordering of the chemicals and reagents. Procurement is typically a disconnected process from the experimental workflow, which often results in tedious administration work and delays in reagent availability for experiments; so one of the potential bottlenecks is to focus on improving procurement and be proactive rather than reactive.

For example, SciNote is integrated with protocols.io, a collaborative research platform that is taking the process of scientific discovery to the next level by allowing researchers to share protocols, communicate possible corrections, and optimize research methods. Also, it is integrated with Labviva, an AI-powered digital marketplace for life science reagents, instrumentation, and services. This way you can find relevant information to choose the right reagents, instrumentation, and service for your work and have a seamless purchasing experience through their eProcurement platform.

20.3.4 Data Analytics Tools

Once you have done your experiments and have collected the data in an ELN, it is time to do data analysis. Even if experiments are executed perfectly, and all the data has been recorded, you still need to make sure how data was trimmed and what were the parameters used for data analytics software; otherwise, the experiment will not be reproducible. Oftentimes, scientists use default settings of tools, which change with versions of software, trim the data "as it has been always done," and forget to log the code of the script they have used for the analysis.

Typical data analytics platforms are R, Matlab, and a good old Excel. If you are using R, Matlab, or similar, they can often pull the data straight from the cloud-based ELN. The benefit of this is that calling directly from an ELN will create a link that will assure traceability. Try to avoid downloading data files and reuploading as you might lose track of which file was used to generate the end result. If you are using Excel, make sure that the ELN is integrated with it so that you can track exactly which version it was that was used and what were the formulas that produced the graphs in the end.

Link between raw experiments and bioinformatics is often missing too. In many, one person would perform the experiments in the lab and another would do the data analytics. Having an ELN can serve as a good bridge between them as it will leverage the exchange of information in a more transparent way.

20.3.5 Other Types of Integrations

Integrations above are the ones most requested and it is where labs start with their integration journey. You may, however, consider integrations with tools that are often not directly connected to everyday work in the lab but can be easily implemented and have significant impact on data reproducibility and productivity in your lab.

20.3.5.1 Scientific Search Engines and Literature Management

Every scientific study starts with literature search. Therefore, you should consider having literature connected to the ELN, through direct integration, file uploads, or just simply through proper referencing and linking to external sources. It is important to document this process as a part of the rest of your experiments. Also, having literature management included as a part of this process will help you later when you will be preparing a scientific paper, patent, or a report.

20.3.5.2 Data Sharing

Is the data stored in your system (e.g. an ELN) easily shareable within the team? What about externally? Is it stored in a way that one can make sense of it without the need of explanation? Can you have control over which data is shared and with whom?

Although data is sensitive and can be shared only with specific people, easy access to this data by these people can reduce the number of meetings and e-mails exchanged in order to get everyone in sync. This can be crucial for people with a lot of meetings and those who are traveling a lot as they can have the data readily available whenever they have time to review it.

Sharing data publicly is also a big part of science, and it is often not done to the fullest extent. Scientists often cite "as has been done before" just because it is inconvenient to find the right protocol and transfer it to an appropriate format for sharing. The ability to share research with just a few clicks enables other scientists and collaborators to learn from your work much faster.

Even if your data is confidential, you will most likely have to report the findings to someone within the company or to a client. For example, most ELNs can create some sort of reports, and even if not, a lot of this can be automated just using export functionalities of ELNs.

20.3.5.3 Publishing

Writing a scientific paper is often a long and tedious process, especially if multiple contributors are involved. Using tools to help with this, such as Authorea and SciNote Manuscript Writer, can help coordinate the efforts, streamline content creation, help with formatting, and simplify the submission process, which can save a lot of time and get work published faster.

In order to take full advantage of these tools you have to have them integrated and the data digitized. An example of this is using a protocol in an ELN, publishing it in protocols.io, and having the digital object identifier (DOI) number ready for submitting as a part of your material and methods section.

Even though publishing papers is a goal for academic researchers, scientists in industry often communicate with each other reports that are very similar to research papers, as they contain the same elements, so even if you work in a private company, automating publishing of knowledge within the company can save a lot of time.

20.3.5.4 Upgrading Plans

If you fully implemented the ELN in your lab, sometimes the easiest way to generate more value is to simply upgrade to a more advanced plan. ELN vendors usually offer advanced plans with more features and integrations. It is a good idea to get the detailed list of advanced features and see how well they map to your requirements based on the data workflow in the lab. If you are not sure if the plan will work for you, ask vendors to walk you through the features and typical use cases. It is their job to help you. Ask for a trial period so that you can try them out before committing.

20.4 Budgeting

The returns on investment into software in life sciences are typically very high and very fast as the experiments and scientists' hourly rates are so expensive. Even very marginal improvements in efficiency and reproducibility can save thousands of dollars of research money. Despite that it is important to have software costs under control as once we have committed on this path, they will surely increase over time. It is important that the cost of license (as well as any future upgrades) work with your budget and the financial plan your lab is following. Regardless of whether you are using grant money or company funds for your research, make sure to put software costs as part of the expenses. Expect that the cost of software licenses might increase at about 20% per year.

If you do not have enough money in your budget, but you see a clear added value for a software plan, vendors typically offer various discount schemes for you, although they might not be something they actively promote on their web page. A few things you can ask for:

- Loyalty program (sometimes vendors will offer significant discounts if you refer them to someone).
- Commitment for multiple years (if you are sure you will use the same vendor for at least couple of years, you can negotiate a better deal by signing a multiyear agreement with them).
- Contribute with use cases and marketing materials (vendors love to publish stories and use cases from their customers).

If you are planning to do integrations on your own, keep in mind that they are custom development, which has significantly higher initial costs, but it also comes with maintenance costs, as APIs, data formats, and your demands change over time. Although the cost of an integration can vary from a few hundred dollars to tens of thousands of dollars, the cost largely depends on how well you understand your needs. The better you can specify which data you want to transfer between systems, less time developers will need to create and maintain the integration. If an integration is well designed and built, it should cost about 10% of the original cost to maintain it per year.

20.5 Continuous Improvement as a Value

Continuous adoption of new software will put strain on your team. It is also impossible to get everything right the first time. So once you made mistakes and implemented software that later needs to get replaced, people might get frustrated. It is important to approach this in bite-size chunks and only move to the next stage once everyone is onboard. This can be a long process so you will need to have patience and follow your digitalization roadmap.

However, if people see that this is what makes them more competitive or that it makes their work easier, they will see the adoption of new systems as a challenge rather than a problem and a part of their job. Make sure you show them how new implementations helped improve your work and celebrate small victories.

References

1 Vasilevsky, N., Brush, M., Paddock, H. et al. (2013). On the reproducibility of science: unique identification of research resources in the biomedical literature. *PeerJ* 1: e148.
2 Vines, T., Albert, A., Andrew, R. et al. (2014). The availability of research data declines rapidly with article age. *Current Biology* 24 (1): 94–97.
3 Pryor, G. (2012). *Managing Research Data*. London: Facet.

Part VI

Vision of the Future and Changing the Way We Do Science

With continuous improvements in mind, we will conclude the book with insightful expert comments on the subject of the future of science. Many of the described technologies are already becoming important for the scientific purpose, and here we identify those that might shape the next 5 to 10 years and change the way we do science.

21

Artificial Intelligence (AI) Transforming Laboratories

Dunja Mladenic

Department of Artificial Intelligence, Jozef Stefan Institute, Jamova 39, 1000 Ljubljana, Slovenia

21.1 Introduction to AI

Time is now and now is the time: We cannot pretend that it is not happening, digitalization is spreading in our work and daily life. There are good and bad sides of it, but to take the best out of digitalization, we should be ready to use it. This may involve raising awareness of the general public and also recognizing the potential of digitalization in specific domains and areas such as laboratories. Moving from paper to digital is much more than just changing the medium for storing the data.

We would like to be not only ready for today but also shape our tomorrow. Understand the opportunities and challenges of digital transformation and predict at least some of the consequences.

This chapter first discusses opportunities, needs, and challenges related to the role of artificial intelligence (AI) in transforming laboratories. As AI is a very broad field, different subfields can be involved in transforming laboratories. Here, we focus mainly on machine learning, text mining, and complex systems analysis, primarily having in mind data and processes present in the laboratories in general. Depending on the specific nature of the laboratory, importance of security, complexity of data sources, and other characteristics, AI can bring a whole variety of methods and automatization possibilities, including robotics, sensor data analysis, planning, and others.

21.1.1 Opportunities

The data comes in a variety of forms, from database records and well-organized facts to images, sensor measurements, written comments or observations, longer texts, sound and video recordings, activity on social networks, satellite images, etc. Handling the volume, variety, and velocity of arriving data is commonly referred to as handling big data. Technology and AI methods are developing to enable storing, handling, and real-time analysis of rich data. There are still many research challenges including multimodal and cross-lingual data analysis, data fusion on heterogeneous

Digital Transformation of the Laboratory: A Practical Guide to the Connected Lab, First Edition.
Edited by Klemen Zupancic, Tea Pavlek and Jana Erjavec.

data streams, and complex systems analysis including artificial and biological systems. Nevertheless, we already have AI methods and tools available for real-time big data analytics that can be utilized in digitalized laboratories and to some extent also to support the process of transforming laboratories.

Today's society and especially businesses are recognizing data as assets. Businesses collecting and buying data for their own usage or selling them to the third party are very much present. Moreover, individuals are starting to recognize that their data have value; for some commodities that come labeled as free actually we are paying by providing our personal data, such as zip code, traces of phone location, search keywords, and entertainment preferences. Even by simply being an active member of the society without explicitly providing any data, we are leaving digital traces by phone usage, media consumption, credit card usage, using the Internet, etc. On the other hand, humanity is developing new ways to perceive the world and ask questions about it. The opportunity here is to become aware as individuals, professionals, and a society of the ongoing transformative processes supported by digitalization and consciously steer them to our benefits and benefits of the future generations.

21.1.2 Needs

Living and working in the world of data and information, we also have needs for coping with the information overload. We are exposed to media and marketing proactively reaching for our attention and often infiltrating their messages in our way of thinking and approaching life. Very often instead of making our life easier, the new product and services are making things more sophisticated and complicated. This is even more evident with the new generations that are already changing their beliefs about life, their ways of communication, and information consumption. The attention span is shortening, and smart mobile devices are becoming an integral part of life and socialization. Spending a day outdoors with friends without looking at the mobile devices may soon become rarity. Talking to a living person on the other side of a line is already becoming a rarity, especially in customer services.

21.1.3 Challenges

The question is how we can use machines and AI to support us in simply being human, expressing and communicating from our creative and intuitive nature, rather than becoming mechanical and adjusting ourselves to limited views of what digitalization and automatization can offer to us. To address this challenge, we propose to reach out for common knowledge and wisdom. When someone is using a hammer, it is clear that it cannot be a hammer to blame for the outcome, as it is a tool. Maybe less evident is the usage of our mind as a tool. We tend to identify with our thoughts, feelings, and emotions, ignoring that it is our responsibility to choose which thought to nourish and follow. This is very evident in yogic philosophy, a yogi "*takes life as a gift, he takes the body as a gift, he takes the mind as a tool...He meditates to clean his mind so that mind can be sharp and intelligent and intellectual and intuitive.*" [1]

To get the best out of AI, we suggest to see it as a collection of tools or even as a potential. To apply AI in our lab, we can check what is available and choose something off-the-shelf or articulate what is needed in collaboration with AI experts to see if we can get user-tailored tools or solutions. Maybe it will turn out that some problems cannot be addressed by today's methods, but our efforts may inspire researchers to develop new methods and approaches. In any case, AI methods, like a hammer or our mind, are tools, and it is on us humans to use them consciously and for the right purpose.

21.2 Artificial Intelligence in Laboratories

Digitalization of laboratories is happening in some areas faster than in others, depending on the needs and opportunities to make a change from paper to digital. AI can serve as a catalyzer in the process, offering a whole range of additional services including sanity checking, outlier detection, data fusion and other methods in the data preprocessing phase, different methods for data analytics and modeling, monitoring of data consumption and other dynamic processes in the laboratory, and decision support to domain experts (Figure 21.1).

21.2.1 Data Preprocessing

Laboratories can have different motivations to go digital, and regardless of the motivation, having standardized processes and data representation is advisable. Even if at the moment we are not interested in combining data from different sources or exchanging data with other laboratories, this may become important in the future. Researchers are reporting on data analysis results benefiting from performing analysis on data from multiple laboratories, for instance in risk classification on cancer survivors [2]. It can also be that digital images can help to provide an estimate of some process we are observing, which then requires adequate online measurement techniques, e.g. using smartphone cameras to estimate biomass concentration during cultivation in shake flasks [3].

Once the data is digitized, AI methods can help in data preprocessing. The methods today can handle a variety of data including measurements and results of testing, textual descriptions, images, and voice recordings. AI methods can contribute to the data preprocessing phase in different ways depending on the nature of data, the processes of obtaining and storing data, or some other data properties. For instance, we can have a tool based on machine learning and statistical data analysis to highlight unusual values or suspicious sequences of values, identify inconsistency over different data sources and data modalities, and enable reducing the number of tests by automatically filling in the missing values or pointing out where more data should be collected.

An example scenario where AI methods can help is in organizing and sharing laboratory notebooks. Using digital media instead of paper enables sharing of collaborative writing, automatic processing, cleaning, and connecting different parts

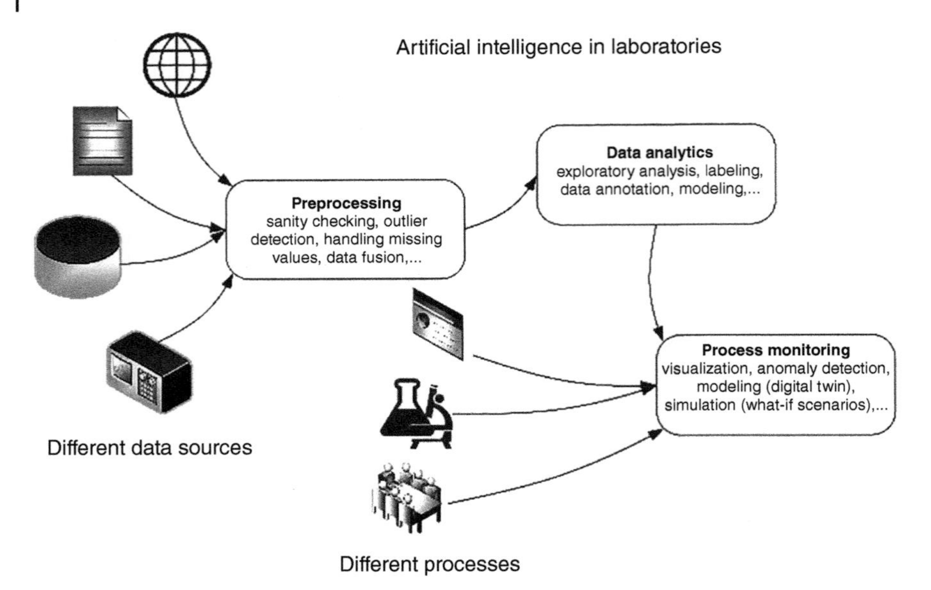

Figure 21.1 Artificial intelligence can support laboratories on different points, from data preprocessing and fusion of data from different sources to data analytics and monitoring of different processes. Source: Courtesy of Dunja Mladenic.

of the written notes. One can use advanced search over a collection of laboratory notebooks, find similar past notes (also across different laboratories potentially written in different languages, if needed), automatically organize the notes according to some predefined ontological structure, bring additional annotations, or automatically predict values of missing metadata. However, all this potential is useless, if it does not support laboratory workers in their work. A study performed in the domain of physics and astronomy to understand better potential barriers in adopting digital notebooks in laboratories has identified that compared to paper notebooks, digital notebooks may suffer from limited configurability, lack of user control, and not being physically attached to the equipment they refer to Ref. [4]. Most of the identified barriers can be addressed in the development of ICT systems for laboratory notebooks.

21.2.2 Data Analytics

Computerization of laboratory work is present for decades in the area of healthcare. Digitalization of laboratory results and medical journals has significantly contributed to efficiency of storing, analyzing, and utilizing medical knowledge. Moreover, as pointed out in [5] "the early computers were also used for evaluation of laboratory results as well as for digital filtration, imaging, pattern recognition and dose planning". AI methods today enable efficient organization of large amounts of heterogeneous data supporting efficient search and retrieval. On the top of that, depending on the data modality, the user can be provided with powerful data exploration tools including rich data visualization, automatic outlier detection, data modeling, and prediction.

Moreover, data analytics can be applied in different phases of laboratory work from monitoring and directing data collection, data preprocessing, storing, and modeling to searching historical laboratory data (e.g. test results and notebooks) and enabling scientists to share data and models across problems and laboratories. AI and in particular machine learning methods have been used for classification and prediction on medical data for decades. We used to get paper records giving symptoms and medical diagnoses usually for no more than 100 patients. We type the data in a computer to create training examples and run machine learning to create models for predicting the probable diagnoses, given the symptoms. Today, there are millions of patient records that are generated in digital form, many medical images stored in databases, and millions of research papers in biomedical knowledge bases. The challenge is to articulate the right questions, the prospective research hypothesis to be tested, and the appropriate approach.

An example scenario of rich text data analytics to enrich workflow of health professionals is support for exploratory data analysis based on automatic classification of reports, related research papers, and even media news. With today's capability of text analytics, we can connect content over different natural languages and enable the users to analyze media coverage and health news bias around the globe [6]. Moreover, the data can be enriched linking it to similar situations in the past (which can apply to laboratory scenario if appropriate amount of digitized data is available), finding related notes or articles (potentially in a cross-lingual setting), and providing related additional data sources (general, such as weather or specific, such as related outcomes from other laboratories around the world). Additionally, usage of domain-specific vocabulary and ontologies can help in search and modeling, taking descriptions of data and processes to the next level, going beyond the surface form of the words to semantics, and reasoning (if we encode the available knowledge in the form of knowledge graph).

21.3 Process Monitoring

The processes in laboratories involve activities and data that may benefit from being monitored and modeled. Businesses and industry are increasingly using modeling of complex real-world systems to cover scenarios such as better understand the properties of the systems, simulate hypothetical scenarios and answer what-if questions, and detect anomalies resulting from interaction of many data streams in real time. Here, we are talking about modeling dynamics and outcomes of a process or a set of possibly interconnected processes. Historical data can be used to build a reference model that can be adjusted, given the current context and trends of the monitored process.

As AI methods are utilized to construct a digital twin of a machine, logistic processes, or a production plant, they can be also used to construct a digital twin of some processes in the laboratories. This would enable monitoring for interdependencies, possible anomaly detection, and simulation of future developments of the process, giving the professionals the possibility to ask what-if questions.

With the capability of real-time modeling and monitoring of the incoming data, we can analyze digital laboratory notebooks over time within the same laboratory or across different laboratories. In a similar way, as machine learning methods have been used in decades of research publications to predict the next big thing in science [7], one can analyze and monitor processes and data flows in laboratories.

21.4 Discussion – Human in the Loop

Digitalization and integration of AI in different sectors and wider society is unfolding. International governmental organizations such as UNESCO, OECD, and UN together with European Commission and national governments are actively participating in shaping the future in the light of digitalization. For instance, OECD has appointed special international expert groups addressing different aspects related to the role of AI in the society. OECD AI Policy Observatory (see Website: OECD.AI) makes even more evident the significance of Artificial intelligence in our changing society and the role of data. Namely, data is becoming a new factor in the production process, a public good and thus we need a regulatory framework to facilitate the free flow of data while ensuring trust.

We are witnessing how AI is changing our lives by improving life quality (e.g. in healthcare, efficiency of farming and productions systems, climate change mitigation/adaptations, and security), but we also recognize that it is bringing potential risks (e.g. opaque decision making, biased suggestions, intrusion in privacy, and usage for criminal purposes). As explained in the European Commission White paper on AI "*To address the opportunities and challenges of AI, the EU must act as one and define its own way, based on European values, to promote the development and deployment of AI.*" [8] The aim is to enable scientific breakthroughs and ensure that new technologies are at service of humanity improving our lives while respecting our rights. Establishing a sound regulatory framework is important as trustworthiness is a prerequisite for uptake of digital technology as it is becoming an integral part of our lives. High-Level Expert group of European Commission issued guidelines on trustworthy AI listing seven requirements: human agency and oversight; technical robustness and safety; privacy and data governance; transparency; diversity, nondiscrimination, and fairness; societal and environmental well-being; and accountability.

Today, we recognize that sustainable economic growth and societal wellbeing is increasingly drawing from value created by data. "*AI is one of the most important applications of the data economy.*" [8] As digitalization and AI are spreading in the labs, we should be prepared and look in advance to defining the role of AI in the lab and redefining the role of humans, while leaving space for changes. It is important to also consider the responsible use of AI in laboratories in alignment with EU legislation.

Regardless of how much and in which processes in the laboratories we use AI, we should remember that AI can cover some of the intelligence, but there are other dimensions that humans bring to the process that cannot be covered by AI. We can say that intelligence is our capability to analyze some situations, learn from mistakes we have made or have seen others made, predict consequences of some action before taking it, and develop a strategy to achieve a goal. Intelligence is connected to clear, focused, and selective thinking and needs our guidance to avoid getting lost in details or fantasies. According to yogic philosophy, intelligence is one of the three creative forces needed to manifest success in life: consciousness, intelligence, and energy [9]. For instance, knowing what and how we want to manifest, we need resources/energy to really do it; knowing what to manifest and having resources, we need a strategy. Humans play a crucial role in the process by consciously deciding what is the goal of some work or laboratory experiments, intelligently developing a strategy for achieving the goal, and involving resources to implement the strategy and reach the goal.

References

1 Bhajan, Y. (1975). UCLA lecture on self courtesy. https://www.libraryofteachings.com/lecture.xqy?q=date:1975-01-14&id=cb9fe022-9f0f-aa54-c125-a887735c7091&title=UCLA-Lecture—Self-Courtesy (accessed 15 February 2020).

2 Chen, Y.-C., Ke, W.-C., and Chiu, H.-W. (2014). Risk classification of cancer survival using ANN with gene expression data from multiple laboratories. *Computers in Biology and Medicine* 48 (1): 1–7, Elsevier.

3 August, E., Sabani, B., and Memeti, N. (2019). Using colour images for online yeast growth estimation. *Sensors* 19: 894.

4 Klokmose, C.N. and Zander, P.-O. (2010). Rethinking laboratory notebooks. In Proceedings of the 9th International Conference on Designing Cooperative Systems COOP-2010, Springer-Verlag.

5 Davila, M. (2008). Computerization of laboratory work 1: an outline. Report KTH, Philosophy and History of Technology.

6 Costa, J.P., Fuart, F., Stopar, L. et al. (2019). Health news bias and its impact in public health. In Proceedings of the 22nd International Multiconference Information Society – IS 2019 (SiKDD-2019), Ljubljana: Institut "Jožef Stefan," 21–24.

7 Grobelnik, A.M., Mladenic, D., and Grobelnik, M. (2019) The next big thing in science. In Proceedings of the 22nd International Multiconference Information Society – IS 2019 (SiKDD-2019), Ljubljana: Institut "Jožef Stefan," 9–12.

8 WHITE (2020). PAPER on artificial intelligence – a European approach to excellence and trust. COM 65 final, 19 February 2020, European Commission.

9 Singh, S. and Mladenic, D. (2018). Three creative forces to manifest success. In: *Expressing Leadership and Success*. Waldzell Leadership Institute.

22

Academic's Perspective on the Vision About the Technology Trends in the Next 5–10 Years

Samantha Kanza

Department of Chemistry, University of Southampton, Faculty of Engineering and Physical Sciences, University Road, Southampton SO171BJ, UK

There is no doubt that digitalization is taking over and that more and more processes are becoming digital in both the home and the workplace. However, as discussed in Understanding and Defining the Chemical Laboratory's Requirements: Approach and Scope of Digitalization Needed: Stages of Digitalization section of this book, creating a fully digitalized laboratory environment is challenging, and there are still certain adoption barriers surrounding the use of technology to capture scientific research in the laboratory that remain prevalent. We need to revolutionize our laboratory environments, and with the advent of new hardware and software technologies that are rapidly becoming (or are already) available at reasonable costs, there could be a significant change coming to the laboratory as we know it.

22.1 Hybrid Solutions

A key issue in digitalizing the laboratory has always been that paper holds many advantages over electronic devices for ease of data entry and portability. However, in recent years Smart Paper and Pen solutions have started to emerge, creating a hybrid solution that combines the affordances of paper with the benefits of an electronic solution (search, robustness, and backup capabilities). Two current vendors at this time are the Bamboo Spark Smart Notebook [1] and the reMarkable tablet [2]. At present these devices offer electronic notebook capabilities in that you can write on them like paper and electronically save and sync your documents, search them, and if desired convert handwritten notes into a typed format. There are also countless electronic lab notebook (ELN) and laboratory information management system (LIMS) vendors and offerings that have versions for electronic tablets such as iPads. It is not a giant leap to consider that in a few years there will be ELN solutions that either can run off a hybrid notebook or even that there will be specially created hybrid ELNs that act more as a true replacement for a paper lab notebook. These may well be more widely accepted than current ELN solutions as they would

Digital Transformation of the Laboratory: A Practical Guide to the Connected Lab, First Edition.
Edited by Klemen Zupancic, Tea Pavlek and Jana Erjavec.

still retain the usability advantages of paper, although an obvious barrier will be cost and durability of hardware.

22.2 Voice Technologies

Introducing voice to the laboratory environment has a huge potential to revolutionize the way in which we do science and to make vast improvements in the laboratory setup, with respect to the capacity to digitize information, perform experiments, and access vital information inside and outside the laboratory. One of the major barriers to adoption to using certain pieces of technology in the laboratory is ease of access, for example a scientist does not necessarily want to take off his/her gloves if he/she is mid experiment to type something into a computer, whether to access or record information. This is one of the reasons the paper lab notebook has remained so popular in today's digital age, as it is quick and easy to use, and there is typically less of a concern about damaging a pen used to record data into a lab book than an expensive laptop or tablet. But imagine if you could record experimental information through voice? Imagine if you could speak your observations out loud and have them recorded, which would remove the entry barrier to needing to type it out. Similarly, if you could ask an assistant a question such as what amount of this chemical am I meant to be weighing out? Or what are the hazards associated with this chemical? Then you could gain access to this information immediately without needing to look it up in another book or on a computer.

22.3 Smart Assistants

Logically following on from using voice comes the idea of smart assistants. This is not a new technology, we already have smart assistants for the home (e.g. Google Home) and on our mobile phones (e.g. Siri or Cortana), surely a smart lab assistant is the next logical step. Certain projects are already working on this idea including the Talk2Lab initiative at the University of Southampton [3] and the Helix Lab Assistant. These assistants could be linked up with LIMS/ELN systems, inventories, health and safety data, and equipment and instruments, the sky's the limit (Figure 22.1).

22.4 Internet of Things

There is also great potential for internet of things (IoT) devices to make their way into the laboratory. Sensors can be used to measure an inordinate number of things nowadays, and these could be used to monitor experiments, laboratory conditions, specific pieces of equipment, and a whole range of things. These sensors could be linked to voice-activated smart lab assistants to provide a fully immersive controllable responsive laboratory. People have already wired up their homes to have their lights, heating, security, and appliances controlled by IoT and smart assistants; observing and

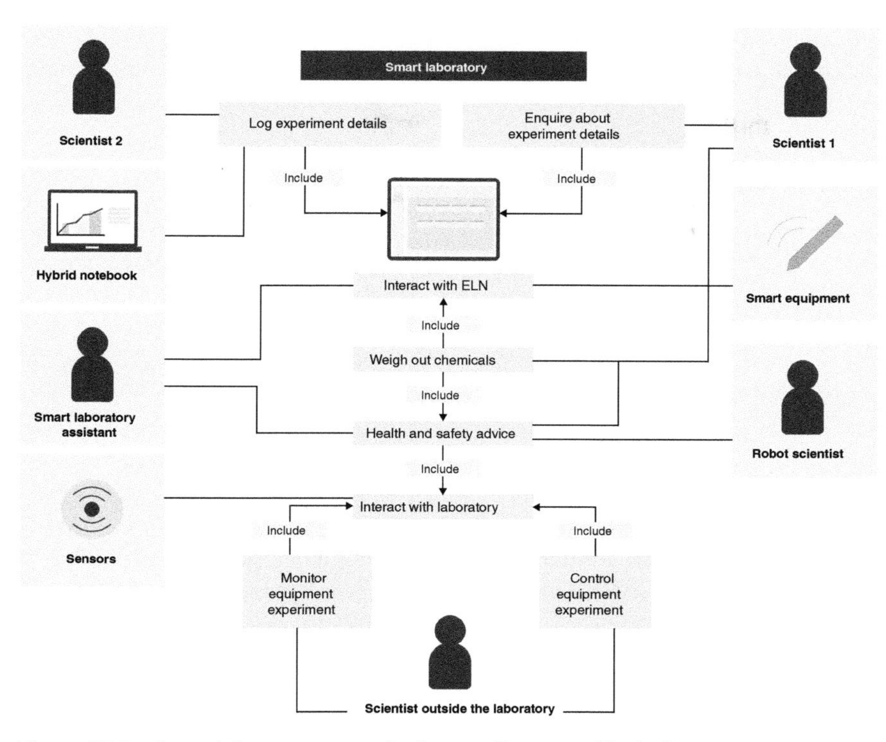

Figure 22.1 Smart laboratory example. Source: Courtesy of Janja Cerar.

interacting with their houses from afar both to control appliances and monitor different aspects of the house. Translating this to the laboratory could mean automated experiments, and or experiments that can be controlled and monitored from afar; video feeds could be used not only for monitoring the laboratory but could also be combined with image analysis to alert the scientist when they needed to return to the laboratory to proceed with the next bit of the experiment or if there are any issues. Further, there is already a plethora of digital equipment that can be linked into laboratory systems; therefore, linking these all together and adding a voice control layer seems like a logical next step. The base technology already exists to create a Smart Lab environment, and researchers are working on the software; therefore, making this advancement is only a matter of time.

22.5 Robot Scientists

Much work is being conducted in the area of intelligent robots, and these are being used in many different areas such as Agri-Tech to pick fruit and treat crops, and in chemistry where research is already underway to create a robot scientist that can perform experiments [4]. Two significant barriers to this technology flooding into laboratories currently are cost of equipment and the intricacy of the algorithms required. A great deal of complex artificial intelligence (AI) has been created for these robots to enable them to perform the required movements to complete these actions, and as the technology becomes more common place, and cheaper mainstream versions of the robots are created, it is well within the realms of possibilities that many more laboratories and scientific research facilities will be frequented by robot helpers.

22.6 Making Science Smart – Incorporating Semantics and AI into Scientific Software

In addition to finding novel ways of bringing the laboratory into the twenty-first century and making it smarter, research is also currently focused on improving the systems used outside the laboratory. AI and Machine Learning is the buzz of 2019, and businesses and academia alike are on a mission to revolutionize everything with the magic of AI. While AI is of course not magic, intelligent algorithms hold great potential to create smart systems (AI is described in detail in the previous Chapter 21).

Additionally, researchers seem to finally be coming around to the fact that AI needs Semantics and vice versa. One of the big challenges facing AI is lack of suitable data, and in an almost equal standard to AI, making data "FAIR" [5] is another passion of the scientific research community. Semantics puts the I in FAIR by providing interoperable data formats that can be read by machines as well as human beings. The combination of these two powerful technologies holds great potential to create better smarter systems.

There are not many Semantic ELNs out there currently, but researchers are working toward adding metadata, annotations, and intelligent data linkage into these systems. Additionally, many ontologies have been developed for different scientific domains, and standards such as the Open Biomedical Ontologies (OBO) [6] and Universal Medical Language System (UMLS) [7] have been formalized to promote reusable and extendable standards within scientific ontologies. As noted in Understanding and Defining the Chemical Laboratory's Requirements: Approach and Scope of Digitalization Needed: Write-up Stage, there is both a lack and a need for more knowledge management in science. Smart research data management systems are the future; systems that can infer related experiments and related projects, can recommend literature related to your work, and that have automatic semantic indexing and searching capabilities.

Similar to voice technologies, IoT, and smart assistants, a lot of these technologies already exist, but some of them are not quite good enough yet and have yet to be combined into ELNs, LIMS, or other scientific software offerings in a way that has made a notable difference to a scientist's working environment.

22.7 Conclusions

A combination of these technologies has the potential to revolutionize laboratories, and realistically many of them already exist. They have yet to make their way into the laboratory to form a truly connected laboratory of the future for three main reasons. Firstly, some of the finer details of the right ways to integrate the technologies have not quite been ironed out yet, for example the incorporation of Semantic Web technologies into ELNs and LIMS systems. Secondly, some work has been conducted to consider the potential of some of these technologies in the laboratory (voice, IoT, and smart assistants), with some preliminary implementations, but work now needs to be done to fully integrate these in laboratories, including user studies, training of voice-based and smart systems to understand scientific terminology, and potential development of new sensors that are geared toward laboratory specifics. Lastly and potentially the hardest issue to overcome is the perspective change required. Everyone from the day-to-day laboratory users to the supervisors, line managers, CEOs, and everyone in between who has some connection to the laboratory needs to get onboard with the idea of using these technologies and adopting new practices to digitalize their laboratories. Ultimately, revolutionizing a laboratory and letting technology change how we do science requires more than just good technology, it requires a change of attitude and organization.

References

1 Wacom (2019). Wacom global. /en-my/products/smartpads (accessed 24 September 2019).

2 reMarkable (2019). The paper tablet. https://remarkable.com/?gclid=Cj0KCQjwzozsBRCNARIsAEM9kBNNQAqdjn8cOL_Z4cPMXpYUmscy-5lHxHud6l71Km1wWGJnV8Vic-waAgnfEALw_wcB (accessed 19 September 2019).

3 https://ieeexplore.ieee.org/document/9094640.

4 Sparkes, A., Aubrey, W., Byrne, E. et al. (2010). Towards robot scientists for autonomous scientific discovery. *Automated Experimentation* 2 (1): 1.

5 Wilkinson, M.D., Dumontier, M., Aalbersberg, J.J. et al. (2016). The FAIR guiding principles for scientific data management and stewardship. *Scientific Data* 3: 160018.

6 OBO (2019). OBO foundry. http://www.obofoundgry.org/ (accessed 17 October 2019).

7 National Library of Medicine (2019). Unified medical language system. https://www.nlm.nih.gov/research/umls/index.html (accessed 17 October 2019).

23

Looking to the Future: Academic Freedom Versus Innovation in Academic Research Institutions

Alastair Downie

The Gurdon Institute, University of Cambridge, Tennis Court Road, Cambridge CB2 1QN, United Kingdom

23.1 Introduction

One might conclude, reading through the chapters of this book, that private/commercial research laboratories are better resourced, better equipped, and ahead of the game compared to academic laboratories. And one might presume that this is because laboratories in the private sector enjoy high levels of investment, generate large profits, and can afford to speculatively embrace the very newest technology before their benefits are proven, while public sector institutions must operate with less funding and demonstrate a greater level of fiscal responsibility. In my view, the reasons are more complex and have not so much to do with funding but more to do with management structures and executive overview.

Most managers will acknowledge that it is easier to make good business decisions from a position that is emotionally and structurally distant. In industry, where the goals of the organization outweigh the needs of the individual, a clear, hierarchical structure enables the CEO, detached from the concerns and preferences of individual researchers, to maintain a holistic overview of the entire organization and its performance. This strong management position ensures that processes are consistent and aligned with a unified vision, and that the organization is able to respond quickly and decisively to new opportunities in a competitive and constantly changing landscape.

23.2 Corporate Culture Versus Academic Freedom

In contrast, a research-intensive university is like a business with 500 CEOs all pulling in different directions, where the culture and tradition of academic freedom allows the needs of the individual to precede the goals of the organization. Principal investigators define their own goals, strategies, and processes, and the institution, because of its complex structure and somewhat marginalized position, is unable to fully support all of these independent activities and interests.

Digital Transformation of the Laboratory: A Practical Guide to the Connected Lab, First Edition.
Edited by Klemen Zupancic, Tea Pavlek and Jana Erjavec.

Another challenge for universities is that academic researchers tend to identify themselves as members of a global community of peers working together in a chosen discipline, rather than a collegiate community within a specific institution. "Drosophila people," for example may care more about feeling aligned with other Drosophila laboratories around the world, than with immediate colleagues in the university that provides their accommodation, infrastructure, and services.

Academic freedom is a precious thing that must be maintained and protected of course – it is a vital force for research innovation – but its absolute superiority is an obstacle to essential and effective transformation within the organization. Without high-level oversight and some degree of process management, it is very difficult for any organization to create a platform for strategic development or to take advantage of opportunities and innovations offered by commercial partners.

Take, for example electronic lab notebooks (ELNs). Essentially a replacement for simple paper notebooks, one might think that identifying and deploying a successful, organization-wide solution should be an easy and pain-free business decision. And apparently it is so, in industry, where there are consistent, efficient, and managed workflows and where research notebook systems have been in use for almost 20 years. In academic research institutions, however, the diversity and constantly changing nature of activities, interests, and collaborations makes the task much more difficult. Within any research group one might find bench scientists working alongside computational biologists, alongside engineers and physicists, and all of these roles have different workflows and processes, and different requirements of a documentation system.

The market is cluttered with specialist research notebook tools for all disciplines, and each vendor tries to differentiate its product by adding more specialist features and integrations. This appeals to a small population of researchers who have a need for these specialist features, but for other researchers it simply adds undesirable bloat and complexity. There is no perfect research notebook product on the market that will satisfy all researchers, all of the time. In a managed environment a best-fit product can be selected and supported, and everyone will adapt and recognize the benefits of common processes; but in most universities this approach is not available, and research groups, by their own choice, have been left to fend for themselves.

23.3 Spoiled for Choice, but Still Waiting for the Perfect Solution

Principal investigators in universities are overwhelmed by grant applications, publication deadlines, teaching commitments, and much more. According to a survey [1] undertaken in Cambridge in 2017, more than 80% of researchers would like to switch to a digital notebook system, but principal investigators are intimidated by the bewildering choice, the commitment required and the difficulty of disengagement, and the prospect of disruption during implementation – they simply do not have the time to research or implement new technology that will require their research groups to adapt to new methods. Consequently, only a small number of PIs have subscribed to commercial products; some have developed informal DIY systems using familiar and readily available productivity software; most have procrastinated and are waiting for guidance or for a solution to reveal itself.

Young researchers on the other hand, in academic laboratories, tend to be enthusiastic experimenters and adopters of new products and methods on a personal/individual basis, and many are taking advantage of free individual user licenses offered by vendors. Inevitably, this will lead to a proliferation of different products even within single research groups and a proliferation of locations where critical data is stored in the cloud. Perhaps, this is appropriate or even the only possible outcome, given the diversity of activities undertaken in university laboratories.

Flexibility within an organization does not promote agility. In fact, the "everyone for themselves" nature of academic interests and working practices has become a ball and chain for many institutions – trying to marshal and direct researchers to adopt common workflows is, as the saying goes, like trying to herd cats. So how then might we improve the capacity of the academic research community to embrace change and strategic opportunities, without interfering with its freedom and flexibility?

23.4 Building a Single, Shared Infrastructure for Research Data Management

It is certainly possible to identify some processes within the academic research workflow that could be harmonized across the entire community – across all disciplines in all institutions in the country. All researchers need to document their work, for example. They also need to store their data safely and, when the time comes, they need to publish their data and make it accessible to the whole community. These are basic activities and common goals for every researcher, in every discipline, and in every university, and they could be supported by a single, joined-up, and national infrastructure for research data management.

The features provided in a typical ELN can be divided into three categories: features that are essential for everyone; features that are useful and relevant to everyone; and specialist, discipline-specific features that are useful to a limited number of researchers only. I propose that a single, universally relevant, discipline-agnostic research notebook platform could be delivered to the whole academic community as part of a national infrastructure, providing all of the essential features to researchers in all disciplines. It is true of course that most disciplines have their own unique needs and expectations, and this is where I believe there is a great opportunity for commercial partners to develop interface skins and bolt-on tools to cater for those specialist needs. Researchers would benefit from all the choice that the marketplace would offer; commercial partners would benefit from immediate access to a much larger client base; and underneath that layer of diverse, innovative, specialist features, all critical content and data would remain in a single, basic software platform, providing consistent workflows, quality, experience, and a shared culture and language for every researcher in every academic institution in the country.

Crucially, separation of the basic documentation functions and the underlying database from commercially developed, advanced interface designs and tools will remove many of the current anxieties that are barriers to uptake. Researchers will be able to experiment with innovative interface designs and specialist features quickly, easily, and without risk of disruption, divergence, or difficulty of disengagement. As easily as switching on/off designs in a modern web publishing platform, they will be able to identify and develop their own ideal processes and workflows, confident in the knowledge that their content remains safe and unchanged in the national research data management infrastructure.

Data storage is a looming problem for all researchers. The volume of data that is stored in academic research institutions is growing, and the rate at which it is growing is accelerating. Funding bodies have imposed retention periods that exceed the typical life expectancy of any storage hardware, and compliance requires commitment to a cycle of investment in storage infrastructure. All institutions across the country are currently locked into this cycle – even some individual departments within those institutions – at very great expense they are all designing, building, and maintaining their own local, independent, and unique storage systems, despite all sharing exactly the same goal. There are also tens of thousands of repositories around the globe, and more being built every day, all with the same purpose: to curate and disseminate published data and tools for the benefit of the research community. But the community cannot be best served by resources that are distributed across such a large range of systems. I believe that there is no need for such a proliferation of independent-but-identical storage infrastructures, and that most of this data, both live and published, could be stored in a single, national infrastructure – a small network of very large data centers distributed around the country – accessible to all researchers in all academic institutions, via ORCID credentials that will remain persistent irrespective of location and career progression. Data would remain in that single infrastructure throughout its entire lifetime, and researchers would move around it, operating, analyzing, sharing, and publishing the data remotely.

In addition to the extravagant multiplication of independent infrastructures and the enormous financial burden on institutions, I believe the growing volume of research data will soon cripple current, familiar methods of managing data in files and folders. Best practice in research data management is widely promoted and recommends robust file-naming conventions and well-organized directory structures and compliance with the FAIR principles (that all data should be findable, accessible, interoperable, and reusable). But it is not uncommon for modern research directory structures to be 30+ levels deep and populated by large datasets with necessarily complex and cryptic filenames. Looking forward, the process of finding old data created by colleagues who no longer work in the lab, by searching for filenames or rummaging through folders and speculatively double clicking on files, will become increasingly hopeless. Unless we change our approach, these files will become lost – unfindable – submerged in a rising ocean of unmanaged data.

In a national data management infrastructure with a repository-style interface, metadata and persistent identifiers (attributed immediately to all live and published datasets) would ensure easy discovery and retrieval of data even many years after its creator has left the laboratory, either directly through the repository interface or via persistent links in the integrated documentation platform described above. And, as with the documentation service, the storage service would be built to enable commercial partners to develop a marketplace of optional, bolt-on, discipline-specific designs and specialist features.

A national approach – building and sharing a single, joined-up platform for research data management – will bring many advantages: economy of scale; local complexity reduction; consistent experience and processes for the entire research community; and a common, familiar, and collaborative culture and language for everyone. Perhaps the most significant of all, however, would be the creation of a framework for rapid and easy deployment of new tools and resources to the entire academic research community, created by both in-house developers and commercial partners, independent of any individual institution. Researchers would retain all of the academic freedom they need to pursue their research interests, with great confidence in rock-solid national infrastructure, services, and support, while cherry picking their favorite designs and features from a menu of risk-free, easily deployed innovations offered by commercial partners.

23.5 A Journey of a Thousand Miles Begins with a Single Step

To put this idea into the context of a "connected lab", my view is that academic research laboratories are unlikely to find common ground, voluntarily, that will foster and enable institutional agility or the ability to take advantage of innovative technologies in any smart or managed way. If principal investigators had more time and resilience to experiment with innovative solutions within their own laboratories, I can only imagine how many different, independent solutions would evolve to meet the same need across any single institution. Far from being connected, this vision of the future seems disjointed, messy, and unmanageable. A suggestion can be to think in a direction of creating a hub that connects all institutions together; connects commercial developers to the entire community; and enables researchers to identify and connect with those innovations that best suit their individual needs – quickly, easily, and without risk.

It seems unlikely that a national research data management infrastructure will arrive as a result of gradual evolution of our current working practices; rather, it will have to be purposefully built over a period of many years by a bold and forward-thinking governing body, and this is a very big ask. There will be substantial cultural, technical, and practical challenges to overcome along the way and some difficult decisions relating to redistribution of funding for the greater good. It is a simple and ambitious idea, but surely only one of many ideas that will emerge as the community becomes engaged with the issues described above.

The breaking point for our current data management practices is still some distance away in the future, but I believe that the time to start a meaningful conversation about the future is right now. There is, now, within academia, an enthusiastic and energetic community of researchers who are passionate about improving data management, and who have the connections and the opportunities to raise awareness of the issues described above, and to start exchanging ideas about possible solutions. I appeal to all readers to join that community; to join the conversation; and to start shaping the future of academic research data management.

Reference

1 https://doi.org/10.5281/zenodo.268403

24

Future of Scientific Findings: Communication and Collaboration in the Years to Come

Lenny Teytelman and Emma Ganley

protocols.io, Berkeley, CA 94704, USA

For the past three centuries, there have been two ways for scientists to share their knowledge. If you are confident in your results, you publish a paper. If you want to share preliminary findings prepublication, that would be in letters or in person at large or small meetings. However, over the past decade, there is a strong signal for web-enabled acceleration of sharing of both final and preliminary results. This is an early but clear signal that is likely to be transformative over the next 10 years.

24.1 Preprints: Reversing the Increased Time to Publish

A paper from Ron Vale in 2016 showed an alarming trend – over the past 30 years, the time to first publication for biomedical graduate students has been increasing [1]. This is not a consequence of journals taking longer to peer review or publish but is rather a change in expectations: papers have been getting longer, with more figures, data, and coauthors. The omnipresent "Reviewer 2" who requests every next-step experiment to be performed for the paper to be worthy may be partially responsible, but journals, funders, and scientists themselves have also led to this change in what constitutes a publishable unit. It is somewhat surprising that despite automation, kits, and a general acceleration in the capacity to do high-throughput science with sophisticated tools and equipment, the time to publish has increased.

With recognition of these hurdles and delays to publication, the good news is that biologists are finally starting to adopt preprints, with bioRxiv being the main preprint server. Our biomedical community is a full 20 years behind the physicists in embracing preprints, as physicists have done with the arXiv server that was launched in 1991 [2]. Still, better late than never. The estimate for the average time to publish from submission to the paper going live is nine months [3]. This is likely an underestimate of the true time to publish as it counts the lag from submission to publication at a given journal but does not account for the fact that many articles cycle through rounds of submission, review, rejection at one journal, and then resubmission to a new journal – rounds that can take years for a single paper. Sharing a

Digital Transformation of the Laboratory: A Practical Guide to the Connected Lab, First Edition.
Edited by Klemen Zupancic, Tea Pavlek and Jana Erjavec.

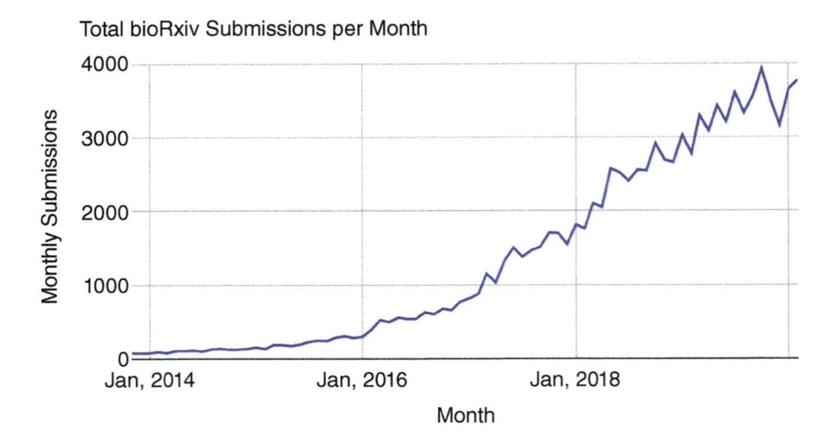

Figure 24.1 Monthly submissions to bioRxiv. Numbers based on: http://api.biorxiv.org/reports/content_summary. GoogleSheets for above plot: https://docs.google.com/spreadsheets/d/1Z96GgBEpzpqNXzMAFmnvY_s9HpdP4fijxFF6b4WuQDs/edit?usp=sharing. Source: Figure created by Emma Ganley.

preprint when the researcher has a paper ready to submit to shorten the time from the start of a Ph.D. to others reading the resulting manuscript by a year is a dramatic acceleration.

We are still far from preprints becoming the dominant norm in biology as only 30 000 out of 1.4 million biomedical papers published in 2019 are in bioRxiv. However, the exploding adoption – as seen from the graph below of monthly submissions to bioRxiv – suggests that we are not too many years away from the point where most authors share a preprint long before their paper is published (Figure 24.1). Following on the heels of bioRxiv, many other preprint servers have also been launched (many are hosted by the Center for Open Science), some research discipline specific. One of the most recent additions is medRxiv, which was launched in June 2019, and is also seeing quick growth and uptake by the community, most recently evidenced by a burst of activity around the COVID-19 outbreak. The benefits of rapid sharing of research seem to become clearer in the midst of a global health emergency – even traditional publishers have agreed to make relevant research openly available (even though we believe this should be standard for all scientific research).

[Numbers based on: http://api.biorxiv.org/reports/content_summary. GoogleSheets for above plot: https://docs.google.com/spreadsheets/d/1Z96GgBEpzpqNXzMAFmnvY_s9HpdP4fijxFF6b4WuQDs/edit?usp=sharing]

24.2 Virtual Communities

Beyond preprints, science is entering a phase of online collaboration and troubleshooting that used to be possible only at conferences. The web has enabled pre- and postpublication discussion that was a dream of many just 15–20 years ago.

Adoption of this was anemic at first, and some rushed to pronounce impossible what is quickly becoming the norm today.

Back in the late 1990s, innovative publishers began pushing for online discussion. Vitek Tracz is a publishing visionary who created a research discussion portal BioMed Net. It was too early and did not gain wide adoption [4]. A few years later, Mr. Tracz created Biomed Central, becoming the first fully open access publisher, and all journals added a commenting section to each article. The Public Library of Science did the same across all of their journals a few years later. However, most articles across both of these publishers remained without comments, and few had any real extensive discussion [5]. The paucity of comments was mistaken by some to mean that scientists only communicate via papers. However, the reality is that the real barrier to such discussion was the burden of creating separate accounts for every journal; discussion at conferences happens because people are already there, and the same applies to virtual communities.

It is now possible to self-organize your own virtual community; some early adopters practice fully open research, making the details of all of their current research and results open in almost real time. The resource http://openlabnotebooks.org offers a platform where scientists can share their lab notebooks live online. One user, Rachel Harding, has discussed the benefits to this approach, especially when working on rare disease projects; she estimates that time to publication of results is reduced down to only about a month (as compared with longer if needing to develop a full story to the point where it would be ready to submit even as a preprint) [6]. Science Twitter is now a prime example of online conferencing. Robust debates and discussions are common among scientists on Twitter, both around preprints, published papers and techniques and ideas. There are even digital conferences and poster sessions that run directly on Twitter [7]. ResearchGate has millions of scientists, and its Q&A section is buzzing daily. On protocols.io, some public methods now have hundreds of comments with troubleshooting that includes results, images, and brainstorming. There are also dedicated online discussion communities for specific techniques or disciplines (seq-answers, bio-portal, and nanopore). The online journal club PubPeer has many daily comments, with researchers often flagging papers that have inconsistencies or data manipulation. The nonprofit hypothes.is is used by many journals to allow line-by-line commenting and discussion on any article, visible to all, without having to create a new account with that specific publisher. Hypothes.is also partnered with COS to allow annotation of content on the preprint servers that they host [8].

The 1990s dreams of the innovators are now a growing reality. This new era of preprints and dynamic online discussion is also spurring innovation that merges both trends. Overlay journals such as Prelights are adding review and commenting layers to preprints. bioRxiv announced their "Transparent Review in Preprints" functionality, which also uses the open annotation potential of Hypothes.is within groups, in October 2019, allowing for preprint review [9]. PREreview is a new nonprofit platform dedicated to changing the peer review and journal club culture to focus on preprints, providing rapid constructive peer review before submission to a journal, and possibly instead of it.

It is hard to predict what impact these trends will have on traditional journals and conferences. At the very least, they already supplement traditional channels in a way that massively speeds up the communication and makes it more inclusive across the world, regardless of travel and subscription or publishing budgets.

24.3 Evolving Publishing Models

Some publishers and journals are experimenting with their procedures for peer review and publishing, primarily with a view to speeding up the process, expediting the dissemination of new research, but also to try and alleviate the burden on the scientific community. In 2018, eLife started a trial whereby if a revision was invited after peer review, the journal was agreeing to publish the final manuscript regardless of how the authors chose to respond to the reviewers' comments. eLife plans to continue experimenting with peer review as discussed by Professor Michael Eisen, upon his appointment as editor-in-chief in March 2019 [10].

Taking a slightly different approach, ReviewCommons was announced in December 2019 as a collaboration between ASAPbio, EMBO, and several other publishers [11]; the strategy and idea is to provide journal-agnostic peer review for submissions at preprint stage. If the authors wish, the reviews received can be posted to the preprint making use of bioRxiv's functionality. And then, with peer reviews in hand, the authors can decide whether to first revise and which participating journal they wish to submit both manuscript and reviews to.

Publishing models have also evolved and expanded significantly in very lateral directions in recent years with the growing recognition that there are research outputs other than the traditional style journal articles that deserve both some form of publication and archiving, and recognition in their own right: data, code/software, and protocols/methods. Recognizing these outputs, their reuse and value to the community is significant as some researchers who might be expert at honing a new microscopy technique may not also have the knowledge to apply an approach to identify new biological insights, but this does not make their scientific advances less worthy.

Although there is no clarity yet about the future of traditional journals and publishing (what an open science advocate's ideal might be seems to still be quite removed from what the average researcher is ready for), it does seem as though we are moving toward a new era of dynamic peer review that takes place after the results are public as a preprint in some form, a time where researchers are assessed on the sum of all of their research output and not just the number of formally published articles in certain journals of perceived high prestige.

24.4 Funders Are Starting to Play a Role in Facilitating and Encouraging Rapid Sharing and Collaboration

Funders used to call for proposals, select the grantees, distribute the funds, and then wait for progress reports. This is changing with many funders asking how they can accelerate science, beyond the impact of the money directly. The National Institutes of Health now has a section on Reproducibility for researchers who are preparing their grant applications [12]. Funder mandates for sharing data and code are becoming more common. Going a step further, a number of funders have been investing energy into boosting rapid collaboration and sharing among grantees, long before even the preprint is ready.

An exemplary funder dedicated to encouraging rapid sharing is the Gordon and Betty Moore Foundation. In February 2019, they announced a set of highly risky grants to bring genetic tools to marine protist research; their strategy was to fund multiple groups globally with a mandate that grantees openly share their method developments, the successes along with efforts that did not work. This is a first acknowledgement from a funder of the high levels of trial and error in method development and the need to ensure that all efforts are shared [13]. To support this requirement and help nucleate a collaborative and openly sharing community, the program officers from the Moore Foundation set up a group dedicated to this work on protocols.io and encouraged sharing there. They also conducted online lab meetings, sponsored physical meetings of the grantees, and more. Their approach and lessons learned from it are described in detail in the article' Strength in numbers: Collaborative science for new experimental model systems' [14].

Equally exciting are the efforts of the Chan Zuckerberg Science Initiative (CZI). Their mission includes not just funding new research but accelerating the progress of science itself. As part of it, one of their core values is Open Science and, at the time of writing, their website states, "The velocity of science and pace of discovery increase as scientists build on each others' discoveries. Sharing results, open-source software, experimental methods, and biological resources as early as possible will accelerate progress in every area" [15]. To that end, they are actively investing in open source software for researchers and open science platforms, including bioRxiv and protocols.io. The foundation seems to be the first to require their fundees to share preprints prior to publication. And similar to the Moore Foundation, the program officers at CZI have put a lot of effort into getting grantees to collaborate and discuss their research efforts both in person and virtually. CZI even employs a significant number of software engineers to help build the tools for grantees to facilitate sharing and collaboration.

A new funder, Aligning Science Across Parkinson's (ASAPs), currently describes its mission as, "ASAP is fostering collaboration and resources to better understand the underlying causes of Parkinson's disease. With scale, transparency, and open access data sharing, we believe we can accelerate the pace of discovery, and inform the path to a cure" [16]. Again, this funder places a strong emphasis on collaboration and openness. And although ASAP only recently issued its first call for proposals and has not made awards yet, the proposal itself is stunning, setting a new precedent, as it asks applying investigators to detail their "collaborative history" and demonstrate past adherence to open science practices [17].

Funders are perfectly positioned to support and guide researchers into adapting new research practices and creating new baseline standards. In Europe, Plan S from cOAlition S has set the stage for the expectation of immediate open access to research from any of the funders in the coalition. It will be really interesting to see if any larger more established funders will follow these leads, and how things will progress from there.

24.5 Conclusion

These are truly exciting times for science communication and research practice. While technology has leapt ahead in leaps and bounds, tools to support researchers in both how they perform and record their science and how they store and make public their approaches and results are now starting to catch up. A need for more accountability has been recognized within science as a whole. Can we access tomorrow the methods and results that were created today? Who can access these research outputs, just the researchers and a select few, or are they globally available? Have we recorded sufficiently how the research was undertaken so that others can repeat, and can we correct any errors introduced along the way? Immediate, dynamic scientific research is standing on a springboard, poised to leap and dive into fully open research and science. Simultaneous efforts to move away from current mechanisms of research assessment are needed to ensure that these efforts are recognized and an individual's science is assessed for career progression.

References

1 Ronald, D. Vale (2015). Accelerating scientific publication in biology. *PNAS* 112(44): 13439–13446. https://doi.org/10.1073/pnas.1511912112.

2 Paul Ginsparg (2011). ArXiv at 20. *Nature* 476: 145–147. https://doi.org/10.1038/476145a.

3 Bo-Christer Björk and David Solomon (2013). The publishing delay in scholarly peer-reviewed journals. *Journal of Informetrics* 7 (4): 914–923. https://doi.org/10.1016/j.joi.2013.09.001.

4 Kim Giles (2003). Elsevier waves goodbye to BioMedNet web portal. *Nature* 426: 744. https://doi.org/10.1038/426744b.

5 Euan Adie (2009). Commenting on scientific articles (PLoS edition). Nascent publishing blog. Accessed 20th Nov, 2020. http://blogs.nature.com/nascent/2009/02/commenting_on_scientific_artic.html.

6 Rachel J. Harding (2019). Open notebook science can maximize impact for rare disease projects. *PLoS Biology* 17 (1): e3000120. https://doi.org/10.1371/journal.pbio.3000120.

7 Event announcement (2019). Accessed 20th Nov, 2020. https://www.rsc.org/events/detail/37540/rsc-twitter-poster-conference-2019.

8 Heather Staines (2018). Hypothesis and the Center for Open Science Collaborate on Annotation. Hypothesis blog. Accessed 20th Nov, 2020. https://web.hypothes.is/blog/cos-launch.

9 CSHL news article. Transparent review in preprints. Accessed 20th Nov, 2020. https://www.cshl.edu/transparent-review-in-preprints.

10 eLife podcast. Episode 58: Meet Mike Eisen. Accessed 20th Nov, 2020. https://elifesciences.org/podcast/episode58.

11 EMBO press release (2019). Review Commons, a pre-journal portable review platform. Accessed 20th Nov, 2020. https://www.embo.org/news/press-releases/2019/review-commons-a-pre-journal-portable-review-platform.

12 NIH Grants & Funding, Policy and Compliance documentation, "Resources for Preparing Your Application". Accessed 20th Nov, 2020. https://grants.nih.gov/policy/reproducibility/resources.htm.

13 Smriti Mallapaty. "For risky research with great potential, dive deep", Nature index news blog, Accessed 20th Nov, 2020. https://www.natureindex.com/news-blog/for-risky-research-with-great-potential-dive-deep.

14 Waller, R.F., Cleves, P.A., Rubio-Brotons, M., Woods, A., Bender, S.J., et al. (2018). Strength in numbers: Collaborative science for new experimental model systems. *PLOS Biology* 16 (7): e2006333. https://doi.org/10.1371/journal.pbio.2006333.

15 Accessed 20th Nov, 2020. https://chanzuckerberg.com/science.

16 Accessed 20th Nov 2020. https://parkinsonsroadmap.org.

17 Tweet. Accessed 20th Nov, 2020. https://twitter.com/lteytelman/status/1184536640313856007

25

Entrepreneur's Perspective on Laboratories in 10 Years

Tilen Kranjc

BioSistemika LLC, Ljubljana, Slovenia

Almost all technological innovations have made their path through laboratories throughout the phases of research and development. Yet, the technology that is used during the R&D process is often not on par with the subject of discovery or development. Despite this common misalignment, there are certain laboratory technologies and products that fundamentally changed the way we do our research: Laboratory device integration, electronic laboratory notebook (ELN) and laboratory information management system (LIMS), cloud platforms, process automation, and laboratory automation are already used in many laboratories. But what's next? Are there more advanced laboratory technologies on the horizon? Certainly, yes, and they will do two things: fundamentally change how we do research and simplify the process of adopting new technologies.

25.1 Data Recording

The basis of scientific methodology is good practice in data recording. Despite all the technological advances, paper and ink remain the scientist's best friends. This controversy has been discussed widely in this book; therefore, we are going to focus on the practical aspects of using the paper-and-ink method for data recording.

Why is paper-and-ink still considered the most practical method for recording the data in laboratories? They are certainly cheap, and the blank canvas gives the scientists more freedom with structuring the data. On the other hand, paper-and-ink method is highly error prone, time consuming, and unshareable. Finally, it has been used for centuries and therefore deeply rooted in science practices.

We have seen in the previous chapters that digital data recording is the ultimate goal as we know it; however, the change will not happen overnight. One interesting approach, seen in other industries, is to bring the digital data recording closer to the user by building a bridge between paper-and-ink and digital records. One such approach is digitization of paper-and-ink records, although such electronic records are not fully utilized. But, the consumer market demonstrated some advanced

Digital Transformation of the Laboratory: A Practical Guide to the Connected Lab, First Edition.
Edited by Klemen Zupancic, Tea Pavlek and Jana Erjavec.

technologies in writing and voice recognition that soon might be more often used in laboratories.

25.2 Recognition of Voice and Writing

Lately, we have gotten used to controlling devices with voice commands through smart assistants, such as Alexa and Siri. These smart hubs can control a whole ecosystem of devices, such as TVs, lights, heating, and door locks. The breakthrough of such assistants can be attributed to the developments in voice recognition technologies in the past few years. There are some attempts to bring such assistants to the laboratories as well. LabTwin released their digital lab assistant in 2019 [1]. It claims to be able to lookup information, access previous experimental data, take notes, and record data, all with voice control. It is an application that works from a smartphone. Another such solution is HelixAI [2], still in development, but claims similar functionalities. No information is available about how they integrate into the laboratory ecosystem of devices. Such digital lab assistants will likely find their way into more laboratories around the world, inferring from the popularity of digital assistants in our homes. But besides that, they might be of special interest for scientists with disabilities or vision impairment. Integration of voice control with laboratory instruments would allow them to actually use the instruments (through voice control), which they might have difficulties with now [3].

25.3 Data Recording in the Future

How much data will we actually have to record manually in the future? The ultimate goal to this question should be "none." Technologically this is possible today, if the instrument software has application program interfaces (APIs) that an ELN or LIMS can access and understand.

An API would allow the other developers to more easily integrate the instrument with other laboratory software solutions, such as ELNs. Many experimental parameters could be transferred automatically to the instrument, as well as the instructions about what to do with the data. The ELN or any other central platform would hold a full record of the experiment, including the environmental parameters from sensors in the room. The scientists would not need to record any data.

25.4 Experimental Processes

One of the main benefits that digital transformation brought to the industry is the ability to iterate, simulate, and collaborate during the design process. This significantly reduced the number of failed experiments and helped to identify the designs that are more likely to be successful. The usage of digital tools in engineering, electronics, and construction is today the standard practice. Such good practices are rarely used in research planning, mostly due to a lack of digital collaborative tools. While scientists are drawing layouts of multiwell plates and pipetting plans, these remain low-level planning tools. But tools for research project management, simulations, data aggregations, and predictors are still rarely used, although some of them are already available.

25.5 Research Project Management

For research project management scientists could use widely available general project management tools. In practice, such skills are rarely taught during a typical PhD curriculum, and therefore only a few scientists actually keep a good track of their project. With ELNs becoming more widely used in the research groups this might change, since modern ELNs have some basic project management features. Regardless, project management is one of the most important transfer skills, and therefore, it should become a common practice during the PhDs, since 80% of the graduates will seek a job outside of academia, where such skills are demanded.

25.6 Experimental Planning

Good experimental design is crucial to achieve a proper quality level in science, as well as to keep the costs under control. Despite such big value propositions, the scientists are often under pressure to start the laboratory work as soon as possible and therefore often reduce the experimental design process to the bare essentials.

The digitalization in other industries revolutionized the design phase. Computer-aided design tools, collaboration and sharing, simulation tools, etc. empowered the engineers to design better products for less money. During the design phase, the engineers iterate over ideas and concepts. Then they can use rapid prototyping tools to quickly validate their ideas. Finally, the product is only built after the designs are validated and signed off.

While this approach cannot be directly translated into scientific methodology, we can learn some good practices. We can already start today by taking enough time to design the hypotheses, strategy, and experiments; then discuss it with the team; and only after thorough discussion actually going to the bench and do the work. This might seem to delay the work, but in reality such a process will both save you time and money. This is because you will think about different scenarios and be more systematic. Systematic approach helps you get high-quality data, cover more edge cases, and include appropriate controls.

Giving more importance to the design phase will also bring other innovations to this field. We can hopefully see computer-aided design tools for life sciences that span beyond plasmid design. Such tools would help you outline the hypothesis, fill in the background data, and plan the experiments to confirm the hypothesis. Such a tool would also ask you to define the controls and perform optimization or validation. Moreover, it will facilitate better team collaboration and integrate with ELN.

There are some tools already out there that perform parts of this, such as GeneIO [4] and FastFinder [5].

25.7 Virtual Reality

The cost of experimental work can be further controlled by simulating the working environment for training purposes. Therefore, the new laboratory staff would be trained in a virtual environment and could try/interface with the equipment before actually getting to the lab. Such approaches are already being tested by some device manufacturers. They even use the virtual reality gear to create a virtual environment where you could interact with the equipment in a 3D space. One of the best known examples is IKA, a manufacturer of laboratory equipment, who spun-off an augmented reality company Realworld One [6]. They are developing tools to help with training the laboratory staff by the help of virtual reality.

25.8 Smart Furniture

One obvious place where data can be presented to scientists is the bench itself. Tabletop displays are not new and have been used in medicine for a while now, but the technology has not made it to the lab yet. It has a great potential to emulate paper-based lab notebooks, since you could have all the data displayed right next to the work you are doing. Besides that, the screens are not taking away additional space on the bench and can be cleaned as any other laboratory surface, without damaging the hardware.

An alternative to this would be to have an over-the-bench projector that would project the relevant information to the bench. This would be easier to install as it would not require to replace the existing benches. The drawback is that your hands would be in the way of the projected image, which would cause shadows and potentially make you miss important information.

Both implementations would benefit from detecting equipment and consumables on the bench. This would enable automatic redistribution of information to the visible parts of the table as well as annotation of the equipment and consumables. An example of this is that a sensor would detect a microtiter plate on the bench and could display a pipetting plan by shining illuminating appropriate wells underneath the plate according to the plan. Such technology is already available on tablets, such as PlatR [7], but has not been integrated into the bench itself.

Such systems typically have cameras built in which enable gesture control. This is great for touchless operation, which is sometimes needed, especially if the hands of the scientist are full or have to remain extra clean. Additionally, having cameras built in the bench could serve as a barcode scanning tool or just to simply capture the parts of the process.

Although these systems are not yet available off-the-shelf, individual elements are already available and can be implemented as individual modules. For sure, smart furniture will play an important role as one of the enablers of digitalization.

25.9 Experiment Execution

The experimentation in a life science laboratory today is much different than it was 30 years ago. Besides the advancements in instrumentation, one major difference is the amount of experimental work the scientists do. In the hypothesis-free scientific approaches, the scientists need to perform thousands of experiments simultaneously. This gave rise to laboratory automation technologies which are being used in such experiments, mostly because the workload is too big. Most commonly such tools are used for pipetting, the so-called liquid handling robots. The scientists still have to load the samples and run the analysis. While fully automated labs exist, they are still rare. Such partially automated laboratories can also be called attended laboratories, because they still require humans to do some tasks. Eventually we can expect very little human interaction in the lab. But until then the scientists will likely still have to do some tasks manually. What we will likely see is the use of laboratory assistants, as discussed above, providing smart voice control and assistance.

Unattended laboratories do exist today, although they are rare. The inspiration comes from fully automated warehouses and factories. The unattended laboratories today use mostly the technology used in automated factories, simply because the special equipment does not exist yet. We will look at some trends in laboratory automation in the following section.

25.10 Laboratory Automation Trends

Laboratory automation has undergone big developments in the past decade, mostly due to the rise of the so-called omics experiments. Omics approaches mean that the analysis is done in a hypothesis-free way. For example, in genomics the scientists analyze the whole genome, instead of selecting a few genes to look at. A less-known omics approach is high-content screening, where, for example human cells are grown in a multiwell plate, and each gene is individually deleted to see how it affects the cell. The readout is often with a microscope. This means multiplexing the experiment 20 000 times, including cell growing, treatment, microscopy, and analysis. One such experiment would take years to do manually, while it can be done on a robotic platform in a few weeks. Robotic platforms are also less error prone than manual work and provide more consistent results.

Automation of the process described above means coupling different robots together into a streamline workflow. The multiwell plate with cells grown in it needs to be first mounted into the liquid handler, where the reagents for gene deletion are dispensed, dried, and then the cells suspension is dispensed into each well. The plate then needs to be transferred to the incubator, then back to the liquid handler for the fixation and staining, and then transferred to the automated microscope, where it needs to be mounted and imaged. The transfers between the liquid handler, incubator, and microscope are therefore usually done by robotic arms. Plates waiting for analysis (which is often the bottleneck) are held in a so-called plate hotel, which is a place to store the plates. The robotic arm can put the plate into the hotel, take it out, mount it to the microscope, etc. An experimental workflow, including all this equipment and automation, is called a workcell. The downside of workcells, as we know them today, is that it is a complete machine that usually cannot be used for other purposes. This means that we need a new (usually expensive) liquid handler if we want to use it for other assays. Lately, some labs started using mobile robots, which is a robotic platform that can move around the lab, with a robotic arm and plate hotel mounted on it. With such a robot, the workflow components do not need to be colocated anymore and can be more freely used in other workflows.

The coupling of laboratory instruments into workcells requires that the equipment is somehow modular and programmable through APIs. While many devices today are automation ready, the workcells are still customly built and require many custom solutions. One big limitation is a lack of automation standards, which is being addressed lately by SiLA [8] (standardization in lab automation). This standard is well accepted by the manufacturers, but most of them still have not implemented it.

Laboratory instruments are designed to protect the manufacturer's interests and are therefore relatively closed. Some scientists therefore adapted the open-source mentality, widely present in the software community. A rising number of open-source laboratory equipment is being designed; however, scientists still need to build it themselves if they want to use it. There are a few exceptions though, for example an open-source liquid handler OpenTrons, [9] an open-source qPCR device openQPCR [10], and a collection of tools from GaudiLabs [11]. Besides the instrument the kit includes full documentation, plans, blueprints, and code. Interestingly, the price can also be 10 times cheaper than their commercial counterparts. Using an open-source robotic arm and with some hacking knowledge the scientist can build her own workcell for a fraction of the commercial cost.

25.11 Cloud Laboratories

Cloud computing revolutionized the IT sector, because it provided masses the access to high-end infrastructure without worrying about maintenance, security, and administration. This gave rise to small businesses that could otherwise not afford to maintain their own data centers. The trends are showing that the laboratories will likely undergo a similar transition in the future.

A cloud laboratory is a platform where scientists could perform certain experiments in an automated laboratory. The scientist needs to design the experiment through a web interface of the cloud laboratory, ship the samples, and then wait for the results that are delivered through e-mail. Meanwhile, at the laboratory, the samples are processed in an automated workcell which provides reliable and reproducible execution of the experiment. Scientists do not need to own any expensive laboratory equipment and might not even need their own laboratory.

There are many benefits of using a cloud laboratory. Most of the research equipment today is underutilized and rarely shared between different organizations. Some instruments require high maintenance and specialized training to use, which usually requires a dedicated person. Automation platforms are still relatively expensive, and therefore many labs do not use it. This means more manual work, more errors, and performance bottlenecks. In a cloud laboratory, a scientist can perform as many experiments as he/she wants.

Laboratory-as-a-service has been available for some years already; however, the real cloud laboratories, where the scientists can directly interact with the platform, are only a few. The best known is Transcriptic, [12] founded in 2012 and backed by Google Ventures. They offer biomarkers, cell-based, and biochemical assay workflows that can be designed and executed through their web application. They managed to sign up some big names, including Ginkgo Bioworks [13] and Eli Lilly [14].

The price of an average experiment on Emerald lab, [15] another cloud laboratory provider, is 25 USD [16]. This means that the experimentation becomes accessible to masses that might not have been in a laboratory before. While this lowers the barrier to begin with experimentation, it also allows people to use the technology for malicious purposes, which will have to be addressed in the future.

25.12 Data Analysis Trends

Data analysis has been a topic of many controversies in science, and therefore, it is not surprising that it has also undergone some of the largest advancements. Scientists nowadays have tools available to simplify their data analysis, either using statistics software (i.e. R [17], SPSS [18], SAS [19], and Minitab [20]), specialized data analysis suites (i.e. Orange [21] and Knime [22]), and data analysis platforms (i.e. Chipster [23], Spotfire [24], and Expressions [25]). One of the main features of data analysis suites and platforms are analysis workflows, where users can create a data processing workflow made of multiple steps and tools. Such an approach maintains the traceability of raw data and reproducibility of results. This is important today, because studies suggest that most scientific papers have flaws related either to experimental design or data analysis.

Well-documented data analysis workflows will have to be included in the research papers sooner or later, which will lead to a wider adoption of advanced data analysis tools and require advanced training in data science for life scientists. The side effects of this will be easier sharing of data analysis protocols, more discussions about it, and easier replication of published studies. The online platform protocols.io allows scientists to share their research protocols and there will be likely something similar for sharing data analysis protocols.

Up until now we mostly wrote about research data. Besides that, laboratories produce a lot of other data that are not even regularly recorded. A typical example is environmental data, such as temperature and humidity. Then there are others, such as instrument uptime, instrument/bench/equipment usage, status of equipment, and reagents use. Such data can help lab managers and management to track the utilization of the equipment and accordingly schedule maintenance or upgrades. It also helps to determine when the laboratory is at the limits of capacity and needs to be upgraded, or whether it can be downsized. In practice, very few laboratories track such data today, although they are important key performance indicators. Due to a lack of sensors and resource management systems in laboratories tracking, such data is also time consuming. With a wider adoption of such systems and sensors, we can expect that the laboratories will also adopt new methodologies to gather business intelligence.

25.13 Artificial Intelligence

Artificial intelligence (AI) is mostly discussed in terms of gaining new insights from data. Besides that AI can also be used to optimize the laboratories and enable more reproducible data analysis.

Some data analysis steps are still best done by humans. Highly trained scientists can do a very good job at segmenting images, pattern recognition, and separating background noise from real data. In the past few years, machines came very close to human performance and sometimes even exceeded it. This is due to the advancements in machine learning, particularly deep learning. Deep learning made a breakthrough in computer vision and voice recognition, the technologies which are already built into consumer products. The same technologies can be used in data analysis workflows, for example to segment the microscopy images, find patterns in EEG waves, and decide whether the cell in the images is dead or alive. The tasks that were done manually up until now can be today automated. This further eliminates subjectivity from the data analysis workflows and allows more reproducible science.

AI is a complex decision-making system, which can be employed in a general laboratory system. When the experimental work becomes more automated and unattended, the scientists will spend much more time building experimental designs and workflows. AI can help with building dynamic workflows, where the next experimental steps are decided based on the data. A simple approach would be to automatically identify whether the data from the experiment are good enough, or the experiment needs to be repeated. Or, the AI algorithm could automatically decide which proteins from proteomic study should be further evaluated with more targeted approaches and also automatically start the experiments. This goes in line with cloud laboratories, where experimental workflow optimization will be crucial for the performance, especially when a large number of experiments are done simultaneously and the decision-making process would just take too long.

Today, we can see that AI is slowly being introduced into our work routine, mostly to help with everyday tasks. In the future, when laboratories will become automated in sophisticated ways, we will see AI algorithms protruding into the deepest levels of automation and other laboratory systems. The algorithms will predict errors before they happen, design experiments around hypotheses, suggest data analysis approaches, and write reports (such as AI Manuscript Writer of SciNote). Such AI will be incomparably more advanced than today's AI algorithms, so there is much to look for in the next 10–20 years.

25.14 Data Visualizations and Interpretation

One of the disruptions we expect to see in life sciences is how we publish the data. The scientific papers are published in the form of a printed report. A big limitation here is that the reports cannot be interpreted by machines, and therefore, the data cannot be used to link the studies together and distill new insights. A step toward this would be to include raw data, machine-readable analytical workflows, and machine-readable data analysis workflows with each report. Then the scientists would still need to manually connect the data together. The next step is to design some sort of standardization for publishing raw data. For example, a Western blot is often used to find the effects on the levels of a protein before and after treatment with a substance. If scientists want to use such data in biological models, they first need to find the right protein identifiers and unify the substance names. Just by adding proper protein identifiers (e.g. UniProt [26]) and substance names (e.g. IUPAC [27] or SMILES [28]), it would be much easier to link the data together. Making scientific publications more machine readable should be the top priority when setting the standards for scientific publishing.

With human-readable printed reports remaining the de facto standard in scientific publishing, there is a need to present the data in an easier-to-comprehend way. Big data is difficult to visualize; therefore, scientists found new ways to distill the large databases into nice figures. A very popular way to present data is infographic, which adds relevant graphics to the graph to make it easier to understand. Another very popular tool used lately are dashboards. Dashboards are used to track the processes in real time. They are widely used in other industries, where real-time data is distilled into a few graphs to show the status of a few key performance indicators, such as sales, production rate, or new user registrations. In life sciences, dashboards are less common. The trend among lab device manufacturers is to install multiple instruments (e.g. qPCRs in a diagnostic lab). All instruments in one lab would then be monitored through a cloud platform in a form of dashboard. This is a relatively new trend; therefore, we are likely to see more device manufacturers to start building such platforms.

25.15 Databases

One of the big revolutions in data sharing in science was the introduction of public databases. Best known today are databases of genetic and protein sequences, providing genome data to everyone for free. There are many other databases where scientists can upload their raw data, for example data from DNA sequencing or microarray data. Such databases make data available to many other scientists to build upon.

These databases provide useful data for other scientists; however, they are unable to link the data from multiple studies. As we discussed above, the main reason is the lack of standardization in organizing the data. Making the data semantic using standard terminology and descriptions of relationships between data points can help the machine to better link the data together. It takes a lot of manual curation to describe the data appropriately; therefore, these tasks are rarely done in such public databases, mostly due to limited resources.

Example A scientist wants to search for all data related to a certain gene. Commonly genes were changing their identifiers and names through time, and current search engines do not keep track of these changes. Certain genes have more than 10 different names, and only a thorough investigation of the literature reveals all of them. Then there are also many publications that refer to protein name (instead of gene name). Including these in the search takes another research into all protein identifiers and names. Although many search engines automatically extend the search with additional names for the same gene, in practice this rarely covers the whole landscape.

25.16 Conclusion

The laboratories are certainly ready for a major technological and cultural disruption. While emerging startups and initiatives fight to find a spot in the laboratories around the world, many people think about the disruptions that will happen in the long term, e.g. next 30 years. The mantra we hear often nowadays is "data is money." One may take this mantra sarcastically – the laboratories are data factories; however, most of the laboratories complain about the lack of funding for science. So how will this be addressed in the next 30 years? One aspect that has not been thoroughly investigated yet is the monetization of research data. The monetization does not necessarily mean selling data. More likely there will be some other business models used, involving public–private partnerships – after all most research institutions are funded by the taxpayers money. Public ledger technologies, such as blockchain, can help transparently track (mostly indirect) contributions of research data to commercial products.

There are probably many other technological advancements we are about to experience in the next 30 years. The reality is that technological disruptions in laboratories come with a certain delay, compared to the other industries. Smart factories are becoming a reality today, and they already proved their value. In the scope of the "data revolution" that we are part of, the laboratories, or the so-called data factories, might soon catch up and technologically exceed the typical brick-and-mortar (smart) factories.

References

1 https://www.labtwin.com
2 https://www.askhelix.io
3 Austerjost, J., Porr, M., Riedel, N. et al. (2018). Introducing a virtual assistant to the lab: a voice user Interface for the intuitive control of laboratory instruments. *SLAS Technology* 23 (5): 476–482. https://doi.org/10.1177/2472630318788040.

4 https://www.biosistemika.com/geneio/
5 https://www.ugentec.com/fastfinder
6 https://realworld-one.com/
7 https://www.biosistemika.com/platr/
8 https://sila-standard.com/
9 https://opentrons.com/
10 https://www.chaibio.com/openqpcr
11 http://www.gaudi.ch/GaudiLabs/
12 https://www.transcriptic.com/
13 https://www.ginkgobioworks.com/
14 https://www.lilly.com/
15 https://www.emeraldcloudlab.com/
16 Lentzos, F. and Invernizzi, C. (2019). Laboratories in the cloud. Bulletin of the Atomic Scientists. https://thebulletin.org/2019/07/laboratories-in-the-cloud (accessed 2 July 2020).
17 https://www.r-project.org/
18 https://www.ibm.com/products/spss-statistics
19 https://www.sas.com/en_si/software/stat.html
20 https://www.minitab.com/en-us/
21 https://orange.biolab.si/
22 https://www.knime.com/
23 https://chipster.csc.fi/
24 https://www.tibco.com/products/tibco-spotfire
25 https://www.genialis.com/software-applications/
26 https://www.uniprot.org/
27 https://iupac.org/
28 https://en.wikipedia.org/wiki/Simplified_molecular-input_line-entry_system

Index

Digital Transformation of the Laboratory: A Practical Guide to the Connected Lab, First Edition.
Edited by Klemen Zupancic, Tea Pavlek and Jana Erjavec.

b

c

d

e

k

l

m

n

o

p